Innovation and Discovery in Russian Science and Engineering

Series editors
Carlos Brebbia
Wessex Institute of Technology, Southampton, United Kingdom

Jerome J. Connor
Department of Civil & Environmental Engineering, Massachusetts Institute of Technology, Cambridge, Massachusetts, USA

More information about this series at http://www.springer.com/series/15790

Valentina Svalova
Editor

Natural Hazards and Risk Research in Russia

Springer

Editor
Valentina Svalova
Russian Academy of Sciences
Sergeev Institute of Environmental
Geoscience (IEG RAS)
Moscow, Russia

ISSN 2520-8047 ISSN 2520-8055 (electronic)
Innovation and Discovery in Russian Science and Engineering
ISBN 978-3-030-06307-8 ISBN 978-3-319-91833-4 (eBook)
https://doi.org/10.1007/978-3-319-91833-4

This Springer imprint is published by the registered company Springer International Publishing AG part of Springer Nature.
The registered company address is: Gewerbestrasse 11, 6330 Cham, Switzerland

Contents

Chapter 1
Technogenesis and Natural Disasters

Victor Osipov and Nadezhda Rumyantseva

Abstract The paper scrutinizes technogenesis in the context of human industrial activity in natural resources development and the creation of social and economic infrastructure on the Earth. The rapidly growing antagonism between technogenesis and the society requires the elaboration of a new strategy of modern civilization development based on keeping the balance between the human economic activity and the environment protection.

Some problems arising from technogenesis development are analyzed, and it is shown how the environmentally friendly economic activity may be managed. As an example, the seismic and engineering geological zoning of Moscow is described, which was aimed at "fitting" items of the technosphere to the natural conditions and at optimizing the interaction between the engineering structures and the environment. The problems in surface and groundwater conservation, assessment of natural risks upon land development, as well as industrial and domestic waste treatment are discussed. The cases cited witness to the necessity of developing the scientifically grounded policy of human technogenic activity based on the principle of technogenesis-environment coevolution

1.1 Introduction

Various processes developing inside the Earth, on its surface, and in the adjacent atmosphere layers cause manifestation of hazardous geological phenomena. The range of developing natural hazards is very wide, varying from disastrous earthquakes to the global climate warming and the hazard of Earth colliding with other cosmic bodies.

Natural disasters in the modern world appear to be the source of severe social shocks, causing large-scale suffering of people and fatalities and leading to huge

V. Osipov · N. Rumyantseva (✉)
Sergeev Institute of Environmental Geoscience RAS, Moscow, Russia
e-mail: sci-secret@geoenv.ru

V. Svalova (ed.), *Natural Hazards and Risk Research in Russia*,
Innovation and Discovery in Russian Science and Engineering,
https://doi.org/10.1007/978-3-319-91833-4_1

economic losses. In the general problem of social security, natural disasters are more and more often considered as one of the principal destabilizing factors preventing sustainable development.

Above eight million people were killed by natural disasters in the last century. At present, the annual economic losses from catastrophes amount to 190 billion USD per year [5]. In 2015, the global damage reached the critical value equal to 380 billion USD. Some countries (Japan) spend up to 5% (8% at the most) of the annual budget for the elimination of natural disaster consequences.

In Russia, the economic losses from natural disasters with the consideration of indirect damages reach 2% of the gross domestic product. The cases of the most disastrous catastrophes that occurred in Russia in the second half of the twentieth and in the beginning of the twenty-first century are the Caspian Sea level rise and flooding its shores in 1978–1995, which caused the damage equal to 6 billion USD; the Sakhalin earthquake (May 1995) destroyed completely the Neftegorsk city, and resulted in more than 2000 fatalities, and the damage higher than 200 million USD. The mean perennial losses from floods and wildfires in Russia amount to hundreds of millions USD. The flood in the Far East was the most disastrous, with its total damage having amounted to 258 billion RUR [5].

The number of natural disasters grew sharply in the second half of the twentieth century (Fig. 1.1).

An increase in the number of natural and technogenous disasters is caused by a number of global processes taking place in the social, natural, and technogenic spheres. These include the growth of population and industrial production in the world, urbanization, environment degradation, and the global climate warming.

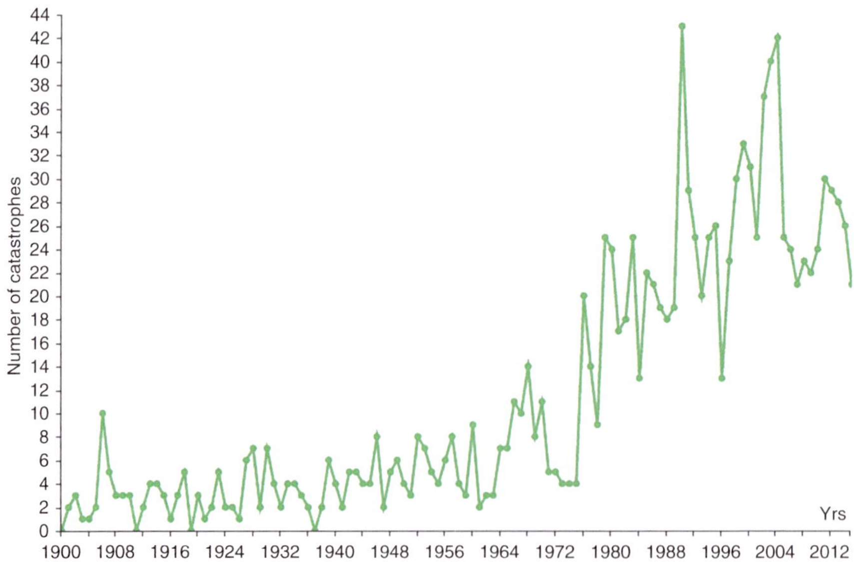

Fig. 1.1 Number of natural disasters registered in EM-DAT in 1900–2003

1.2 Global Climate Change and Its Consequences

The problem of global climate change stands out among the factors promoting the growth of natural calamities number. This hazard is based on two main reasons. The first one is the solar-Earth interactions resulting in the recurrent changes of lasting periods of relative cooling and warming. The second reason is technogenesis resulting in the increasing volume of greenhouse emissions to the atmosphere (carbon dioxide, methane, and nitrogen dioxide), favoring the consumption of solar energy in the near-surface layers of the atmosphere and the rise in its temperature.

The paleoclimate reconstructions reveal a number of megacycles of natural temperature change in the Earth's history caused by the recurrence of lasting periods of relatively low and high temperatures. Each megacycle includes, in its turn, less lasting climatic macro-, meso-, and microcycles. Thus, we may speak about different periods in cyclic temperature changes in the Earth that vary from tens and even hundreds of million years to some hundred years.

The global temperature change on the Earth during the last 150 years is of particular interest: the global climate warming is developing at a rate of approximately 0.02 °C per decade against the background of the general cooling trend. Up to the year of 2015, the temperature on the Earth has risen by 1° as compared to the temperature in the preindustrial epoch. Most researchers are sure that besides natural factors, the Earth climate is being significantly affected by technogenesis in the last 150 years, which resulted in the extraordinarily intense rise in the temperature.

The technogenic activity of a man and atmospheric emissions from industrial works, transport, energy-production facilities, and social infrastructure contaminate the atmosphere with fine particles and greenhouse gases. Above 32 million tons of emissions from stationary and mobile sources enter annually the Earth's atmosphere. The atmosphere in the industrial cities is being contaminated most intensely.

During the last 200 years (1800–2010), the total emission of carbon dioxide (tons per year) from all sources has risen on the Earth from 280 to 390 fractions per 1 mln air fraction; it may reach 800 fractions to the end of the twenty-first century. The methane concentration rose from 0.8 to 1.65 in the same time period and nitrogen oxide from 285 to 310 fractions per 1 million air fractions.

The temperature rise on the Earth starting from the second half of the twentieth century is accompanied by a rapid growth of disastrous phenomena, thawing Arctic ice caps, and degradation of permafrost. Sharp climate changes induced lately a series of natural disasters, i.e., the abnormal heat in Europe, extremal heat and wildfires in the European Russia in 2010, and abnormally cold winter in 2012.

The amount of hydrological disastrous events rose most intensely (Fig. 1.1). Only in 2015, more than 30 large-scale droughts and disasters that stroke 98.6 million people were registered on the Earth. In Russia, a number of catastrophic floods are observed lately, i.e., on the Lena River, Yakutia, in 2001, in Krasnodar Krai in June 2012, and in the Far East in 2013.

A number of major accidents of global threat were caused by natural disasters. For instance, this is the destruction of Fukushima nuclear power plant in Japan. The accident took place on March 11, 2011, after the earthquake with magnitude 9 (against the calculation magnitude 7). The tsunami wave have flooded four of six blocks and disabled the reactor cooling system. This induced a series of hydrogen explosions, melting the active zone, and nuclear emission to the atmosphere.

No less dangerous tragedy took place in December 2004: an earthquake of magnitude 9 in the Indian Ocean induced a huge tsunami, which waves reached the Indonesia coast and a number of other coastal countries. About 280 thousand citizens of these countries were killed.

The role of technogenesis in the climate cooling is most vividly manifested in the state of Arctic region and permafrost degradation. According to the calculations, the total area of close to surface perennial frozen ground in the land of the Northern Hemisphere may reduce in the twenty-first century from 20 mln km^2 to 5.3–12.8 mln km^2 depending on the anthropogenic impact scenario (Fig. 1.2).

Intense thawing of polar ice caps is the second important consequence of the warming in the northern area. The perennial ices are considered to have twice reduced in Arctic during the last 50 years.

Transition of the frozen ground to the thawed state leads to the surface deformations and promotes the development of hazardous natural phenomena over vast areas of the Eastern and Western Siberia, such as thermal karst, thermal erosion, solifluction, seasonal heaving, ground subsidence, etc. It should be noted that salinized frozen ground are thawed and lose the bearing capacity already at negative temperatures.

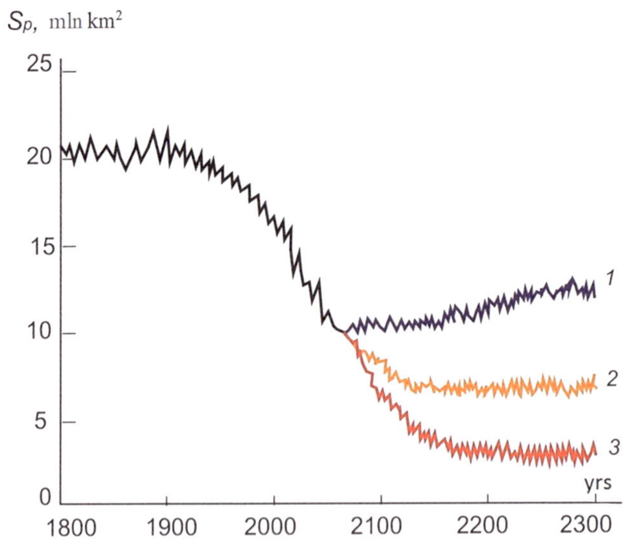

Fig. 1.2 Predicted changing area of perennial frozen ground distribution on the Earth in twentieth to twenty-first centuries for mild (1), intermediate (2), and severe (3) scenarios of anthropogenic impact [2]

a) b)

Fig. 1.3 Territory affected by thermokarst in Western Siberia: space image (**a**) and explosive thermokarst funnel at Yamal Peninsula (the helicopter view) (**b**)

Thermal karst is one of the most widespread processes related to the permafrost degradation. Figure 1.3a shows the rate of affection by thermal karst of the territory in one of the northern regions in the Western Siberia. Recently, thermokarst formations of another kind, i.e., of pneumatic or gas-explosive origin, were found there (Fig. 1.3b) [1]. The funnel formed from the explosion is about 60 m in diameter on the surface and a depth over 50 m (Fig. 1.3b). A parapet of ground thrown to a distance up to 120 m is observed around the funnel. The funnel shape testifies to its formation as a result of powerful gas explosion.

1.3 Strategy of Struggle Against Natural Disaster

The irreversible growth of the number of hazardous events causes the necessity to adopt a new strategy of struggle against natural and technogenic catastrophes based on the risk assessment and elaboration of methods of mitigating their consequences. Implementation of this strategy will permit us to pass to the economic planning and development with the account of existing hazards, to increase safety of the society, and to reduce social and economic losses. The strategy should be based on "the global prevention culture" proceeding from the scientific prediction of the future disasters. It is better to prevent the disastrous phenomena rather than to eliminate its consequences. The international experience proves that the expenses for prediction and providing readiness to the emergencies are up to 15 times lower than the prevented damage.

The development of methods of predicting disasters based on deterministic and probabilistic approaches appears to be one of the most important items in this strategy. These approaches take into account the vulnerability of the technosphere and permit predicting possible social and economic losses as well as assessing risks. Risk maps appear to be the most important document for elaborating of strategic plans on the timely engineering protection in order to minimize the consequences of future catastrophes.

Institute of Environmental Geoscience RAS (IEG RAS) is involved in the research on compiling the risk maps for many years. In collaboration with the Extreme Situations Research Center, we have compiled more than 30 risk maps for the country as a whole and its separate regions and cities [4]. The special GIS created for this purposes is widely used for compilation of these maps.

The maps of individual natural risk, which are built both for particular hazard types and for integral hazard produced by several hazardous phenomena, are the maps of most widespread type. For the Russia territory, three maps of the integral natural risk were built to a scale of 1:20,000,000, showing the risk produced by six hazards, i.e., earthquakes, floods, storms, landslides, mudflows, and snow avalanches. The first map (R_e1) shows the losses with fatalities in case the mentioned processes are realized, the second (R_e2) deals with the fatalities and injuries of people, and the third map (R_e3) considers both death and injuries of people as well as economic losses. For example, Fig. 1.4 displays the integral map of individual natural risk (R_e1).

The obtained estimates of individual natural risk from six types of natural hazards vary within the Russia's territory from negligibly small to very high values, exceeding 30 × 10–5 for the fatality (Re1), >100 × 10–5 for the fatality and injuries of different heaviness, and >150 × 10–5 for fatalities, injuries, and economic losses.

The maps of seismic risk built for the entire country and almost for all seismic regions in Russia are the most widespread among the maps of individual risk from particular hazards. The former include the map of natural individual risk (Re3) for the Northern Caucasus, Siberia, Krasnodar Krai, Stavropol Krai, the territory of Petropavlovsk-Kamchatskiy, etc.

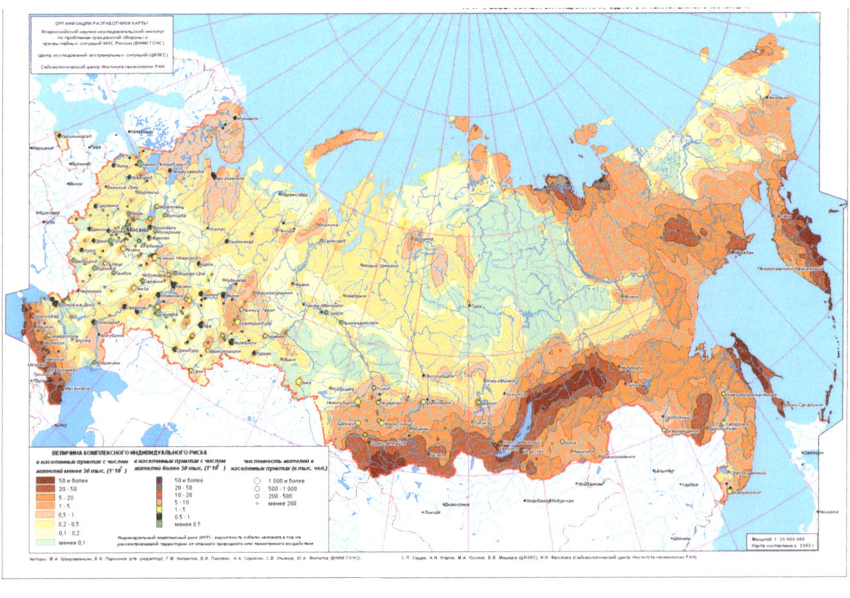

Fig. 1.4 The map of complex individual natural risk for population of the Russian Federation with the account of earthquakes, floods, hurricanes, snow avalanches, landslides, and mudflows

Development of techniques for adapting to ecosystems is the second important aspect of the strategy of struggle against natural disasters and mitigation of their consequences [3]. By adapting to ecosystems, we should understand not the attempts to change the latter according to the society needs but, on the contrary, adjusting the technogenic activity to the existing ecosystems. In doing so, the approved axiom should be followed: a man should not interfere with the natural processes and induce their negative development. Thus, the idea of adapting to ecosystems under the conditions of technogenesis consists in minimizing the transformation of former upon solving issues of ensuring environmental safety of the society. Inscribing our activity to ecosystem processes and the engineering protection of the created technogenic objects appear to be the basic techniques of this idea implementation.

The engineering activity of people with the account of environmental risk based on building the map of zoning of the territory under consideration by the hazardous processes is the vivid example of above statements. For instance, we may mention the work performed at the Sergeev Institute of Environmental Geoscience RAS on the seismic microzoning of the Imeretinskaya lowland in the Greater Sochi area, before the construction of Olympic facilities – 2014 there (Fig. 1.5). In the territory allotted for the Olympic construction, the lowest seismicity zones were outlined, within which the main Olympic facilities were allocated.

As an example of engineering protection and mitigating the nature-society conflict, we may mention the construction of a 24.5-km-long dam in the Gulf of Finland in 2009, which protects St. Petersburg from floods upon the water level rise in the

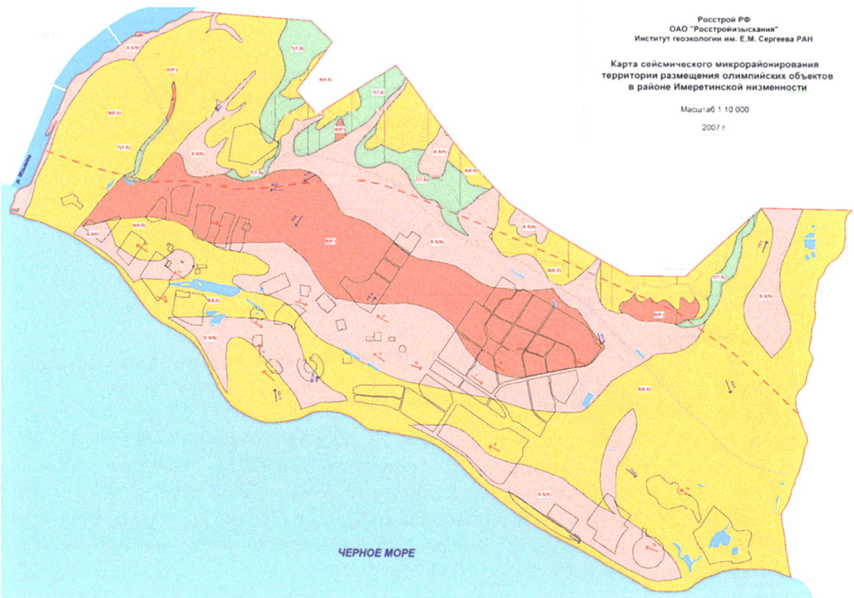

Fig. 1.5 The map of seismic microzoning of the Olympic object territory in the Imeretinskaya lowland

Neva River. The city of St. Petersburg was subjected to flooding 305 times since the day of its foundation. The dam can resist the water rise to 4.2 m, which completely prevents flooding of the city in the future and saves citizens from the permanent threat and huge material losses.

1.4 Conclusion

The global climate warming developing on the Earth is accompanied by a sharp increase in the number of catastrophic natural phenomena causing huge social and material losses. Elaboration of the strategy for society-environment interaction is the most important task for the future generations of scientific community. The main goal of this strategy is to develop in harmony with the environment.

Development of methods of disaster prediction based on deterministic and probabilistic approaches is one of the most significant elements of this strategy. Taking into account the technosphere vulnerability, this permits predicting possible social and material losses and assessing natural risks.

The second important element of the strategy should be the principle of adapting and following the basic axiom: a man should not interfere with the natural processes and induce their negative development. The scientifically grounded critical parameters of technogenesis development are of utmost importance in this field. It is important that a man in its economic activity should adjust to natural processes, i.e., should strive to incorporate technogenesis into the integrated process of society and nature development.

Acknowledgments This study was supported by the Russian Science Foundation, project no. 16-17-00125 “Assessment of risk caused by hazardous natural processes in urban areas.”

References

1. Bogoyavlenskii, V. I. (2014). The danger of catastrophic gas emissions in the permafrost zone. Arctic. Funnels on Yamal and Taimyr Peninsulas. *Burenie i neft' (Drilling and oil), 9*, 13–19.
2. Mokhov, I. I., & Eliseev, A. V. (2012). Modeling global climate changes in XX–XXIII centuries upon the new scenarios of anthropogenic impacts RCP. *Doklady akademii nauk (Reports of Academy of Sciences), 443*(6), 732–736.
3. Osipov, V. I. (2010). Management of natural risks. *Vestnik RAN (Herald of RAS), 8*(4), 291–297.
4. Osipov, V. I., Frolova, N. I., Sushchev, S. P., & Larionov, V. I. (2011). Assessment of seismic and natural risk for the population and territories of Russian Federation. Extreme natural phenomena and disasters. In V. M. Kotlyakov Ed.-in-Chief, IG RAN; Compiled by Sobisevich, A.L., IFZ RAN; – Moscow: IFZ RAN, *Volume 2: Uranium geology, geoecology and glaciology* (pp. 28–48). Moscow: IFZ RAN, Probel 2000 Publ (in Russian).
5. Porfir'ev, B. N. (2015). Economic consequences of catastrophic flood in the Far East in 2013. *Vestnik RAN (Herald of RAS), 2*, 30–39.

Chapter 2
Complications and Prospects of the Hydrogeological Substantiation of Drainage in Deep Tunnels

G. I. Batrak, I. A. Kostikova, I. A. Pozdnyakova, E. A. Karfidova, and L. S. Toms

Abstract Currently, many automobile tunnels are being designed and put into operation in urban areas including Moscow. This is associated with the need of the city to unload the motorways in the absence of sufficient-free areas for the construction of land routes. Experience of operation has shown the expediency of building such facilities on the city territory. Several deep tunnels were commissioned in Moscow including the northwestern tunnel, 27 m deep and 3.1 km long.

The Alabian-Baltic tunnel has been under construction since 2006. A great difficulty in its operation is created by groundwater that seeps into the tunnel. In winter these leaks lead to icing on the road which can trigger accidents. Unfortunately, it has not been possible to completely seal the tunnel to the present day. All this determines the need for water drainage. However, closely located houses, many of which have already suffered from construction, as well as the passing subway tunnel, require careful justification for such solutions.

Currently, the most dangerous for the accident-free operation of the tunnel is the water inflow into it. Breakthroughs of water-saturated sands or the process of forced mass suffusion was observed in several sections. This process could manifest itself both as a sudden arrival of quicksands into the pit and breakthroughs of water-saturated sands through interbeds of clay soils. It was the main reason for the decompaction of the soil massif containing the tunnel, the anomalously large and uneven sediment of buildings, and the formation of dips.

A numerical hydrodynamic model was created within the framework of the work. The model analyzed various options for the organization of drainage, including the zero option (without drainage). It is shown that between the technologically altered level in the aquifer and the bottom of the tunnel, a significant difference in altitude remains, which leads to seepage of groundwater into the tunnel. Analysis of possible technical solutions to reduce water inflows in the tunnel has shown that a permissible option is a forced drainage inside the contour of the wall in the ground.

G. I. Batrak (✉) · I. A. Kostikova · I. A. Pozdnyakova · E. A. Karfidova · L. S. Toms
Institute of Environmental Geoscience, Moscow, Russia
e-mail: gib74@mail.ru

V. Svalova (ed.), *Natural Hazards and Risk Research in Russia*,
Innovation and Discovery in Russian Science and Engineering,
https://doi.org/10.1007/978-3-319-91833-4_2

2.1 Introduction

Currently, a large number of automobile tunnels are being designed and put into operation in the city of Moscow. This is associated with the natural need of the city to unload the motorway in the absence of sufficient-free areas for the construction of land routes and interchanges. A new stage in the development of the underground space of the city began in 2003, with the launch of the Lefortovo tunnel, in which the Institute of Environmental Geoscience was also involved. The depth of its laying is 30 m, and the length is 3.2 km. The experience of subsequent operation showed the possibility of building such facilities in the city. After that, several deep tunnels were built in the city, including the northwestern tunnel, 27 m deep and 3.1 km long.

The Alabian-Baltic tunnel (ABT) has been under construction since 2006 as part of the Big Leningradka project. After numerous postponements of its full opening, it is likely to be fully open in the first half of 2016. However, the greater complexity in its operation is created by groundwater that seeps into the tunnel. In winter these leaks lead to icing on the road, which can trigger accidents. Unfortunately, it has not been possible to completely seal the tunnel to the present day. All this determines the need to consider the possibility of water depletion. However, closely located houses, many of which have already suffered from construction, as well as the passing subway tunnel, and other automobile tunnels, require careful justification for such decisions.

2.2 Analysis of the Influence of the River on the Hydrodynamic Situation

The valley and bed of the Tarakanovka River are still clearly distinguished in the relief of the earth's surface of the urban landscape. In the basin there are nine tributaries of the second order along the left and ten on the right sides of the valley.

In the course of the work, the morphology of the modern earth's surface was analyzed, the inheritance of the river valley was estimated by the modern urban relief, and directions and shortest paths of runoffs (streamlines) and places of concentration of surface wastewater were estimated: the length of the paths and volumes of surface runoff in the cadastre division, proceeding from the assumption that the main collection of rain sewage is carried out on the street-road network.

The river valley continues to play an important role in the accumulation of surface waters in the territory. ABT is laid along the river valley, and the surface runoff from all adjacent territories tends to it. The part is intercepted by a network of rain sewers, and a part reaches a depression in the relief around the tunnel and seeps into the ground. The index of topographic humidity shows how much the conditions of the territory are favorable for the infiltration of atmospheric precipitation down into aquifers. In the identified valley of the Tarakanovka River and especially in the ABT area, these conditions are most favorable. Despite the fact that the Tarakanovka

River itself is currently channeled, its channel filled with loose sand deposits serves as a water collector, which, taking into account the much deeper laying of the tunnel in comparison with the channel, is intensively discharged into it, collecting water from the tributaries buried beneath the technogenic deposits.

2.3 Analysis of Hydrogeological Conditions in the Section of the Tunnel

In the hydrogeological section, in the interval from the surface of the earth to the top of the Callovian-Oxfordian clay, three aquifers are distinguished: Supra Moraine, distributed fragmentarily; Supra Jurassic, widespread throughout the whole territory; and Jurassic in phosphorite sands, distributed fragmentarily. The horizons of the upper part of the section, the Upper Jurassic and the Jurassic, are everywhere separated from the deeper aquifers in the Carboniferous sediments by weakly permeable Callovian-Oxfordian clays. The first aquifer from the surface of the earth in the Upper Carboniferous sediments is the aquifer in Perkhurov limestones, which is widespread everywhere. The Upper Jurassic has a complex structure, due both to the heterogeneity of the geological structure and to the lithological heterogeneity of the water-bearing sediments.

Depending on the geological structure, the water-bearing sediments may or may not include alluvial sands, fluvioglacial sands of the Don-Moscow horizon, Lower Cretaceous sands, Upper Jurassic sands, sandy loam, etc. Variants of the structure of the strata in the section are many; they change often. A frequent and abrupt change in water-bearing sediments is also a result of the filtration heterogeneity, which is confirmed by a large spread in the values of the filtration coefficient from 0.1 to 5 m/day of the same fluvioglacial fine-grained sands.

In the hydrodynamic relation on the section of the tunnel, there are prerequisites for downward filtration, since the groundwater levels of the above-ground aquifer at its distribution areas are set above the groundwater levels of the Upper Jurassic aquifer. The natural hydrodynamic conditions – the conditions for inflow and outflow, the direction and speed of the flow, and the change in the position of the ground surface of the groundwater in the ABT area and the adjacent territory – were violated even before its construction and were associated with the construction of the metro and other underground infrastructure facilities (Fig. 2.1). Large underground objects (the ABT itself, as well as other tunnels of the metro, the Volokolamsk and Leningrad tunnels, the collector of the Tarakanovka River in the immediate vicinity of it) exert a significant influence on groundwater. They change the direction of the natural groundwater flow, the speed of filtration, and the depth of the groundwater level, especially in the areas where they are located.

In regional look, it is possible to change the location of the areas of recharging and discharging groundwater, its volume, and parts of the general groundwater balance in the adjacent territory. The listed possible changes in the hydrodynamic

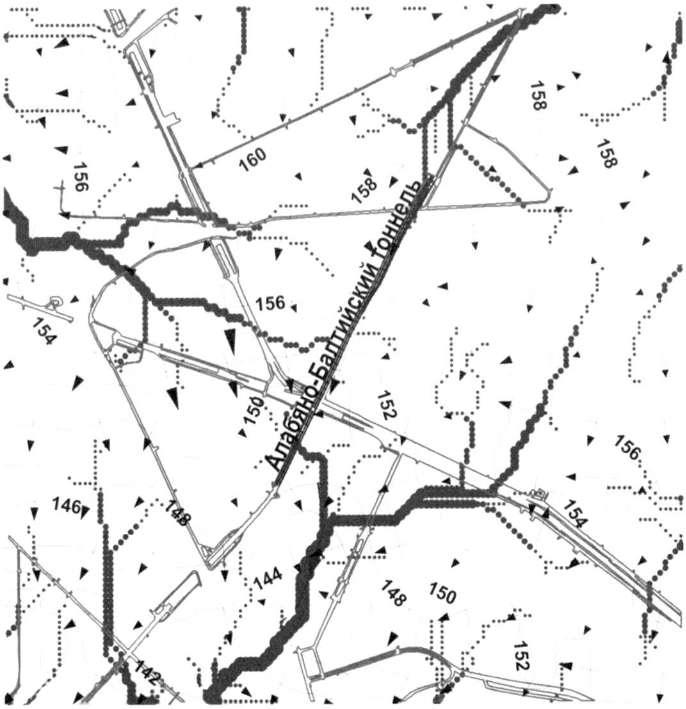

Fig. 2.1 Schematic map of hydrogeological conditions prior to the construction of ABT (isolines are the hydroisogypses of the Upper Jurassic aquifer; vectors are the direction of the flow of groundwater)

conditions along with the heterogeneous structure of the hydrogeological section and the significant filtration heterogeneity of the water-bearing sediments affect the water inflows into the tunnel and, ultimately, its safe operation. They should be taken into account not only on the ABT site but also on the adjacent territory due to its large extent and depth of the site. The tunnel passes through the entire heterogeneous hydrogeological strata and from the hydrodynamic point of view is the deepest drain, to which the flow of groundwater rushes around other underground structures from a large area of the entire water-saturated stratum over the Callovian-Oxfordian clays.

2.4 Results of Hydrogeodynamic Modeling

Within the framework of the work, the region of the catchment basin of the river was investigated. In connection with the complex construction conditions and the strong variability of the geological environment, a numerical hydrodynamic model was developed for the analysis and prediction of the hydrogeological situation, as well as for the hydrogeological justification of the drainage system. The size of the model was chosen in such a way as to characterize the zone of primary recharge of groundwater, both in the prevailing conditions and the predicted period of drainage functioning. Dimensions of the modeling area were 3900 × 3900 m. The basis of the hydrogeodynamic model was the three-dimensional geological model created by the GIS MapInfo. The archive data of the IGE RAS were used from the database, accumulated in the process of large-scale geological mapping and other projects in Moscow. This base was supplemented by the results of recent years' surveys carried out during the construction of the facility. In total, 1478 wells with a total depth of 54,078 m were considered for the study area. Five hundred twenty-two wells with a total depth of 16,125.7 m were used to construct the three-dimensional model. Thirty-three geological sections with a total length of 160.8 km were gathered to construct digital surfaces of aquifers and weakly permeable layers of the geofiltration model (Fig. 2.2).

Since the geofiltration model was created for drainage simulation, the main problem was the detailing and implementation of the water heterogeneity of the water-bearing sediments in the super-Jurassic aquifer that significantly influenced the results of vertical drainage simulation. Since often interchangeable interlayers of uneven grained sand, sandy loam, and loam are not sustained either in a section or in

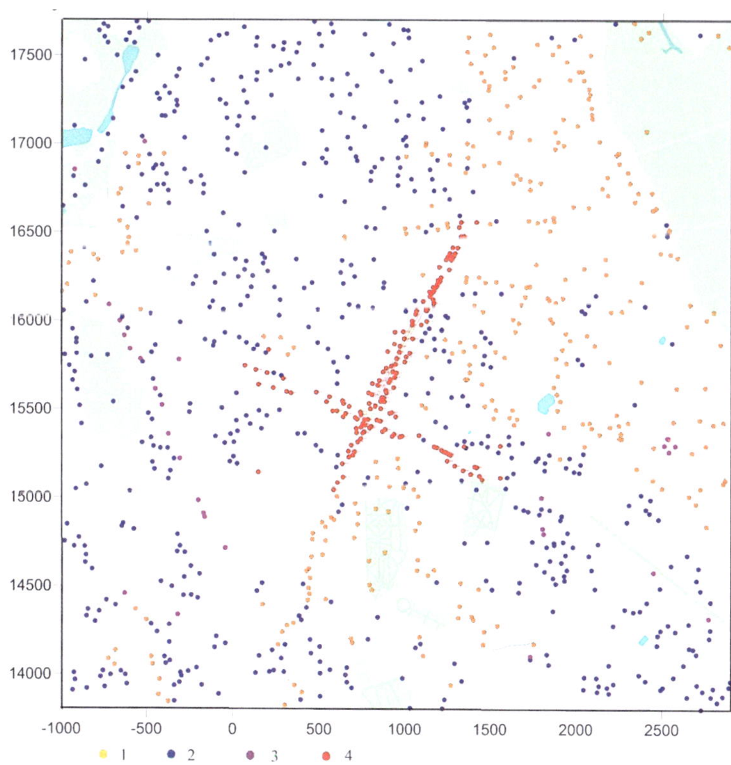

Fig. 2.2 Boreholes over the territory

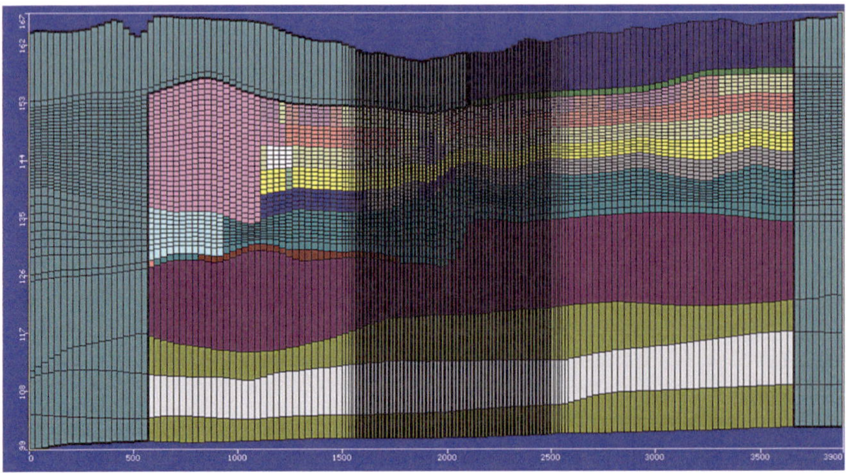

Fig. 2.3 Section of the model

a plan, it was impossible to break the entire thickness into several homogeneous layers with different filtration properties, corresponding to the lithology of the water-bearing rocks. The developed original technique, allowing to take into account the chaotic nature of variability of lithology and filtration parameters of the water-bearing strata, consisted in the partition of the Upper Jurassic aquifer into several reservoirs and recalculation by the known dependence for the center of each model block of filtration coefficients, taking into account the change in power and the filtration coefficient of the prevailing lithological differences. The resulting number of reservoirs for the Upper Jurassic horizon was five, and zones with unique values of the filtration coefficient was more than 40 (Figs. 2.3 and 2.4).

Created in this way, the original model, consisting of 13 layers, beginning with the over moraine and ending with the roof of the Sunday clay, was realized with the help of MODFLOW 2000, using a licensed program for entering, preparing, and displaying the results of Visual Modflow 4.3. After solving the inverse problem, i.e., restoration of the natural and man-caused situation prevailing before the construction of the model, which consisted in the calibration of the geofiltration parameters of the model – the filtration coefficients, the recharge parameters, and the boundary conditions – the model was redivided into 41 layers for more precise design of tunnel structures and other underground structures. The model analyzed various options for the organization of drainage, including the zero option (without drainage). It is shown that between the man-made level in the aquifer and the bottom of the tunnel, a significant difference in altitude remains, which leads to seepage of groundwater into the tunnel. Analysis of possible technical solutions to reduce water inflows in the ABT has shown that a permissible option is forced drainage inside the wall contour in the ground. Legitimate drainage, which could intercept the flow on its way to the tunnel, will require much higher capacity of water-reducing

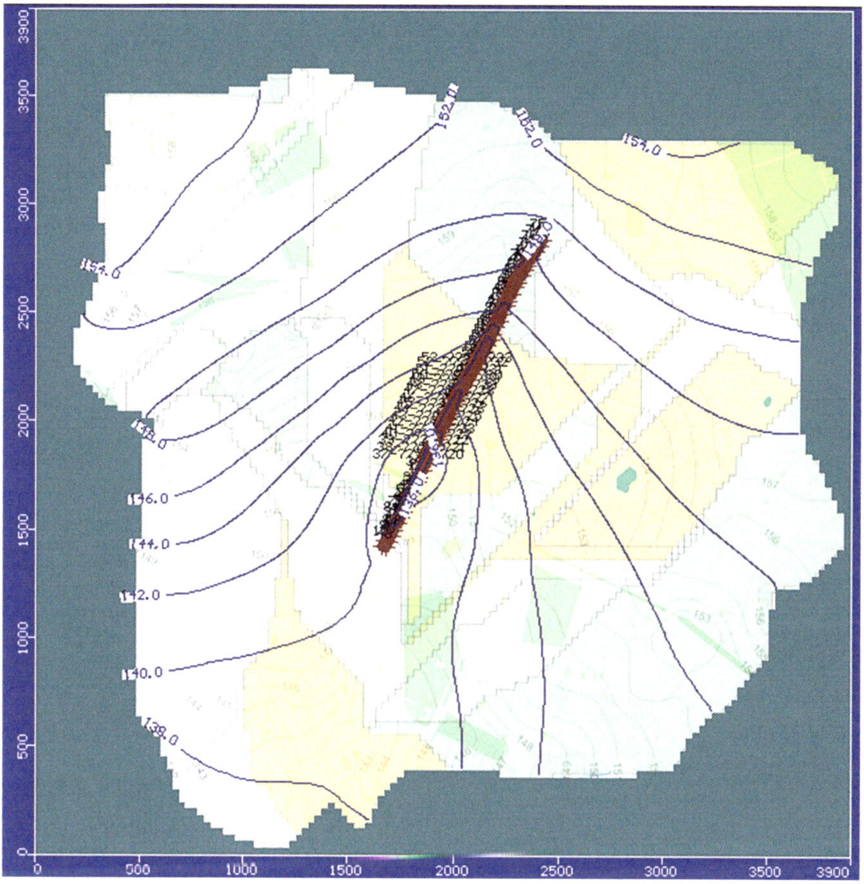

Fig. 2.4 Groundwater levels when drainage operates in steady state regime

installations, lower the level of groundwater in a larger area than with in-circuit drainage, and give more inflow to the drainage system. In addition, this option is also dangerous because such a regional lowering of the groundwater level can cause subsidence and activation of exogenous processes on a large territory (Fig. 2.5).

Another variant of dumping water into the Perkhurov aquifer through vertical bypass wells will not allow monitoring of the drainage process, i.e., when activating negative processes of a geological or technogenic nature, it is simply difficult to cover such wells or reduce the flow through them. With this option, it is also possible to contaminate coal-bearing horizons that have a drinking purpose.

Thus, the main option for in-circuit drainage, which is carried out by drilling wells from the wells of the drainage network, passing through the bottom of the tunnel, was considered as the main one.

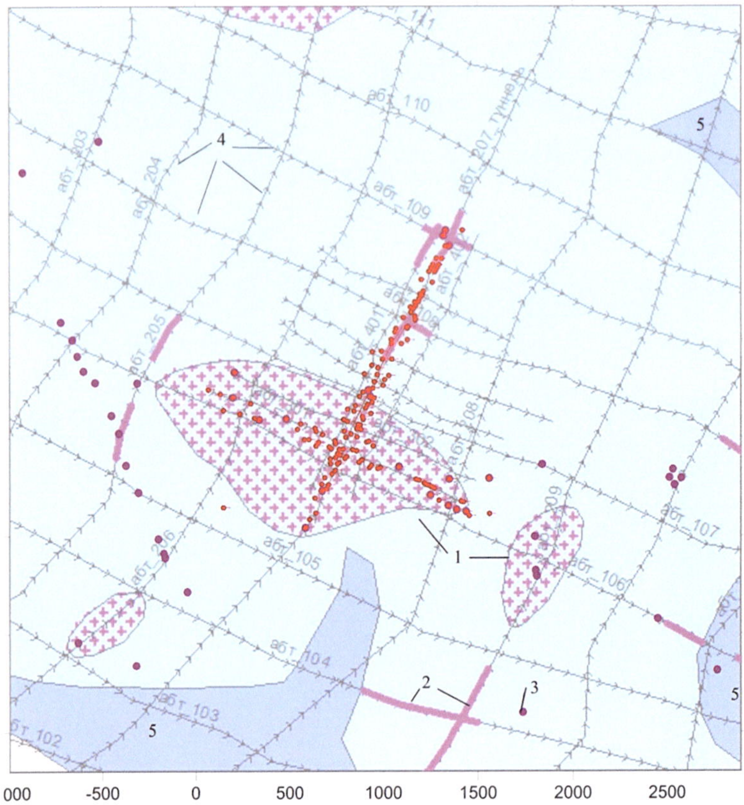

Fig. 2.5 Cross-sectional mesh

2.5 Conclusion

According to calculations, the total inflow into the system of in-circuit drainage for the period of achievement of the hospital is about 2000 m^3/day, provided that 103 water wells equipped in drainage wells are operating. If these wells are used as self-draining (without forced water withdrawal), the groundwater level will decrease substantially, but the drainage norm at the level of the roadway will not be achieved. Nevertheless, a decrease in the pressure gradient in the enclosing structure can lead to a significant reduction in water inflows into the tunnel.

Prior to the beginning of works on dewatering, the organization of monitoring of earth surface subsidence and deformations of buildings must be envisaged. At its carrying out, the basic attention should be given to buildings and constructions which not only fall into the zone of influence or directly border with it but also stand a few meters from it.

Chapter 3
Probabilistic Approach to Forecasting of the Risk Caused by Groundwater Contamination

Irina Galitskaya, Irina Pozdnyakova, Gleb Batrak, Leonid Toms, and Alekcandr Palagushin

Abstract In this chapter, a probabilistic approach to forecasting of the risk caused by groundwater contamination is considered. The general conceptual model submitted for consideration includes two parts: (1) a model of the impact of contaminated groundwater resulting from urban sources; and (2) a model of the impact of contaminated groundwater on the recipients. Examples of the probabilistic approach to forecasting of the risk of groundwater contamination are considered in the territories of the leaching fields of domestic wastewater, the gas station and radioactive waste repository. Stochastic modeling was used for risk forecasting. The probability of water intake contamination, damage to the population from contamination of drinking water, and the value of the risk were calculated.

3.1 Introduction

The safeguarding of security has always been one of the major problems of individuals, society, and the state, but this problem has become particularly topical since the second half of the twentieth century. The development of civilization has led to increases in natural and technonatural hazards, disruption of the ecological balance, and environmental degradation. Significant environmental degradation occurs not only at local and regional levels but also at the global level. It is especially important for the populations of cities (especially megalopolises) to ensure environmental safety from threats caused by environmental contamination. Analysis, evaluation, and forecasting of anthropogenic environmental changes and the hazards and risks associated with them are important elements of sustainable development of urban territories, determining their optimal use and population safety.

I. Galitskaya (✉) · I. Pozdnyakova · G. Batrak · L. Toms
IEG RAS, Moscow, Russia

A. Palagushin
Gidrospetsgeologiya, Moscow, Russia

V. Svalova (ed.), *Natural Hazards and Risk Research in Russia*,
Innovation and Discovery in Russian Science and Engineering,
https://doi.org/10.1007/978-3-319-91833-4_3

In this chapter, a probabilistic approach to forecasting of hydrogeochemical risk (risk caused by groundwater contamination) is considered. Hydrogeochemical risk includes the following factors: (1) the probability of hydrogeochemical hazard formation and its realization; and (2) the damage caused by hydrogeochemical hazard realization. Hydrogeochemical risk assessment should take into account both the risk of groundwater contamination (the so-called risk of exposure) and risk of contaminated groundwater impact (the risk after exposure).

This study presents a generalized conceptual model of hydrogeochemical hazard and risk occurring in a social and natural–technical system, developed on the basis of scientific synthesis of various theoretical models and long-term research. This conceptual model allows us to represent the main stages, the factors and processes of hydrogeochemical hazard and risk formation, and interconnection in the system "DM (decision maker)–technogenic sources–groundwater–human." The developed conceptual representations are the basis of the methodology of forecasting and management of hydrogeochemical risk in urban areas.

Forecasting of risk is performed at a level of complexity that depends on the projection stage, intensity of potential technogenic source impacts, importance of objects under consideration, receptivity of groundwater and objects to external influence, degree of uncertainty, and expected value of risk. According to this, the method of hydrogeochemical risk forecasting is selected. Different methods are used for probability estimation: from a method of subjective estimation of probability, method of analogy and expert methods to complex statistical procedures and numerical stochastic modeling. In the represented example, risk forecasting was evaluated with use of numerical simulation.

3.2 Methodological Principles

The Main Terms

On the basis of the scientific publications of Rowe [8], Petak and Atkinson [6], Mark and Stuart-Alexander [5], Udias [10], Ayala [1], and others, a generally accepted interpretation of risk has ultimately been formed for natural and anthropogenic hazards as a probabilistic measure of losses, which can be defined by multiplication of the probability of the adverse event (P) by the damage value (L):

$$R = P \cdot L \tag{3.1}$$

The necessity of "hazard" and "risk" differentiation is noted. The complexities of terminological base creation are connected with the specifics of the object of research and the research technique, different genesis of hazards and risk (urban, natural, social, and others), and conversion from worldly and ordinary notions to scientific notions.

For the determination terms "hydrogeochemical hazard" and "hydrogeochemical risk" we have assumed as a basis conventional notions of hazard and risk, founded on creation experience of the terminological base, which has been universally recognized and has satisfied the requirements of concrete fields of knowledge [4].

A hydrogeochemical hazard is a hydrogeochemical state of groundwater (a hydrogeochemical situation) posing a threat to people and/or the environment. We consider "hazard" as an objective reality (state, material property, situation, and other), existing in respect of society and nature. A hydrogeochemical hazard can be conditioned as both low and high chemical substance concentrations in comparison with hygienic standards. A hydrogeochemical hazard can be both natural and urban. We shall consider only coupled urbanonatural hydrogeochemical hazard and risk.

Hydrogeochemical risk is the hydrogeochemical hazard measure. It includes the following quantitative factors:

1. Probability of hydrogeochemical hazard formation and its realization
2. Damage caused by hydrogeochemical hazard realization
3. Uncertainties in the damage and probability values

It is important to emphasize that hydrogeochemical risk is considered as an integral factor. Hydrogeochemical risk assessment should take into account both the risk of groundwater contamination and the risk of impact of polluted groundwater on population and infrastructure objects.

Structure of Hydrogeochemical Hazard and Risk

Hydrogeochemical hazards are characterized by more various sources and factors of formation that greatly complicate the model of hazard and risk. The model presented below was created on the basis of our long-term experience in hydrogeochemical hazard research, and we also took into account the experience of environmental contamination risk assessment [3].

There are the following essential elements in the structure of hazard and risk: the subject of the hazard, the environment, and the object of the hazard (Fig. 3.1). The decision maker is the first-rank hazard subject. The decision maker can be one person or group of persons, working out the collective decision. The decision maker brings about urban sources (second-rank subjects)—in other words, the urban hazard. Urban hazard realization can cause a hazardous hydrogeochemical situation. The first-rank object of a hazard in a city territory is a human. The second-rank objects of the hazard are objects of the infrastructure immediately related to a human. The term "environment" includes various environmental components. In addition to these three components (subject–environment–object), an external person (positioner), estimating an actual or potential hazard, is usually entered. In the process of hydrogeochemical hazard formation and realization, it is possible to change the subjects' and objects' role functions. The first- and second-rank subjects can become the objects of hazardous "secondary" impacts; for instance, a factory where leakages have occurred that is waterlogged by aggressive groundwater, which brings about destruction of the foundation and maybe even collapse of the building.

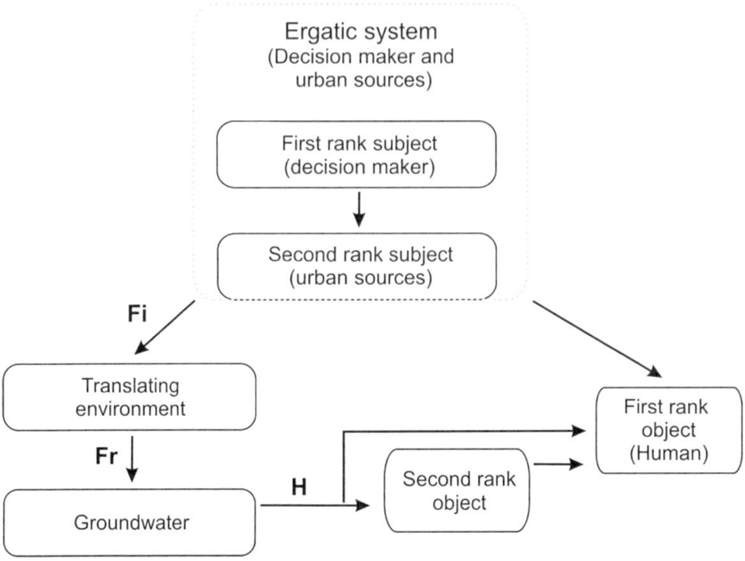

Fig. 3.1 Structure of hazard and risk

For risk analysis, it is reasonable to consider two models. Model 1 is "ergatic system (decision maker and urban sources)–groundwater"; model 2 is "groundwater–human" (Fig. 3.2). In the first model, the probability of hazardous hydrogeochemical situation occurrence is estimated. In the second model, the hydrogeochemical risk to recipients is estimated.

The formation and realization of a hydrogeochemical hazard can be considered as a complex event that forms because of more simple events. The probability of the complex event can be valued on the basis of knowledge of the probability of the simpler events.

Model 1: Ergatic System - Groundwater

The possibility of a hazardous hydrogeochemical situation can appear already at the idea stage. It depends on social factors, economic factors, political reasons, the competence and professionalism of the decision maker, learning about the situation, and the influence of information on the decision maker and the population. Later, at the project stage, the risk can change (increase or decrease)—for instance, in the design, distribution of enterprises, etc. The result of the subject's intellectual and organizational activity at the idea and project stages is an urban source creation at the stage of construction. Thereby, the technogenic impact with specific parameters (intensity, composition, periodicity, etc.) appears at the construction and operation stages. It is important to note that the initial impacts are only potentially dangerous, since their real hazard, in terms of the formation of the hazardous hydrogeochemic situation, appears only in the process of interaction with the environment.

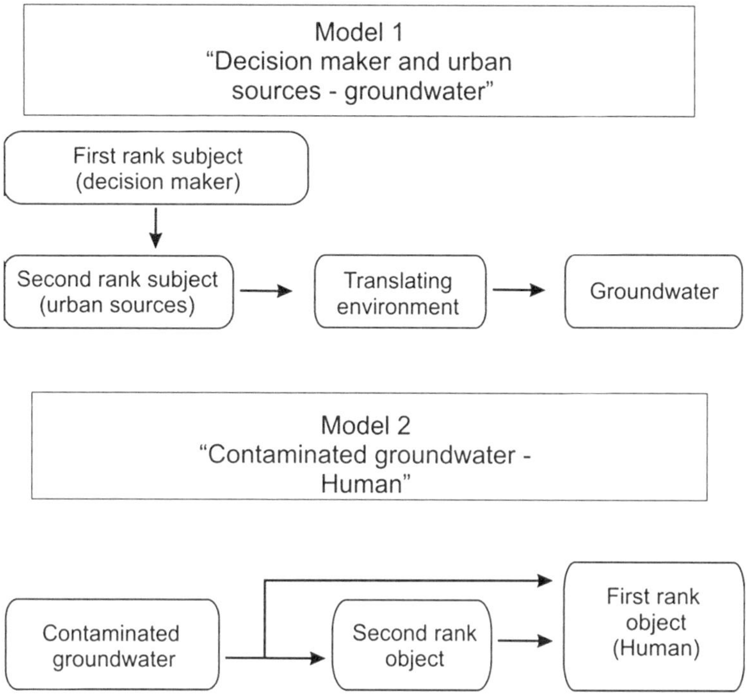

Fig. 3.2 Scheme of hydrogeochemical hazard formation (model 1) and realization (model 2)

Urban sources are characterized by different types of impact, the time of the action, the composition and concentration of contaminants, and so on. The formation of the hydrogeochemical hazard can be conditioned by different types of source impact (direct (chemical) and indirect (physical, biological, and others)), different characters of impact (normal or emergency), and different source actions (deterministic or probabilistic). Two types of urban source stand out in the nature of the action: (1) the operation of the source determines the impact on groundwater (for example, water inlets and drain holes); and (2) the impact of the source is conditioned by different failures. The first type of impact is deterministic and the second is probabilistic (defined by the probability of failure). In practice, for risk analysis, the frequency of the hazardous event is often used for urban sources. However, for probabilistic estimation, the occurrence of the hazardous impact is considered as a Markov process, characterized by the simplest flow of the appearance of the event.

Result Impact Formation (External Factor) The resulting impact's intensity and nature are essential factors of the risk of formation of a hazardous hydrogeochemical situation. It is important to assess the probability of the resultant impact's formation above the maximum allowed value, i.e., capability to bring about hazardous hydrogeochemical situation occurrence. The parameters of the resultant impact depend on the type and intensities of the source impact and conditions on the way to its translations. For instance, in the case of a chemical influence the resulting influence parameters depend on the initial amount of the contaminants, chemical properties of elements and their compounds, forms of migration, redox potential, acid–base balance, water–rock interactions, and microbiological factors on the way to translation. It is possible to select two types of situation: the hazard of the resulting influences either increases or decreases in comparison with the initial influence. An example of the first type of situation is an increase of contaminant concentrations in drinking water in the event of leakage from water supply pipes and filtration through a polluted unsaturated zone; an example of the second type of situation is a reduction of sewage contamination as a result of sorption of polluting substances during their filtration through soils and rocks, as well as their degradation or deposition. In many cases, the resulting impact is defined as a contaminant concentration, arriving from the unsaturated zone. It is necessary to define (1) the probability that the contaminant concentration arriving in the water table will exceed the criterion level; and (2) the probability that the time of arrival in the water table will be less than the criterion time.

Groundwater Stability (Internal Factor) The resulting hazardous impact on groundwater is not a sufficient condition for hydrogeochemical hazard formation, because groundwater can be stable with external impacts. The probability of the occurrence of the hazardous situation depends on its internal characteristics and the abilities to save, restore, and/or adjust its composition and characteristics, and/or execute certain functions (for instance, as drinking water). The probability of formation of the hazard is valued in the failure fixation area (point), which depends on the purposes of the study. In most cases, it is the area where the impact of groundwater on humans and infrastructure objects is valued. In space it may coincide with the area of the contaminant arrival into the groundwater or be at a certain distance. When contaminant transport significantly changes the degree of the hazard, issues of dispersion and physicochemical interactions in the water–rock system are possible. Groundwater stability depends on its receptivity, i.e., possibility of reacting negatively to the impact of the factor, and the state, characterized by remoteness from the critical point (for instance, the maximum allowable concentration). Even small impacts on groundwater close to the threshold of stability can bring about negative consequences. Groundwater stability is defined by specific contaminants; it can also be greatly changed over time. It is conditioned by hydrospheric change (for instance, rock sorptive capacity exhaustion and approximation of the critical point). A stability assessment can be made with use of the time value, during which the "groundwater–rock" system retains the ability to resist the external influence, or with use of the loss value (for instance, deterioration of water quality). The particularity of hydrogeochemical hazard formation is its "latent" nature.

Hydrogeochemical hazard formation can result from synergetic effects. In addition, essential factors of hazard formation are simultaneous impacts on the hydrosphere with absolutely different loads (for instance, hydrodynamic, mechanical, heat, or chemical), though separately each of them cannot bring about a hazard situation.Thus, in the first model the probability of occurrence of a hazardous hydrogeochemical situation is determined by multiplication of the probabilities of several occurrences:

$$P(\mathrm{H}) = P(\mathrm{F}) \cdot P(\mathrm{S}|\mathrm{F}) \cdot P(\mathrm{Fr}|\mathrm{F}\&\mathrm{S}) \cdot P(\mathrm{H}|\mathrm{Fr}\&\mathrm{F}\&\mathrm{S}) \tag{3.2}$$

where *P*(F) is the probability of the source impact F on groundwater during the given period; *P*(S | F) is the conditional probability of the watershed location in a zone of the source impact F; *P*(Fr | F & S) is the conditional probability of the resultant cumulative impact formation above the maximum allowed value Fr in the case of the source impact F and the corresponding watershed location; and *P*(H | Fr & F & S) is the conditional probability of event H, which is a result of groundwater instability in the case of the resultant cumulative impact Fr.

If there are a few types of possible impact, in such a case the probability *P*(Hi) of the groundwater defeat is calculated for each Fi and the resulting probability is calculated by the formula of total probability, taking into account the possible dependence of these impacts. Two cases of geochemical hazard formation are presented in Fig. 3.3.

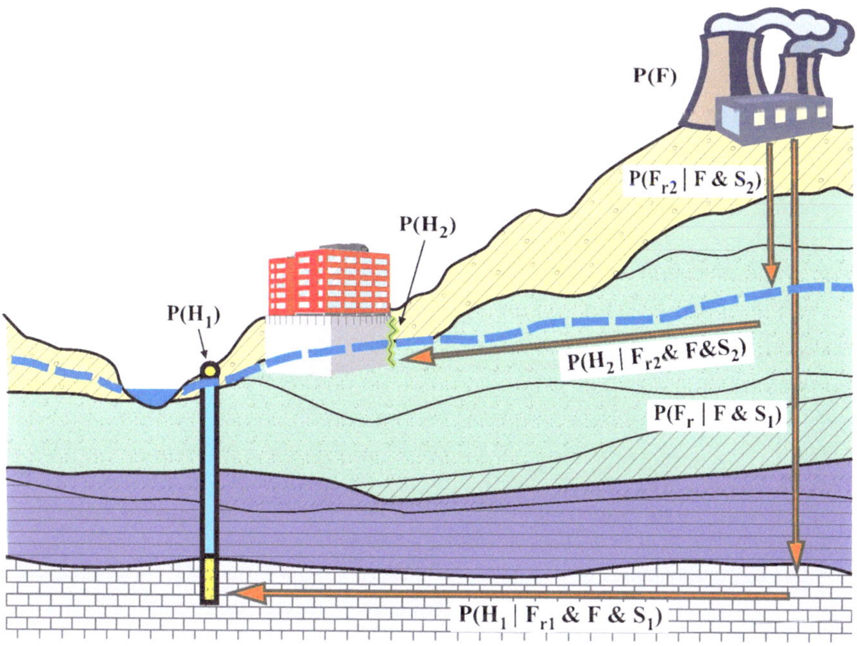

Fig. 3.3 Scheme of hydrogeochemical hazard formation

Model 2: Contaminated Groundwater - Human

The hydrogeochemical risk to recipients is estimated in the second model. Here the impact of contaminated groundwater on humans (the first-rank object) and objects of the material sphere (second-rank objects) is considered. The object choice depends on the purpose of the investigation. Hydrogeochemical hazard realization is the cause of social and economic losses. The social losses from the contamination are determined as worsening health of urban dwellers from the direct impact of the contaminants. The economic losses in this case are determined as the probability of an indirect effect on humans, i.e., an aggressive groundwater impact on basements or other underground engineering structures in the waterlogged areas, leading to their destruction. As mentioned above with regard to model 1, the decision maker "organizes" the risk of hydrogeochemical hazard formation, creating urban sources. In model 2 he also "organizes" the risk of hydrogeochemical hazard realization, placing the buildings in a zone of aggressive groundwater influence, allowing use of insufficient water treatment, and so on.

The hydrogeochemical hazard is characterized by multiple furcated successions (chains) of interactions in technical and natural systems. For instance, leakages from the water supply system or sewers can cause waterlogging and change the chemical composition of groundwater. In turn, these events can increase groundwater aggressivity, cause chemical and mineralogical rock transformations, change filtration characteristics, and cause deformation behavior changes. This generates new breakings of the water supply system or sewer pipes, additional leakages, and foundation deformations, which provoke other processes—manifestations that often turn out to be unexpected. The measures that are taken can break the dangerous synergetic chains but herewith generate new ones.

For model 2 the probability of damage to the object depends on the probability of (1) a hazardous hydrogeochemical state; (2) contact of the recipients under consideration with contaminated groundwater; and (3) the occurrence of vulnerability of the recipients. Contact of the object with contaminated groundwater can be deterministic (location of a building in the territory, waterlogging by aggressive groundwater), and probabilistic (casual use of water from a pit).Overall, taking into account the factors of the risk in model 2, the integral hydrogeochemical risk is expressed by the following dependency:

$$R = P(H) \cdot P(S^*|H) \cdot P(L|H\&S^*) \cdot L \tag{3.3}$$

where P(H) is the probability of a hazardous hydrogeochemical state (refer to formula (3.2)); P(S* | H) is the conditional probability of the contact S*of the recipient with groundwater; and P(L | H & S*) is the conditional probability of damage L in the case of contact of the recipient with contaminated groundwater.

Forecasting of risk is performed at the level of complexity, which depends on the projection stage, intensity of potential technogenic source impacts, importance of objects under consideration, receptivity of groundwater and objects to external influence, degree of uncertainty, and expected value of risk. According to this, the method of hydrogeochemical risk forecasting is selected. Different methods are used for probability estimation: from a method of a subjective estimate of probability, method of analogy and expert methods to complex statistical procedures and numerical stochastic modeling.

At the initial stage (preinvestment) in connection with information deficit, an analog method and expert methods are used. At the stage of investment feasibility, an analytical method using simplified statistical distributions, such as triangular distribution, is recommended. Usage of these methods is based on the following proposition: the probability of a "final" undesirable event can be valued on the basis of the probability of simpler event estimation. The hydrogeochemical hazard formation and realization can be considered as a complex event that forms because of more simple events. The ultimate probability of the complex event is defined as the product of the probability of a primary (triggering) event multiplied by the conditional probabilities of all other events, provided that the probability of each consequent event is calculated on the assumption that all other events have already happened. In this case, the calculation is done by formula (3.2).

At the stage of project risk, the assessment procedure is complicated. Analytical solutions cannot give the correct result in the presence of irregularities in the environmental system; therefore, it is (quite) reasonable to use numerical stochastic modeling when probabilistic blocks are included in the model, which describes the deterministic geofiltration and geomigration processes. Previously an analysis of all components of hydrogeochemical risk and marking of probabilistic and deterministic events will have been carried out.

When assessing the probability of groundwater contamination, first of all it is necessary to analyze this complex event on the basis of consideration of simpler events. It should be ascertained what events can be considered as probabilistic and what events can be considered as deterministic. When we use analytical methods, the values of the simpler event probabilities can be assessed separately and then multiplied; an example of a rather similar approach was presented by Rosen and LeGrand [7]. When numerical simulation is used, probabilistic blocks are included in the model, which deterministically describes filtration and migration processes.

The examples of a probabilistic approach for forecasting of groundwater contamination risk are considered in the territories of the leaching fields of domestic wastewater, the gas station and the radioactive waste repository. Stochastic modeling is used for risk forecasting.

3.3 Simulation of Contaminant Transport with Risk of Groundwater Contamination Under Leaching Fields Via Hydrogeological Windows

These leaching fields of domestic wastewater, constructed in 1892–1898, are located in the immediate vicinity of the Moscow River floodplain and were used initially for treatment of domestic and then industrial sewage until 1985. Sewage water and sludge contain great quantities of organic matter, ammonium and nitrate–nitrogen, heavy metals, and oil products. After filtration throughout the unsaturated zone below the field, part of the wastewater has been collected by the drainage network and dumped into the river, but another part of it has reached the groundwater.

More than 200 wells were drilled for investigation of the geology, engineering geology, and hydrogeology conditions of the area. A study of the geology, engineering geology, and hydrogeology of this site showed that the hydrogeological conditions of the studied area are represented by an upper sandy alluvium aquifer overlying an aquitard composed of Jurassic clays, which overly the lower aquifer in the Carboniferous limestone. Water intake wel located on the other side of the river and a study of groundwater contamination in the well-capture zone is very important.

Recent detailed investigation showed that Jurassic sediment composition varies significantly in the study area from heavy clay with 20–30 m of thickness to sandy clay and loam with less than 1 m of thickness. Moreover, areas were found where Jurassic sediments are composed of sands (Fig. 3.4). We called these sites of direct interaction of upper and lower aquifers hydrogeological windows in the aquitard. So the hydrogeological windows are sites of significantly reduced thickness of clay, or even of clay absence.

The hydrogeological conditions of the site have changed also as a result of leakages of wastewater from leaching fields. The upper aquifer groundwater level depth has decreased to 1–2 m within the river floodplain and to 6–7 m at the alluvial terraces. The soils and groundwater have been contaminated by organic matter, heavy metals, oil, and nitrogen. As a result of groundwater pumping, the lower

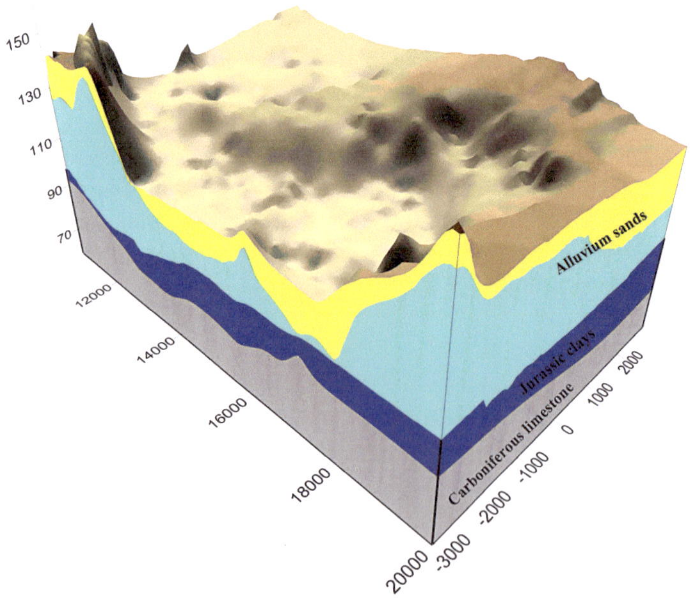

Fig. 3.4 Model of the geological structure of the study area

aquifer groundwater levels have been significantly lowered (from 25 to 75 m), causing downward flow of contaminated groundwater from the upper aquifer through the aquitard into the lower aquifer.

When estimating the probability of groundwater contamination under leaching fields in Moscow we concluded, on the basis of the situation analysis, that the deterministic events are the impact of the source (the contaminated unsaturated zone) on groundwater *P*(Fi) and the object location in a zone of the source's impact *P*(S/Fi) because the well-capture zone (the object) is located in the way of contaminant migration. The probabilistic event is the resultant cumulative impact formation *P*(Fr/Fi & S) because the contaminant concentration entering the aquifer under investigation depends on the number, location, and size of the hydrogeological windows (as considered below).

To estimate the impact of leaching fields on the target aquifer contamination and to analyze the possibility of dispersion of contamination in the target aquifer due to hydrogeological windows, three-dimensional (3D) geological and hydrogeological models of the area were developed. The geological model of the study area was created using a special program (GSI3D version 1.5) for 3D geological mapping and structure modeling. We used about 1500 geological wells and created 56 cross-sections, which allowed us to create geological layers and find hydrogeological windows in the aquitard. The hydrogeological windows in the aquitard found in the studied area, using 3D geological mapping of the area, are shown in Fig. 3.5.

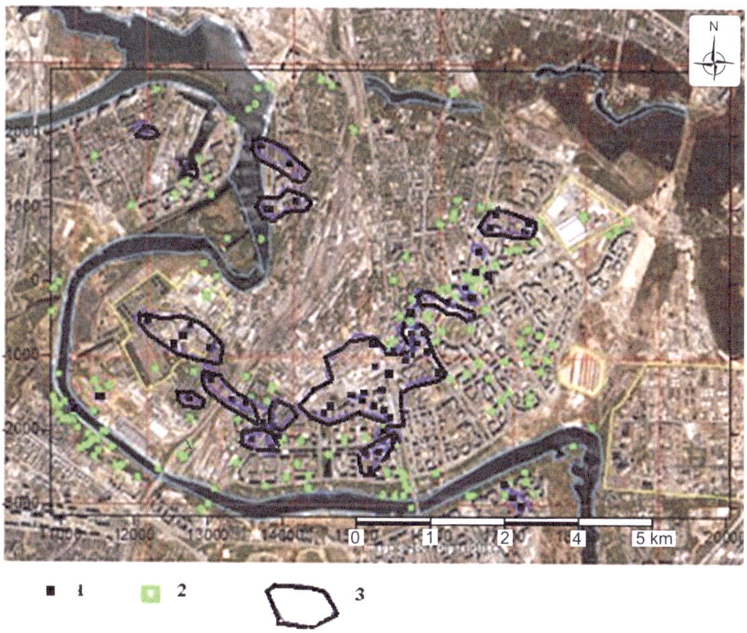

Fig. 3.5 Hydrogeological windows in the aquitard. 1 Well with a hydrogeological window within the aquitard. 2 Well without a hydrogeological window within the aquitard. 3 Possible boundaries of windows obtained by kriging

One can see that there are many hydrogeological windows under the disposal fields. The hydrogeological window size estimated by interpolation varies from 0.02 to 0.30 km^2 and the largest window is in the center of the leaching fields.

The hydrogeological model of the study area was based on the geological model. The steady-state hydrogeological model included three layers: the upper aquifer in the alluvial sands, the aquitard in the Jurassic clays, and the lower aquifer in the Carboniferous limestone. The model area size was about 40 km^2 and the model thickness was about 100 m. The model domain was bounded by the Moscow River to the south, several ponds to the north, and the hydraulic watershed to the east.

Head-dependent boundary conditions were specified along the Moscow River, and ponds and no-flow conditions were defined along the watershed. The model had uniform grid. Each cell size was 100 m (x) $\times$ 100 m (y). For numerical 3D groundwater flow modeling, Processing Modflow Pro version 7.0.26 was used [2].

The calibration of the lower aquifer for the heads was 1 m of the maximum difference between simulated and observed heads. First, we calibrated the model with one averaged value of the hydraulic conductivity for the aquitard or the so-called model without windows in the aquitard to simulate the situation when we had no information about hydrogeological windows in the aquitard. The next goal of development of this unrealistic model without windows was to clearly estimate the importance of hydrogeological windows by comparing the averaged results of simulation of the model with windows and the model without windows.

Therefore, we developed the model by using a stochastic conditional simulation of hydrogeological windows in the aquitard. It was assumed that hydrogeological windows could be a fast pathway for contaminant transport from the polluted upper aquifer (the contamination source) to the target aquifer with freshwater, and these windows could be simulated by using the stochastic approach.

For both models we performed simulation of nitrate - nitrogen transport by using MODFLOW/MT3DMS codes [2]. In both models, the nitrate - nitrogen concentration was specified with an additional infiltration rate under the disposal fields as a dimensionless value equal to 1. For the model with windows, we performed a probabilistic simulation of contaminant transport with an estimate of risk groundwater contamination via hydrogeological windows.

The applied approach of the estimate of the groundwater contamination probability was based on a stochastic simulation of hydrogeological windows in the aquitard. The model of the indicator simulation was used for stochastic simulation of hydrogeological windows in the aquitard. We applied the indicator function $I(x)$:

$$I(x) = \begin{cases} 1 \text{ if window exists in } x \text{ location} \\ 0 \text{ otherwise} \end{cases}$$

One can see that this function has a value of 1 if a hydrogeological window exists in x location and 0 if hydrogeological windows are absent.

Statistical properties of $I(x)$ were calculated using well data (Fig. 3.6). The expected mean and variogram of the hydrogeological window indicator function were calculated using data from 254 wells drilled within the studied area. The

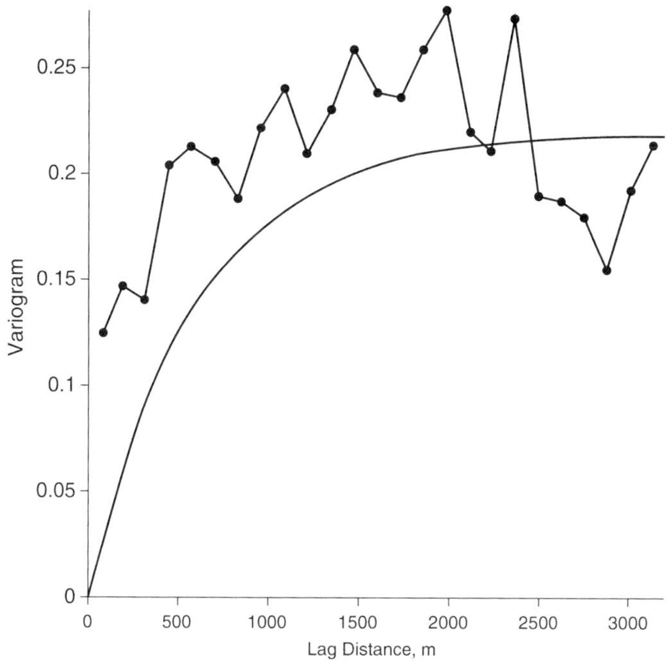

Fig. 3.6 Indicator variogram and its fitting by a theoretical exponential model without a nugget effect

calculated expected mean $E\{I(x)\}$, i.e., the probability of windows, was 0.3 and a spatial correlation of the indicator function in the form of an indicator variogram was identified. The experimental variogram had an isotropic structure. Despite the visibility of a nugget effect, this variogram was fitted by the exponential function with a range of about 500 m.

An approach applied for modeling of risk of the lower aquifer contamination included the following steps:

Stochastic conditional simulation of 50 realizations of aquitard window locations, using SISIM code in the GSLIB library [3]

Transfer of the results of each simulation into the hydraulic conductivity field, using the rule:

$$I(x) = \begin{cases} 1 \text{ then } k(x) = K_{\text{sand}} \\ 0 \text{ then } k(x) = K_{\text{clay}} \end{cases}$$

Running of MODFLOW/MT3DMS codes for each conductivity field and acquisition of concentration fields [2]

Probability analysis of simulated concentration fields

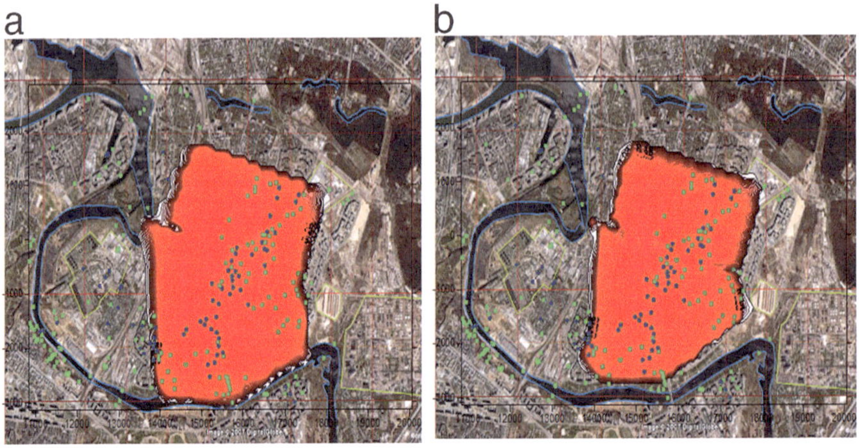

Fig. 3.7 Simulation results: distribution of contamination above the aquitard. (**a**) "No windows" case. (**b**) Expected mean concentration field with windows

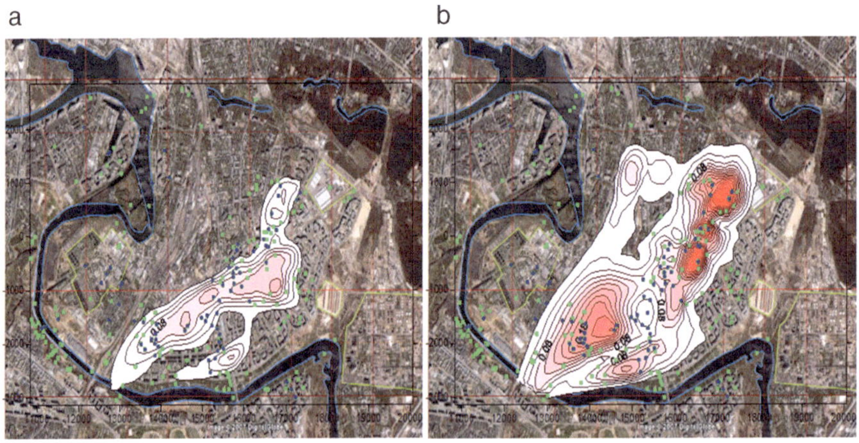

Fig. 3.8 Simulation results: concentration field of nitrate–nitrogen in the target aquifer. (**a**) "No windows" case. (**b**) Expected mean concentration field with windows

The simulated concentration field in the upper aquifer obtained by the deterministic model without windows and the expected mean concentration field with windows obtained by the stochastic model were very similar (Fig. 3.7). Comparison of simulated concentrations in the target aquifer (Fig. 3.8) showed that the maximum simulated concentrations at the end of the simulation period (100 years) in both models are greater than the allowed dimensionless nitrate–nitrogen concentration (0.04), which corresponds to 45 mg/L; the expected mean concentrations in the target aquifer (aquitard with windows) are greater than the concentrations obtained by the deterministic model in the variant without windows; the contaminated area with the expected mean concentration of nitrate–nitrogen in the target aquifer exceeding 0.04 is greater than that in the deterministic model.

The probability analysis of simulated concentration fields obtained by stochastic simulation included analysis of the distribution of the contaminated area in the target aquifer, the distribution of the total mass of contamination in the target aquifer, and the width of the contaminated zone at the boundary or the probability of contamination reaching the water well–capture zone.

The results of probability analysis are shown; in particular, the probability of contamination groundwater formation in the water well–capture zone is about 75–80%. The damage value depends chiefly on the cost of treatment facilities. By multiplying the probability of groundwater contamination by this cost, we could estimate the risk in monetary terms.

Thus, the stochastic simulation shows that hydrogeological windows in the leaky aquitard strongly affect contaminant transport in the target aquifer, including contamination of the water well–capture zone. A stochastic model of the hydrogeological windows' spatial distribution conditioned on the well data was developed. This model allowed estimation of the effect of data uncertainty on forecasting of contaminant spreading in the target aquifer.

3.4 Probabilistic Approach to Assessment of Risk Caused by Groundwater Contamination by Oil Products

The study area is located in the valley of the Moscow River (Fig. 3.9). Two different types of rock—Precambrian crystalline basement rocks, and Paleozoic, Mesozoic, and Cenozoic sedimentary cover rocks—are found in the geological structure. The geological section depth, in accordance with the tasks to be solved, was limited to Upper Carboniferous rocks. The investigated part of the geological section is represented by clays and marls of Voskresenskaya subsuite (C_3vs), limestones of Ratmirovskaya subsuite (C_3rt), clays with interlayers of marl of Neverovskaya subsuite (C_3nv), limestones of Perkurovskaya subsuite (C_3pr), unconformable overlaying Callovian–Oxford clays ($J_{2\text{-}3}k\text{-}o$), and fluvioglacial and alluvial sands and sandy loams, as well as fill-up soils. The geological history of the study area includes a number of important features, which are necessary to take into account when creating a geological model—the basis of geofiltration and geomigration models. Erosion and glaciation often occurred in the Pleistocene epoch, which significantly influenced Jurassic sediment preservation. Within the ancient erosion valleys developed in the territory, Jurassic sediments are absent or eroded (the sites of hydrogeological windows) (Fig. 3.10). It was assumed that hydrogeological windows could be a fast pathway for contaminant transport from the polluted upper aquifer to the lower aquifer with freshwater. The main aquifers in the study area are as follows: the "above-Jurassic" aquifer in the alluvial and fluvioglacial sand and sandy loam, and the Kasimovskii aquifer in the Upper Carboniferous limestone that lies below. The separation layer is represented by Callovian–Oxford clays.

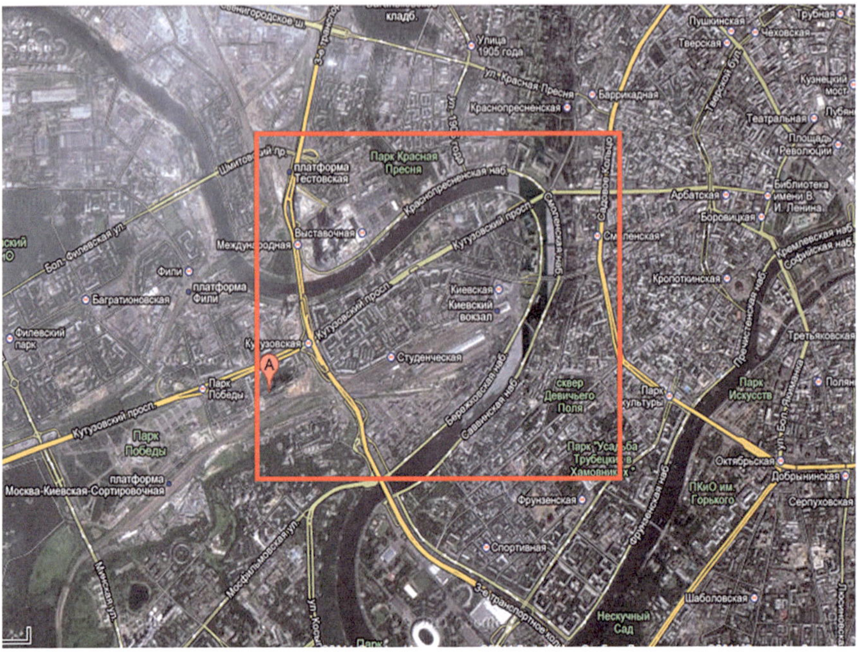

Fig. 3.9 Study area

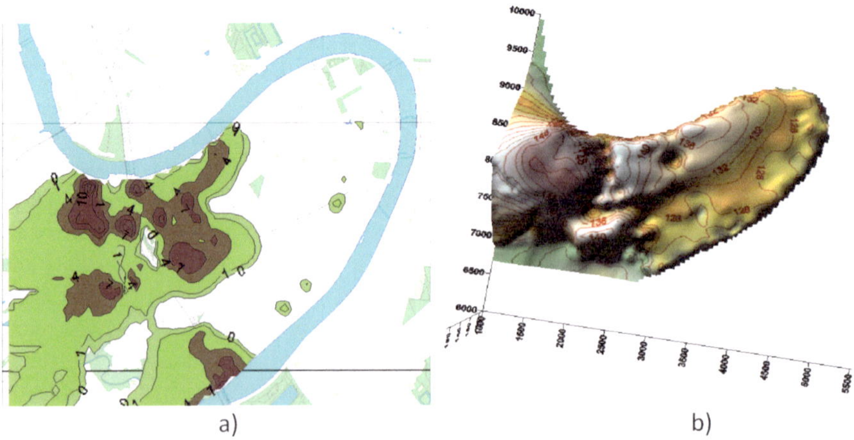

Fig. 3.10 (**a**) Scheme of the Jurassic deposit spread (green and brown areas). (**b**) Study area relief

The main task was assessment of the probability of oil product arrival in water intake wells and monetary risk. The general conceptual model for hydrogeochemical hazard and risk included the following concepts: formation of a hydrogeochemical hazard (model 1), realization of the hydrogeochemical hazard (model 2), and the type of damage. The concept of Upper Carboniferous aquifer contamination (a hydrogeochemical hazard) consists in the fact that as a result of filtration, oil products arrive in the "above-Jurassic" aquifer and from it travel through weakly permeable Jurassic clays to the Upper Carboniferous aquifer. Realization of the hydrogeochemical hazard in this case can occur with use of contaminated groundwater, resulting in potential damage to the health of the population. It is assumed that in this case the major cost of drinking groundwater contamination will be determined by the cost of treatment facility construction.

On the basis of the situation analysis we concluded that certain events are the source of contamination's impact on groundwater *P*(Fi), the object's location in a zone of the source's impact *P*(S/Fi) because the well-capture zone (the object) is located in the way of contaminant migration, and groundwater instability occurrence *P*(V/Fr) because physicochemical interaction of the main contaminant (benzene) with water-bearing limestone is negligible, and dispersion (the main process in our case) was considered as a deterministic process. So the only probabilistic event was the resultant impact formation *P*(Fr/Fi & S), because the contaminant concentration entering the aquifer depends on the location of the hydrogeological windows (Figs. 3.11 and 3.12).

The steady-state hydrogeological model included three layers: the upper aquifer in the alluvium and fluvioglacial sand and sandy loam, the aquitard in the Jurassic clays, and the lower aquifer in the Upper Carboniferous limestone. The model domain was bounded by the Moscow River in the west, north, and east; the Setun River in the south-west; and the hydraulic watershed in the east. The model had a uniform grid. Each cell size was 50 m (x) × 5 m (y). For numerical 3D groundwater flow modeling, Processing Modflow Pro version 7.0.26 was used [2]. We performed simulation of benzene transport by using MODFLOW/MT3DMS codes because benzene is one of the most soluble oil components with low sorption ability. In the model the benzene concentration was specified as equal to 50 mg/L.

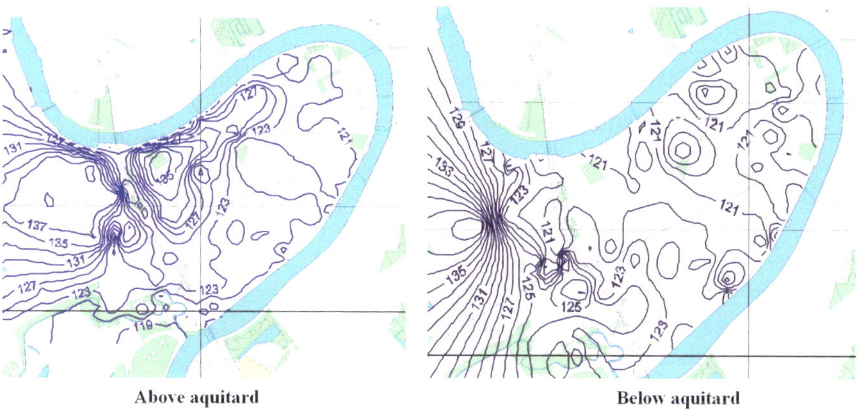

Fig. 3.11 Distribution of heads above and below the aquitard

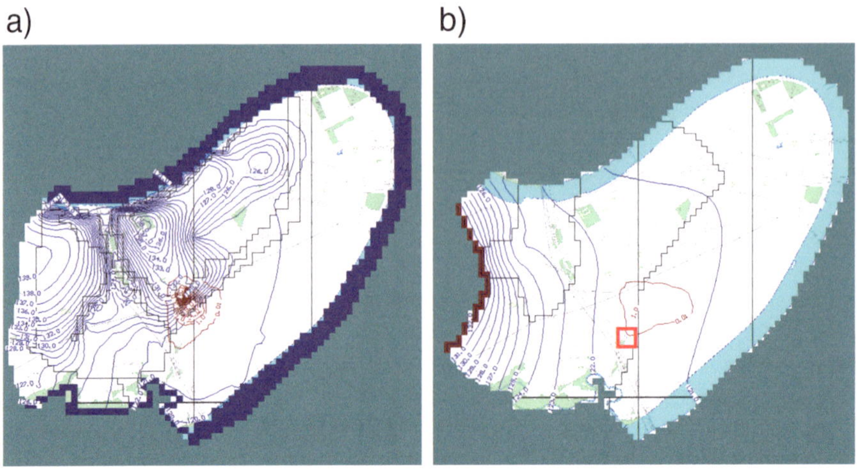

Fig. 3.12 Distribution of benzene (**a**) in the "above-Jurassic" aquifer and (**b**) in the Upper Carboniferous aquifer

The suggested approach applied for modeling of the probability of oil product arrival in water intake wells in the lower aquifer included the following steps:

- Estimation of different variants of Jurassic clay boundary with the use of a random-number generator
- Running of MODFLOW/MT3DMS codes for each variant and acquisition of concentration fields
- Probability analysis of simulated concentration fields

The results of the probability analysis showed that the probability of benzene arrival in the water intake wells in the lower aquifer is about 30%. The value of the monetary risk, with an allowance for the risk to treatment facilities, is 0.3 × \$100,000 = \$3000.

3.5 Probabilistic Approach to Forecasting of Groundwater Radiochemical Contamination Risk

Study Area

The study area is located in the East Ural megazone. Volcanic, magmatic, and sedimentary rock complexes of different ages have developed there. In a significant part of the area, the folded basement is overlain by a cover of Meso-Cenozoic sediments with a thickness of 5–15 m represented by eluvial–deluvial formations of Meso-Cenozoic weathering crust with formations of marine Paleogene and continental sediments of the Quaternary system. Along with plicative structures, disjunctive dislocations of different directions and types are widespread in the area.

The main aquifer complex occurs in the Paleozoic fractured zone and in the lower watered part of the Meso-Cenozoic weathering crust. Groundwater is recharged by direct infiltration of rainfall and snowmelt. The study area is located in the highest part of the terrain, sloping in all directions. This ensures the watershed type of the groundwater regime. The flow is directed in the northwest to the river valley and in the southeast to the marshy floodplain of the river valley. The study area is largely located on the south–southeastern slope of the hill, so the main direction of groundwater flow in the study area is to the south–southwest. This pattern of groundwater flow is controlled by the relief origin and the regional fracture system of the intrusive massif. Groundwater is discharged to the river system.

Methodological Principles

The study area is characterized by the presence of a potential source of groundwater pollution: a repository of medium- and low-level radioactive waste. Strontium 90 is the most hazardous pollutant. The object at risk is the aquifer complex in the Paleozoic water-bearing zone of fractured acidic and medium magmatic intrusions, as it is the main source of drinking water supply for the population and the most poorly protected from pollution.

On the basis of the situation analysis we concluded that certain events are the source of contamination's impact on groundwater $P(\mathrm{Fi})$ and the aquifer recharge area's location in a zone of the source's impact $P(\mathrm{S/Fi})$ because the well-capture zone (the object) is located in the way of contaminant migration. So the main probabilistic event is the resultant impact formation $P(\mathrm{Fr/Fi}\ \&\ \mathrm{S})$ because the strontium 90 migration depends on the parameters in the fault zone and their values are not accurately defined.

To evaluate the contamination risk of groundwater, we used the following approach: first, assessment of the probability of the pollution reaching (which is equal to the level of intervention) the well-capture zone located on the riverbanks; and second, estimation of the value of damage equal to the cost of the treatment works.

The main tasks were the following: (1) assessment of the impact of the fault crossing (in the north-west direction) the area of the radioactive waste repository (Fig. 3.13) on strontium 90 migration; and (2) statistical assessment of the impact of definition accuracy of filtration parameters in the fault zone on the contamination front velocity reaching the water intake near the river.

The techniques of predictive modeling of the impact of the radioactive waste repository on changing of hydrogeological conditions within the study site and the surrounding area included the following stages. The first stage consisted of development of a geofiltration model of the territory. At the second stage, a series of inverse problems were solved in order to clarify the basic parameters of the model (model calibration). The third stage consisted of reproduction in the model of the underground contour of the radioactive waste repository and assessment of its long-term impact on the groundwater.

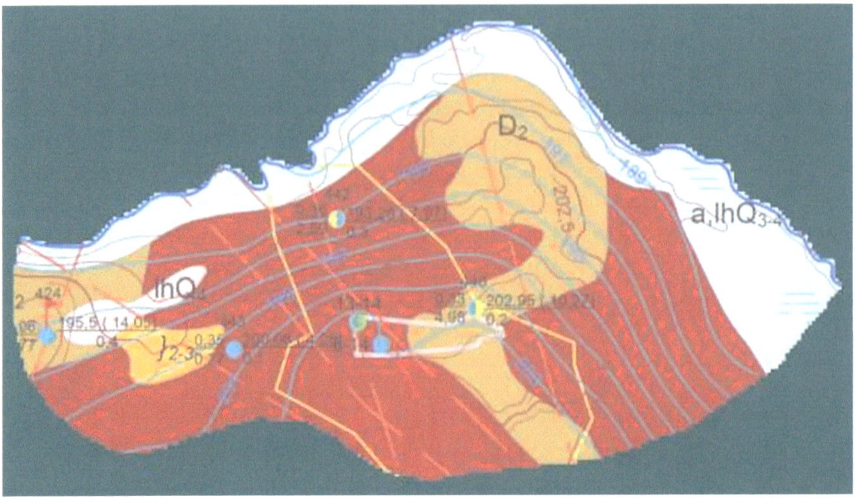

Fig. 3.13 Area of modeling

Geofiltration Schematization

Geofiltration schematization of hydrogeological conditions should take into account the particular tasks solved with the help of the model, i.e., the developed geofiltration model with the further modeling must allow us to completely estimate its potential impact on hydrogeological conditions. For geofiltration schematization, the spatial structure, the filtration medium, the seepage regime, and the boundaries of the groundwater flow were substantiated. Geofiltration schematization was performed by analyzing the hydrogeological conditions, taking into account the specific kind of problem. The hydrogeological features of the study area that must be taken into account were (1) the presence of faults, because the higher filtration coefficients are fixed in adjoining areas according to testing for underground water inflow; (2) the aquifer occurrence in the river bend where the radial flow is spreading; (3) low groundwater levels in combination with the slow flow velocities, which result in strong dependence of the groundwater level depth on the season and the precipitation rate; and (4) the presence of low permeable alluvial deposits in the river bed. Besides, during modeling of changes in radioactive isotope concentrations in groundwater because of leakage, the filtration heterogeneity of water-bearing sediments and the stationary filtration regime must be reproduced in the model.

When substantiating the spatial structure in the geofiltration scheme, one layer was distinguished, including the aquifer, represented from the top to the bottom by sandy loamy layers of permeable alluvial sediments interbedded with loam of variable thickness, and weathering crust sediments, which overlie the fractured intrusive rocks. The groundwater stream in the sandy layers and the fractured rock has a free water table. At a depth of 30–35 m, the fracturing attenuates significantly;

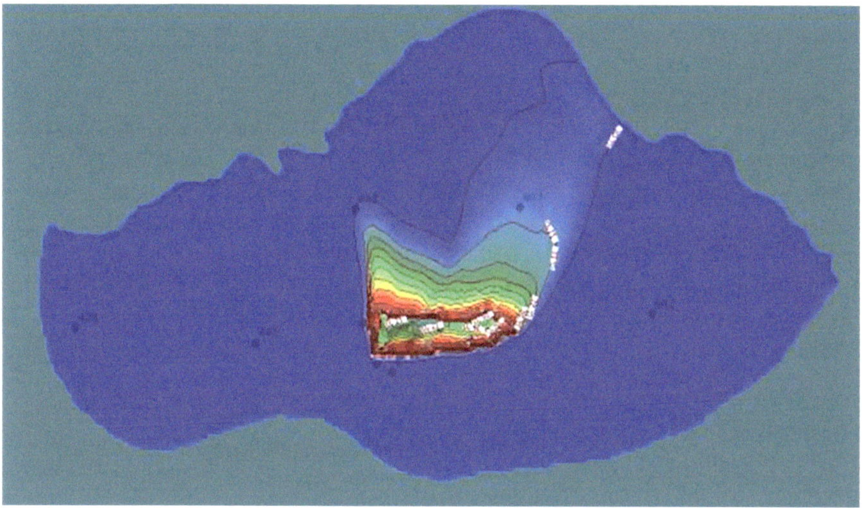

Fig. 3.14 Contamination front reaching the capture zone near the river within 30 years after the start of the migration (in the model without the fault zone)

in terms of the problem under solution, the groundwater flow can be ignored from this depth.

With regard to the hydrodynamic structure of the groundwater flow, the geofiltrational flow was modeled as planned. Its structure was not complicated by concentric local deformations, indicating the presence of inner zones of recharge or overflow. The groundwater flow within the territories adjacent to the study area was characrerized by the tongue-shaped spreading to the north.

The calculation area dimensions were chosen so the model could estimate the radioactive waste repository's impact on changes in the hydrogeological conditions within the territory adjacent to the study area. Given the complex nature of the natural and man-made hydrogeological conditions, the simulation area dimensions in the plan were 4000 m from north to south and 5000 m from east to west (Fig. 3.14). The calculation area in the plan was divided into blocks with a uniform step. A computational grid of 20×20 m steps was used for the whole territory.

Boundary Conditions The filtration area in the plan was bounded in the south, east, and west along the current lines of the natural flow, including the boundary of the groundwater watershed. In the north, the groundwater flow was limited by the river. Thus, there was no need to model the natural flow incoming rate in the filtration area boundaries in the plan. The entire volume of water in the model was formed by the infiltration, which significantly reduced the uncertainty in setting the layer filtration properties.

Parameters of the Model The values of the filtration coefficients were taken on the basis of research. The filtration coefficients range within 0.5–22.6 m/day, and the coefficients of water conductivity vary from 5.9 to 231.5 m^2/day. Elevated water conductivity was observed in the wells located in the immediate vicinity of the fault axis.

Infiltration intensity values in the first approximation were borrowed from publications. The recharge goes into the entire area being affected by the infiltration of precipitation and industrial leakage from water-bearing pipelines. The preliminary value of the infiltration recharge in the modeling area was taken as 68 mm. Discharge takes place to the river system. Upon the modeling, the filtration coefficients and infiltration recharge value were specified at the stage of the inverse problem solution.

The MODFLOW program developed by the US Geological Survey was used for numerical geofiltration calculations. To work with the MODFLOW program, Visual Modflow version 4.6 software (Waterloo Hydrogeologic) was used for input, training, and output of the results.

To calibrate the model for the groundwater level, 15 wells were used, including eight wells in the study area and the rest outside it. Incomplete compliance of the level values obtained in the course of drilling to the model level values is not a drawback of the model, for several reasons. Wells were drilled in different years and different seasons when the groundwater levels were at different depths because of natural fluctuations in precipitation and temperature. In some cases, the accuracy of levels measured while drilling appears doubtful. In general, for a sufficiently large area with significant drops in groundwater levels and with the significant archive data involved, the model and field level convergence is very satisfactory. Obtained in the course of solving inverse problems, the model reflected the current geofiltration conditions satisfactorily. Therefore, the filtration coefficients of fractured intrusive rocks were 0.1–0.5 m/day. The filtration coefficients of weathered crust were 7–10 m/day, and the alluvium filtration properties varied widely. The infiltration recharge was 75 mm/year over the entire filtration area. The resistance of fluvial deposits was $10E^{-2}$–$10E^{-3}$ m/day.

Results and Discussion

The designed geofiltration model adequately reflected the existing natural and man-made conditions. This was primarily due to the fact that the quantity of uncertainties was minimized in the model. Thus, the current flow lines and the river were taken as the boundaries. Accordingly, the infiltration, with the mean value estimated at 15% of the normal rainfall, or 75 mm/year, was the only recharge item of balance. The river was the only discharge area, because the intrusive massif fracturing decreased with the depth, preventing overflow into the underlying aquifers. Numerous lithological varieties in the study area—such as intrusive rocks, weathering crust, alluvial deposits, and fault zone rocks—were tested during the filtration works. There is no information about the resistance of the river bed, but with such a small set of uncertainties, it could be obtained by calibration. As a result, a model distribution level to a high degree was consistent with the observation

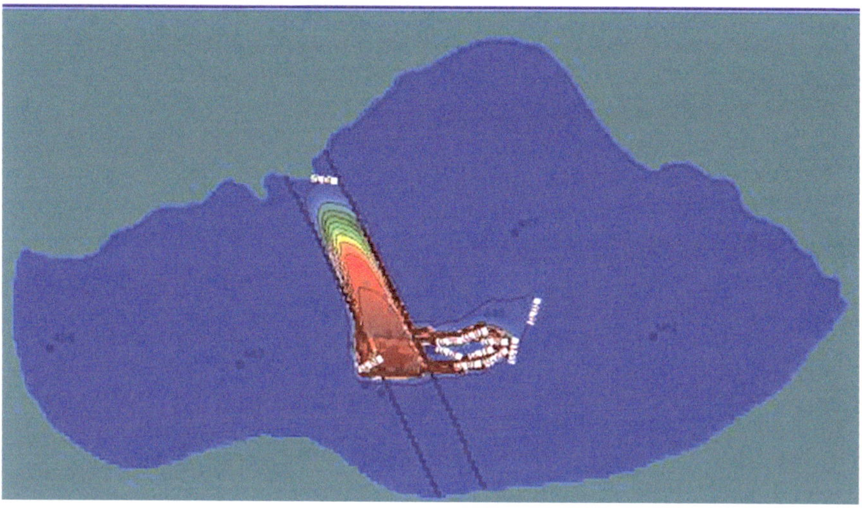

Fig. 3.15 Contamination front reaching the capture zone near the river within 9 years after the start of the migration (in the model with the fault zone)

network data, averaged over a multitude of years to solve the stationary geofiltration problem.

The results of the radioactive waste repository survey allowed us to ascertain that accidental leakage of contaminants can occur mainly through the microcracks by the repository perimeter, while the bottom appears to be securely isolated. Modeling of strontium 90 migration with groundwater showed that excluding the fault zone, the contamination front will go to the northeast to reach the river in 30 years (Fig. 3.14).

Considering the fault zone in the model, strontium 90 will mainly migrate to the northwest, and the contamination will reach the river in 9 years (Fig. 3.15). This shows the importance of tectonic analysis and filtration testing of near-fault and fault zones.

The testing for groundwater inflow, represented by single pumpings, did not characterize the entire fault zone. The values of the filtration parameter dispersion were evaluated on the basis of the assumption that the average value for the fault filtration parameters are normally distributed. The expected value was accepted according to the testing for groundwater inflow. On the basis of these characteristics, the filtration parameters of the fault zone for different quantile was computed (Fig. 3.16). Thus, according to numerical calculations with a probability of 90%, the contamination (equal to the level of intervention) could reach the capture zone near the river in 13.7 years, and with a probability of 10% it could reach the river in 7.5 years.

The calculations of the groundwater contamination halo of strontium 90 were made using the developed geofiltration model. The parameters of strontium 90 sorption (one of the main physicochemical processes in the water–rock system) are largely determined by a combination of acidic–alkaline conditions, competing

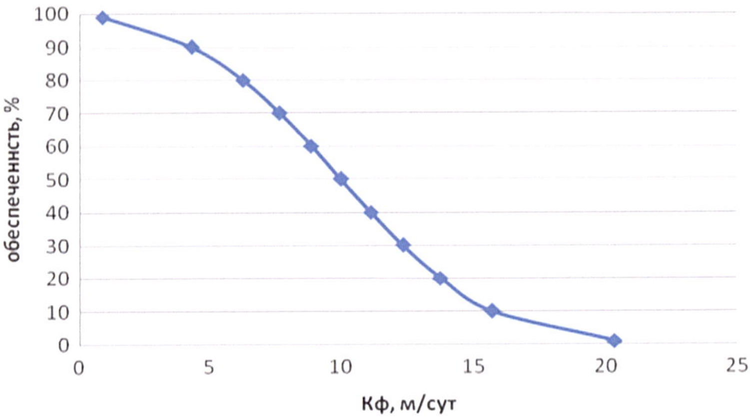

Fig. 3.16 Integral distribution function of the filtration parameters of the fault zone

reactions, and concentrations of radionuclides. Strontium 90, depending on the pH of the aqueous solution, is mainly adsorbed by a cation exchange mechanism.

The Ca^{2+} and Na^{+} present in groundwater reduce the distribution coefficient of Sr^{2+} and raise the radionuclide mobility. In water containing nitrate ions, half of the strontium 90 occurs in the form of aqueous complexes with nitrates, which significantly reduces radionuclide sorption. With no information on the nitrate ion concentration in the leakages in the study area, strontium in the solution is believed to occur in the form of uncomplexed Sr^{2+} ions. In predictive calculations, it was assumed that (1) the initial concentration of strontium 90 in the leaks is 102.7 Bq/L (according to the object analog); (2) the adsorption isotherm is described by the Henry equation, the main parameter being the distribution coefficient Kd; and (3) the value of Kd for strontium 90 is equal to 0.086 L/kg in accordance with the calculated values of the indicator obtained during strontium 90 migration research in fractured rocks in the vicinity of Lake Karachai [9]. The calculations were carried out using the method accounting for the sorption capacity of water-bearing rocks through the effective porosity.

The distribution of strontium 90 concentrations virtually repeated the estimates obtained for the neutral pollutant. The difference was the lower rate of change in the contamination halo, the cause of which was the higher value of the effective porosity of strontium as compared with the porosity of the neutral pollutant. The completed model calculations of strontium 90 migration fully complied with the contamination halos of the neutral pollutant, but with a temporary slowing factor of 1.3. Thus, according to numerical calculations with a probability of 90%, strontium 90 could reach the capture zone near the river within $13.7 \times 1.3 \approx 18$ years (i.e., the probability that the pollution will reach the water intake through 18 years is 0.9), and with a probability of 10% it could reach it within $7.5 \times 1.3 \approx 10$ years. The damage is taken as equal to the cost of sewage treatment plants, which, according to preliminary estimates, would be about $1.5 million. Thus, the risk values, in monetary terms, amount to $1.4 million and $0.15 million in the first and second cases, respectively.

So, the main uncertainty in the risk assessment is the impact of the fault zone on strontium 90 migration. If the effect of the fault zone is insignificant, the predominant movement of the pollution front will be to the northeast. If the effect is significant, the front of the pollutant will move much faster along the fault zone than in the main rock massif. The recommendations for adjustment of the monitoring network were drawn up on the basis of the simulation results. In order to minimize the uncertainty of our knowledge of pre-emptive migration routes, we recommended laying of two observation alignments: (1) from the radioactive waste repository to the river through the maximum flow gradient; and (2) from the radioactive waste repository along the fault zone axis of the river.

3.6 Conclusion

Development of the methodology of the ensuring of population safety, when different type of natural and technonatural hazards activating, is one of the most important problems in modern times, and the relevant decision making should be based on risk analysis and forecasting. The interdisciplinary nature of this problem has led to research in various knowledge domains, which has defined different approaches to classification of risk, methodology, and methods of risk assessment. The results of analysis and synthesis of risk research studies, associated with environmental contamination, have demonstrated the necessity of further development of the theory and methodology of analysis, prediction, and management of these risks. The methodology of assessment and forecasting of technonatural risk is a relatively young research area, intensively developed in various fields of scientific knowledge, and therefore the search for new and improved ways of analysis, forecasting, and risk management as the foundation of population safety is quite legitimate. The main tasks for the future development of the theory of hydrogeochemical risk include (1) improvement of the methodology for forecasting and management of hydrogeochemical risk on the basis of approach integration developed by specialists in different fields of knowledge; (2) development of methodology and techniques for large-scale hydrogeochemical hazards and risk mapping in urban areas; and (3) development of documents governing the procedure of risk assessment and forecasting at various stages of development in urban areas.

References

1. Ayala F. J. (1987). Introducción a los riesgos geológicos. In F. J. Ayala (Ed.), Riesgos Geológicos, IGME, Serie Geologiá Ambiental, Madrid (pp. 3–19).
2. Chiang, W. H., & Kinzelbach, W. (2001). *3D-groundwater modeling with PMWIN* (1st ed.). New York: Springer Berlin Heidelberg ISBN 3-540 67744-5.
3. Deutsch, C. V., & Journel, A. G. (1997). *GSLIB: geostatistical software library and users guide* (2nd ed.). New York: Oxford University Press.

4. Galitskaya, I. V., Pozdnyakova, I. A., Toms, L. S. (2010). *Assessment of hydrogeochemical hazard and risk in the urbanized territories. Global groundwater resources and management.* Selected papers from the 33rd International Geological Congress, Oslo (Norway).
5. Mark, R. K., & Stuart-Alexander, D. E. (1977). Disasters as a necessary part of benefit–cost analyses. *Science 16, 197*(4309), 1160–1162.
6. Petak, W., & Atkinson, A. (1982). *Natural hazard risk assessment and public policy.* New York: Springer Verlag.
7. Rosen, L., & LeGrand, H. E. (1997). An outline of a guidance framework for assessing hydrogeological risks at early stages. *Ground Water, 35*(2), 195–204.
8. Rowe, W. D. (1977). *An anatomy of risk.* New York: Wiley.
9. Samsonova, L. M., Kochergina, N. V., et al. (2004). Modeling strontium-90 delay by fractured rocks in the halo of groundwater contamination in the area near Lake Karachay. *Issues of Radiation Safety, 2*, 30–41.
10. Udias, A., et al. (1999). *Principles of seismology*. Cambridge: Cambridge University Press.

Chapter 4
Landslide and Seismic Monitoring Systems

Alexander Ginzburg, Valentina Svalova, Alexey Nikolaev, German Postoev, and Andrey Kazeev

Abstract The operative monitoring systems of landslides have been developed by Sergeev Institute of Environmental Geoscience RAS (IEG RAS) for a number of large industrial objects and urban areas. The designed monitoring systems of geological hazards for the "Blue Stream" gas pipeline and for Sochi territory are the examples.

4.1 Introduction

IEG RAS has developed Unified Automatic Equipment (UAE) working for Federal Target Program "Development of the Federal Seismic Monitoring and Earthquake Forecast System" to use it at the forecast landslides in the seismically dangerous regions of Russia. The problem of landslide risk reduction on one hand concerns building in landslide-prone areas and on the other hand concerns providing an advanced warning about an activation of landslides which threaten the buildings The complexity of this problem is compounded by a large number of different types of landslides, their occurrence along the slopes and the developmental dynamics during the landslide progression from the formation or activation phase and to the catastrophic destruction phase [1–16].

In preparation for the 2014 Winter Olympics in Sochi, a new road that combines highway with railway was constructed between Adler and Krasnaya Polyana. The combined road further goes on to the Alpika-Servis train station. It passes the valley of the river Mzymta. It is one of the most important objects in transportation infrastructure for the 2014 Olympics in Sochi. Olympic structures such as ski jumps, alpine skiing tracks, etc. are situated in the mountainous part of Sochi.

The goal of research was to identify monitoring technologies and criteria for landslide hazard assessment for effective early warning monitoring of the combined

A. Ginzburg · V. Svalova (✉) · A. Nikolaev · G. Postoev · A. Kazeev
IEG RAS, Moscow, Russia
e-mail: inter@geoenv.ru

V. Svalova (ed.), *Natural Hazards and Risk Research in Russia*,
Innovation and Discovery in Russian Science and Engineering,
https://doi.org/10.1007/978-3-319-91833-4_4

road. The aim of this monitoring system is to support the transportation safety for the 2014 Olympics in Sochi.

The results of the research have shown that as primary monitoring instrumentations should be used horizontal extensometers and inclinometers. They provide the highest accuracy of landslide deformation measurements. The criteria for landslide hazard assessment were developed on the basis of 2010–2013 monitoring results. Mechanisms of landslides in the region have also been taken into account.

4.2 Unified Automatic Equipment

Unified Automatic Equipment (UAE) is assigned to develop wide application diffused informational-measuring and controlling complexes. If such complexes are connected to radio telemetry channels, they provide the possibility to create area monitoring systems of different applications including seismic, geophysical regional monitoring systems, systems of ecological monitoring, etc.

The following devices are included into UAE: remote measuring system (RMS); system for data storage, control, and processing (DSCPS); Cable Connection Terminal for filed conditions, central (CCT-C); and Cable Connection Terminal for filed conditions, remote (CCT-R). Remote measuring system (RMS) is assigned to convert voltage signals coming from the geophysical channels' outputs into analog-digital signals.

RMS does the synchronization of the received information with the Integrated Time System, calibration of the geophysical channels, and calibration of geophysical channel control on the operation of the own hardware and software, forms data arrays, stores data, and transmits the registered and preprocessed data through the serial connection channel of the RS-485 type through CCT-C directly to the DSCPS. General view of RMS is given in Fig. 4.1. RMS operates under temperature range

Fig. 4.1 RMS general view

Table 4.1 RMS technical features

#	Parameters	Nominal value
1.	Number of differential analog inputs	8
2.	Input range, for each channel	$-10 \div +10$
3.	Input differential resistance, kOhm	Not less than 20
4.	Reduction factor of the inphase component, decibel, for each channel	Not less than 60
5.	Interference, decibel, one channel into other (for all)	Not more than -100
6.	Number of the effective bytes of the ADC under sampling frequency of 200 Hz for each channel	18
7.	ADC least significant bit weight range, microvolts, for each channel[a]	2.60 ± 0.10
8.	Absolute error of the whole range, mV, not more than	± 1
9.	Temperature coefficient of ADC, %, °C[b]	± 0.006
10.	Sampling frequency of the ADC, Hz	
	Minimum	Not more than 0.001
	Maximum	Not less than 200
11.	Calibration signal source (DAC) output signal range	
	True output	0 ... 4
	Complementary output	−4 ... 0
12.	DAC bit number	12
13.	DAC least significant bit weight range, microvolts,	1
14.	DAC limits of basic absolute accuracy, microvolts	± 25
15.	Temperature coefficient of DAC, %, °C	± 0.16
16.	DAC output resistance, Ohm	52 ± 3
17.	Load resistance, Ohm, not less	600
18.	Power consumption, Watt	Not more than 1.50

[a]For each RMS parameter is be determined during manufacturing and stated in RMS passport
[b]% of the measured value

from −30 to +50 °C. Humidity is not more than 98% under the temperature of +25 °C. RMS technical features are given in the Table 4.1.

System for data storage, control, and processing (DSCPS) (Fig. 4.2) does the following: reception of information directly from RMS or from RMS through CCT-C, processing of received information, injection of commands through CCT-C or directly to RMS, synchronization of data with common time system, transmission of the processed information through radio communication channel (or through cable communication line) to Information Processing Center (IPC).

DSCPS can operate under the following conditions: ambient temperature up to +50 °C, ambient temperature up to −30 °C, and relative humidity – 98% under temperature +25 °C. DSCPS technical features are given in the Table 4.2.

Cable Connection Terminal for filed conditions, central (CCT-C) (Fig. 4.3) is assigned to the following: to be used in informational-measuring and controlling complexes as a part of area-spread cable monitoring system of several kilometers square for provision of communication through serial duplex channel between

Fig. 4.2 DSCPS general view

Table 4.2 DSCPS technical features

Parameter	Value
Supply voltage	+24
Power consumption from 24 V dc source, not more than	3 W
Number of RS-485 com- ports	2
Time synchronization accuracy, msec	±1
RAM, not less than	6 Mbite
Data rate through serial ports, not more than	
RS-485	115.200 bit/s
Overall dimensions, mm	
Length (with connectors)	264 ± 2
Width	68 ± 2
Height	136 ± 2

DSCPS and CCT-R and to provide galvanic isolation of informational-measuring and controlling complexes communication lines components from the connection line.

CCT-C also can be used as the central monitor in an informational-measuring and controlling complex that provides tracing of the current time and initiation of DSCPS and other units therein according to set schedule. When CCT-C is connected to DSCPS in computer or communication line, it can turn on DSCPS and turn on/off any lines out of schedule under DSCPS commands.

CCT-C can operate under the following conditions: ambient temperature up to +50°C, ambient temperature up to −30 °C, and relative humidity – 98% under temperature +25 °C. Acid or alkali vapors and any other aggressive dirt in the air are inadmissible. Interface type is optically isolated RS-485 and information rate is up to 115.200 bit/s. The number of input lines depends on modification. Line length is up to 1200 m; primary power supply source voltage, 8–30 V; power consumption, Watt, not more than 1,0, and IP rate, IP65.

Cable Connection Terminal for filed conditions, remote (CCT-R) (Fig. 4.3) is assigned to the following: to be used in informational-measuring and controlling

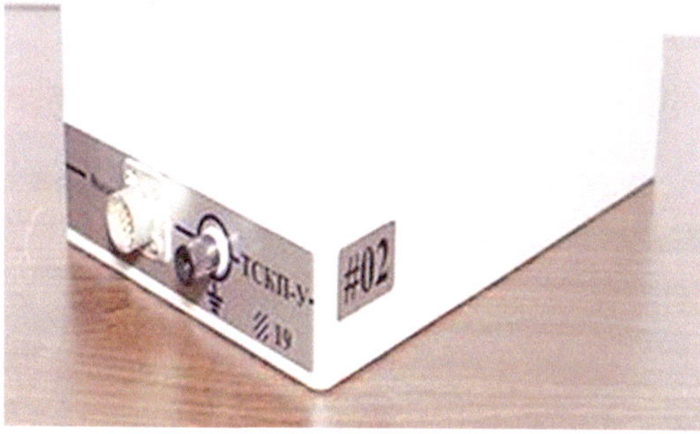

Fig. 4.3 CCT-C and CCT-R general view

complexes as a part of area-spread cable monitoring system with square of several kilometers for provision of communication through serial duplex channel between DSCPS and CCT-C and to provide galvanic isolation of informational-measuring and controlling complexes communication lines components from the connection line.

CCT-C can operate under the following conditions: ambient temperature up to +50°C, ambient temperature up to −30 °C, and relative humidity – 98% under temperature +25 °C. Acid or alkali vapors and any other aggressive dirt in the air are inadmissible. Interface type is optically isolated RS-485, and information rate is up to 115.200 bit/s. The number of input lines depends on modification. Line length is up to 1200 m; primary power supply source voltage, 8–30 V; power consumption, Watt, not more than 1,0; and IP rate, IP65.

Informational-measuring and controlling complex (IMCC) based on UAE. It includes the following units: UAE, DSCPS, CCT-C, CCT-R, and RMS. The number of CCT-R and RMS depends on the objective of the IMCC. Types of measuring channels, their number and placement relatively to DSCPS, and schedule of data acquisition are determined by the main objective of the complex.

Nine cable lines are connected to CCT-C with CCT-R at their ends. The maximum length of cable is 1200 m. Power goes through these cables from CCT-C to CCT-R. The cable line is protected with lightning protection devices that are installed in CCT-C and CCT-R. Information interchange rate for CCT-C and CCT-R is 115 kbit/s according to RS-485 protocol. Nine RMS can be connected to each CCT-R, and up to eight analog-digital converters can be connected to each RMS. Power to RMS is transmitted through CCT-R.

The following restrictions shall be considered in design of IMCC based on UAE: number of analog-digital channels in one system shall not be more than 72; and summary digital information recorded from all analog-digital converters of all RMSes included into one system shall not be more than 64 kbit/s.

There are two operating modes of IMCC that are determined by recording and data acquisition modes: information accumulation mode (information is collected only during visits of the IMCC) and real-time recording mode (IMCCes are combined into regional or local monitoring system by means of radiotelemetering or cable communication lines. In this case information will be recorded in real-time mode).

IMCC operation under information accumulation mode. In that case information will be stored in DSCPS main memory, which is 6 Mbytes. Laptop shall be connected to DSCPS before measuring. A program to configure IMCC shall be run at this laptop to set schedule of inquiries for measuring channels connected to RMS ADCes. CCT-C is also coded with this program (measuring channels turn on/off schedule). Then special recording program shall be run. First it will test IMCC and then synchronize DSCPS time service and control module according to GPS signals. GPS is connected to the laptop. Then the laptop will be disconnected from DSCPS, and IMCC will start autonomous recording of information according to assigned schedule.

According to that schedule, CCT-C will send voltage to DSCPS and to measuring channels indicated by the program. DSCPS inquires all RMSes by turn (with time gap between each inquiry), then compacts information, and forms packets of set format. These packets are stored in RAM. DSCPS synchronizes all RMSes under special commands. DSCPS Operation software is coded in ROM, and it is loaded after DSCPS is turned on or cleared. The software provides DSCPS operation as a main part of IMCC. It controls all connected RMS and stores and transmits measuring data to the laptop.

The software provides the following: initial self-testing of DSCPS and RMS status inquiry; receiving from the laptop and transmission to RMS test frequency for each ADC channel; time control and RMS time synchronization; simultaneous start of measuring in all RMS; RMS inquiry, storing of measuring results; compacting and buffering of data; recoding and starting of RMS after power supply failure that resulted in RMS reinitialization; receiving from a laptop and transmission to RMS any random command in RMS-laptop interchange format and transmission of RMS response to a laptop; and transmission of DSCPS memory data to a laptop, transmission from laptop data packets to DSCPS memory (including program loading), and transmission of the control command to assigned address that provides remote loading and running of tests or running of new version of DSCPS software.

RMS software consists of RMS microprocessor program (DS80C320) and programs for personal computer like IBM-PC that allows autonomous operation of RMS outside RMS-DSCPS complex.

Microprocessor program is placed in microprocessor's ROM or external ROM and provides the following:

1. Initial self-testing of RMS and all ADCes
2. Receiving of commands and their transmission through communication line (RS-485)
3. Transmission of system status and failure reports (system stays operative if some of ADC failed or disconnected)
4. Self coding of test frequency for each ADC channel (from 5 ms to 1 month frequency of measuring or turning off)

5. Measuring, storing, and transmitting of recorded data
6. ADC efficiency control, disconnection of channel in case of ADC failure (the channel can be disconnected under the command of external computer)
7. Operation from DAC (DAC will be connected to one of eight channels under the command, and output voltage will be changed according to assigned program)
8. Time service control: RMS monitors time with accuracy of 1 ms and transmits milliseconds counter together with ADC values from the moment of measuring.

RMS time service can be controlled and adjusted under the commands from external computer or DSCPS. It allows keeping time accuracy as ±1 ms during all time. RMS-DSCPS or RMS-external computer interchange rate is 115.2 or 19.6 k baud (switch is upon a command) and is always performed upon external computer initiative. Each RMS has its own address and responds only to the referenced commands. It is possible to connect up to 254 RMS to one line (some commands can be sent to all RMS at one time. RMS respond is not required in that case). The system provides effective operation even if communication cable line has a malfunction. An interchange is performed with packets, and each packet is accompanied with check sum. In case of interchange error, the packet will be ignored. Measuring results are stored in RMS and are transmitted asynchronously.

An information form DSCPS can be read as follows: A laptop is connected to DSCPS, and information is copied from the main memory to hard disk. Testing and synchronization of IMCC will be performed at the same time. IMCC operation works in data real-time recording mode. To carry out recording in real-time mode, IMCC is combined into regional or local monitoring systems by means of high-speed telemetric connection or cable communication line. In that case information is recorded by controlling computer in the Information Processing Center – IPC– that controls operation of all IMCCes. The number of IMCC included into monitoring system is restricted by hardware and software and can be not more than 16. The maximum interchange rate between IPC and all IMCC is restricted by a throughput of communication system.

IPC controlling computer software is assigned to control IMCC equipment, finding out its failures and receiving data and information storing. It can be installed at IBM-PC (PENTIUM based) with clock speed not less than 166 MHz.

Operation of regional or local monitoring system based on IMCC is determined by configuration file that will be created by special program. This file considers the availability of strategic monitoring points, types of geophysical channels, their parameters, and availability of time synchronization facilities (GPS).

After the start of the program, it reads configuration of the overall local seismic (geophysical) monitoring system and synchronizes time according to Greenwich with accuracy up to decimals of msec and calculates corrections of computer quartz frequency to provide system functioning in case of short-time failure of GPS. Time synchronization is done from GPS every second.

After synchronization it inquires, codes, synchronizes, and starts equipment of some IMCC. The program inquires status of IMCCes equipment (operation status, availability of channels, channel operation status, etc.). In case of any problem, report will be displayed at the screen (this report is also recorded in operation protocol file). Operation program for each measuring channel, frequency of inquiry, and gain factor plus one program of intense movements channel operation including a number of channels are transmitted to central controller of monitoring point. Before starting each IMCC is synchronized with IPC computer time (then synchronization is performed each 10 s). Time of signal transit from IPC to IMCC is also considered in synchronization procedure. After that DSCPS in IMCC starts collecting data from RMS.

IPC controlling computer cyclically inquires data from corresponding IMCC and records them into main memory buffers. After necessary data have been collected, they are recorded into file that corresponds to channel type. Normally these files are stored at other computer and can be read through local network. Nevertheless for short-time experiments, the system can be configured to use local disk of controlling computer. During short breaks of communication (not longer than 10 s), data will not be lost due to DSCPS of each IMCC which has large memory buffer. Operator can carry out calibration of any sensor of the system during its operation. In case of contingency (disconnection of communication with IMCC, its failure, failure of some channels or all RMS channels, or restoration of all above) and under some normal modes like recording of events or sensor calibration, the system will display report at the screen stating Greenwich time of event start, name of IMCC, and channel name plus event description. The report is recorded into buffer of 100 lines size and into protocol file. Operator can check the buffer at any time.

IMCC real-time recording mode (2). In that case IMCC is included into regional or local monitoring system that operates in real-time mode. Operating software of DSCPS is coded in ROM and will be started after turning on or clearing of DSCPS. This software is assigned to provide DSCPS functioning as a main part of IMCC (controlling of RMS, storing and transmission of data, reading of data in attendance mode) to IPC after processing.

Its functions are as follows: initial self-testing of DSCPS and RMS status; receiving from IPC or laptop test frequency for each ADC channel and transmission of these data to RMS; time monitoring and synchronization of RMS time; start of simultaneous measuring in all RMSes; RMS inquiry, storing of measuring results; and data compression and buffering.

If IMCC is used for complex geophysical monitoring systems with sensitive seismic sensors and sensors of intensive movements, DSCPS operational software provides the following: selection of intensive movement channels and storing their data for further transmission of event prehistory to IPC after detection of event; transmission of other channels' measuring results upon IPC request; calculation (in sliding window) of number of the exceeding set threshold recorded by event identification channel; event processing, transmission of event report, its prehistory, and report on time that has passed after event finish to IPC; receiving of program to change output voltage at DAC to perform sensors calibration from IPC and its transmission to RMS upon operator's command; RMS and ADC efficiency monitoring, reporting to IPC on equipment and communication failures, and recoding and restarting of RMS after power supply failure that results in reinitialization of RMS; receiving from IPC of random command in RMS, external computer interchange format, its transmission to RMS and transmission of RMS response to IPC; and transmission of DSCPS memory to IPC, receiving of the packet for DSCPS memory from IPC (including loading of the program) and transmission of control to the assigned address that allows performing remote loading and running of tests or new version of DSCPS main software. DSCPS time service can be control and adjusted upon commands from IPC that includes GPS time signal receiver. It allows keeping time accuracy as ± 1 ms during all time. RMS provides the same functioning as a part of IMCC operating either in autonomous mode and controlled by DSCPS or as a part of IMCC included into regional or local monitoring system; UAE is certificated by GOSTANDARD of RF.

Type approval certificate RU.C.34.004.A. No 13659 for information-measuring and controlling complexes based on Unified Automatic Equipment has been issued according to the decision of the Russian Federation State Committee on standards and metrology on December 25, 2002. The complexes were included into state register of measuring equipment under the No 23981-02B.

Basing on this design, IEG RAS has developed and delivered hazardous geological processes monitoring system to such big industrial enterprises like Russian – Turkey gas pipeline "Blue Stream," Soligorsk Ore Mining and Processing Enterprise (Republic of Belarus), Topsides Induced Monitoring System For Oil, and Gas Offshore Platforms and 2014 Olympics in Sochi objects.

4.3 Russia-Turkey Gas Pipeline "Blue Stream" Monitoring System

The construction of the Russia-Turkey gas pipeline "Blue Stream" was accomplished in 2002 (Figs. 4.4, 4.5, 4.6, 4.7, and 4.8).

The pipeline route of a total length of 1226 km crosses the Black Sea. The pipeline consists of two pipes each with 610 m in diameter. The project volume of transported gas is equal to 16 billion cubic m per year. When approaching the Black Sea, the pipeline crosses the northwestern slopes of the Big Caucasus Ridge, where landslides are widespread. Thirty-five landslides are registered at this section of the pipeline route, seven of them being the most hazardous. The online operating automatic control system of landslide processes was projected for these sites. The following registering devices were installed at each of the seven sites: the seismic acoustic control unit, the inclinometric control unit, and the groundwater level control unit.

Each unit was disposed in a separate borehole drilled in the landslide-prone slope. The equipment of each borehole provides the control of only one parameter. The measurement complex included the gauge of seismic acoustic emission and two units of data registration and collection. A special hydrogeodynamic gauge was used for measuring the groundwater level, and the tree-point extensometer was applied for rock mass displacement measurement. The measured data are communicated to the monitoring center, where they are processed using the special software.

Fig. 4.4 "Blue Stream" scheme

Fig. 4.5 "Blue Stream" during construction

Fig. 4.6 The compressor station, "Beregovaya," view from the sea. Visible glade in which the buried pipeline

Fig. 4.7 The compressor station, "Beregovaya," view from the sea. Visible glade in which the buried pipeline

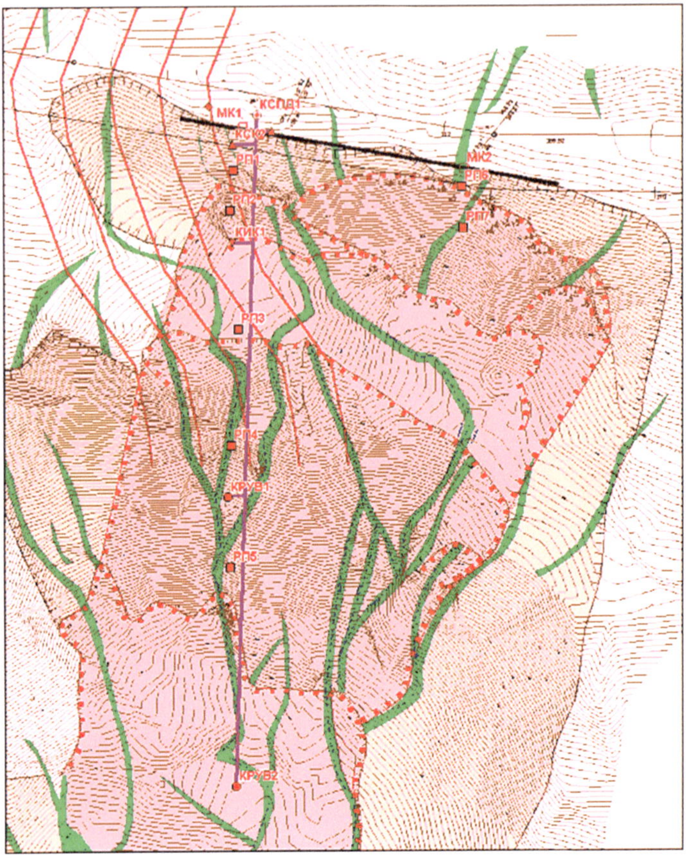

Fig. 4.8 The system of landslide-prone slope monitoring at the Russia-Turkey gas pipeline route with automatic unit for seismic acoustic measurements, unit for deviation control (inclinometer), and unit for groundwater level control

In addition to the surface automatic control, the remote sensing control based on the high-resolution space and aerial survey is used. The remote survey data are also processed using the special software. The developed monitoring system permits to control the conditions of the landslide-prone slopes and thus ensure the safety of pipeline operation at the site of high geological risk.

4.4 Landslide Monitoring System for Objects of the 2014 Olympics in Sochi

The landslide hazard is the main geologic hazard along the combined road from Adler to Krasnaya Polyana. The largest part of the area along the road is landslide-prone area, and numerous landslides historically occurred on the bank slopes.

Fig. 4.9 Sea cluster. Imeretinskaya Depression. Olympic objects during construction

At present, monitoring systems are installed at a number of Olympic structures (Figs. 4.9, 4.10, 4.11, 4.12, and 4.13).

Also monitoring systems were used during constructing the roads. They include an automatic monitoring of main parameters which characterize the state of the landslide area at each moment. Numerous factors contribute to such an active development of landslides in the region, such as a high degree of bedrock weathering on slopes. Abundant rainfalls and saturation of cover sediments also lead to the formation of numerous cracks and sliding. Nine landslide sites have been detected along the route of the combined road during geological investigations (Fig. 4.14). The stages of works are shown in Fig. 4.15. The results of monitoring are in Fig. 4.16.

It was determined that the most widespread type of landslides within the study area is a debris slide. Several sites, especially ones at the beginning of the route, exhibited block-type landslides of compression-extrusion. The investigations have shown that on landslide slopes with relatively long-time displacements and with periodic changes in displacement conditions, the acceleration of a displacement velocity up to hazardous levels may lead to the initiation of the sliding in new areas near upper and side borders of an active landslide. This increase of the size of a landslide can result in a catastrophic destruction.

Two general methods of observations were accepted in the automatic monitoring system and set along the combined highway and railway: (1) extensometric arrays (providing automatic measurements of displacements over the surface of landslide-prone slopes) and (2) inclinometric measurement in drills (supplying measurements

Fig. 4.10 Ski slope

Fig. 4.11 Bobsleigh track

Fig. 4.12 Bobsleigh track

Fig. 4.13 Ski jump

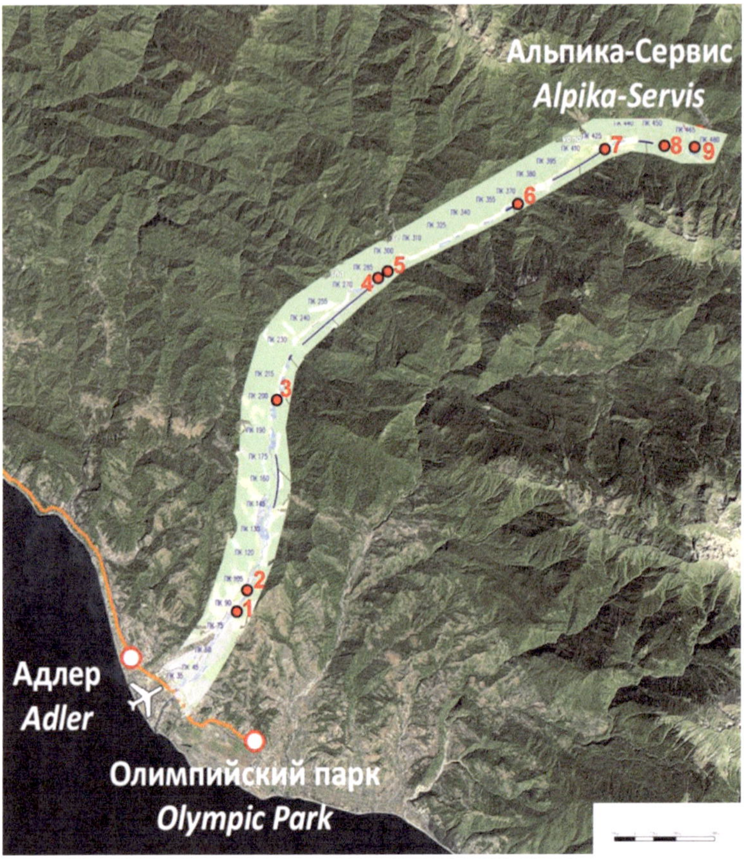

Fig. 4.14 The location of landslide sites along the combined road from Adler to Alpika-Servis (Krasnaya Polyana)

of sliding parameters vs depth, some in manual and automatic modes). The results of research have shown that the most useful parameters related to the characterization of an active landslide state and sliding dynamics, including the progressing development of a landslide during activation, are landslide displacement velocity, depths of slip surfaces, and propagation of active displacements within the territory.

Landslide hazard criteria were proposed for the constructions of the road based on the monitoring data of an active landslide at one of the study sites along the railway from Adler to Krasnaya Polyana. These criteria are based on measurements of displacement velocities and distribution of landslide deformations (including new volumes of ground masses involved along the margins of active landslides) with area and depth [1, 3]. As a result, several monitoring methods as related to the landslide hazard were recommended along the Adler-Krasnaya Polyana railway: automatic observations of displacements over the slope surface using extensometers and inclinometers (during site visits and in partly automatic mode).

Fig. 4.15 The stages of works at railway Adler-Krasnaya Polyana construction

4.5 Conclusion

Unified Automatic Equipment (UAE) and landslide monitoring systems have been developed and constructed by Sergeev Institute of Environmental Geoscience RAS (IEG RAS). The systems were successfully used for a number of large industrial objects and urban areas such as the "Blue Stream" gas pipeline and for Sochi territory in connection with 2014 Winter Olympics.

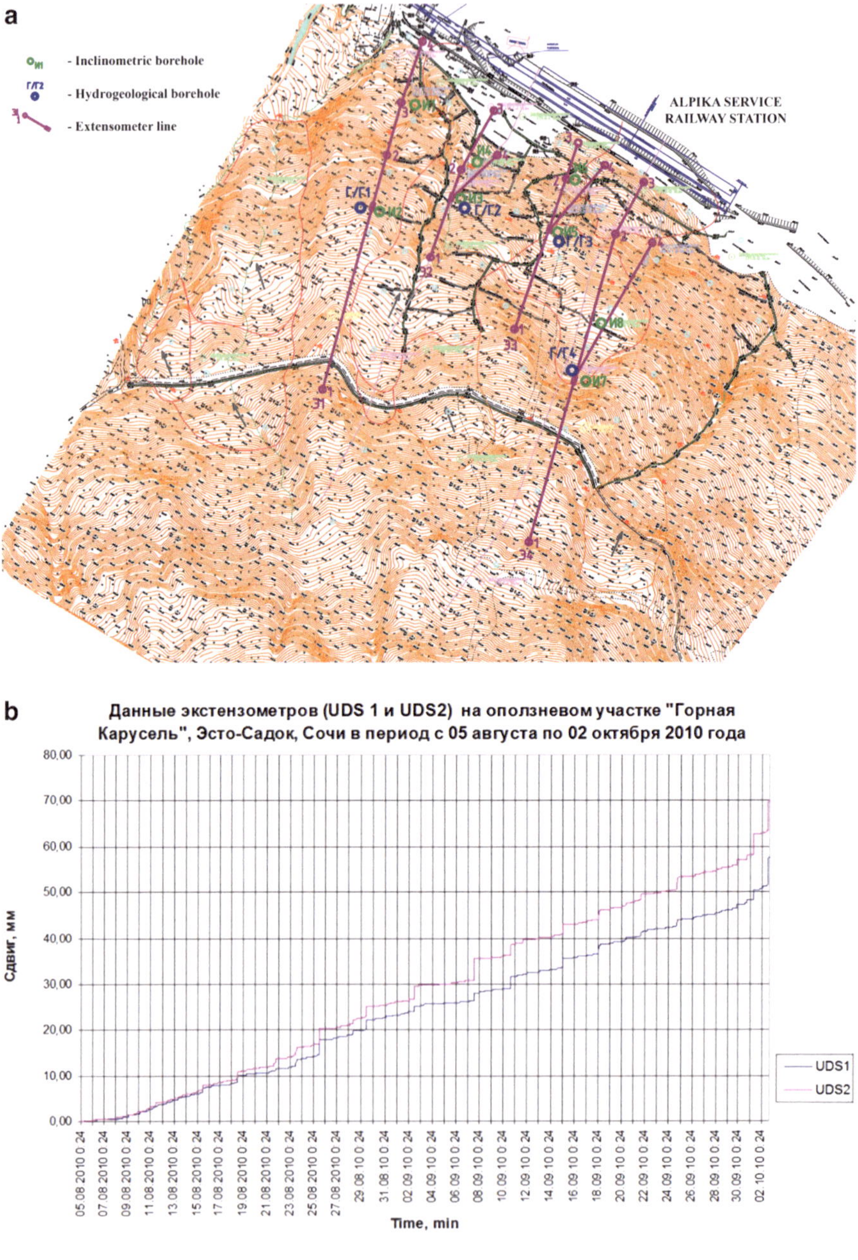

Fig. 4.16 Results of monitoring at railway Adler-Krasnaya Polyana. (**a**) Scheme of monitoring; (**b**) extensometer data; (**c**) inclinometer data

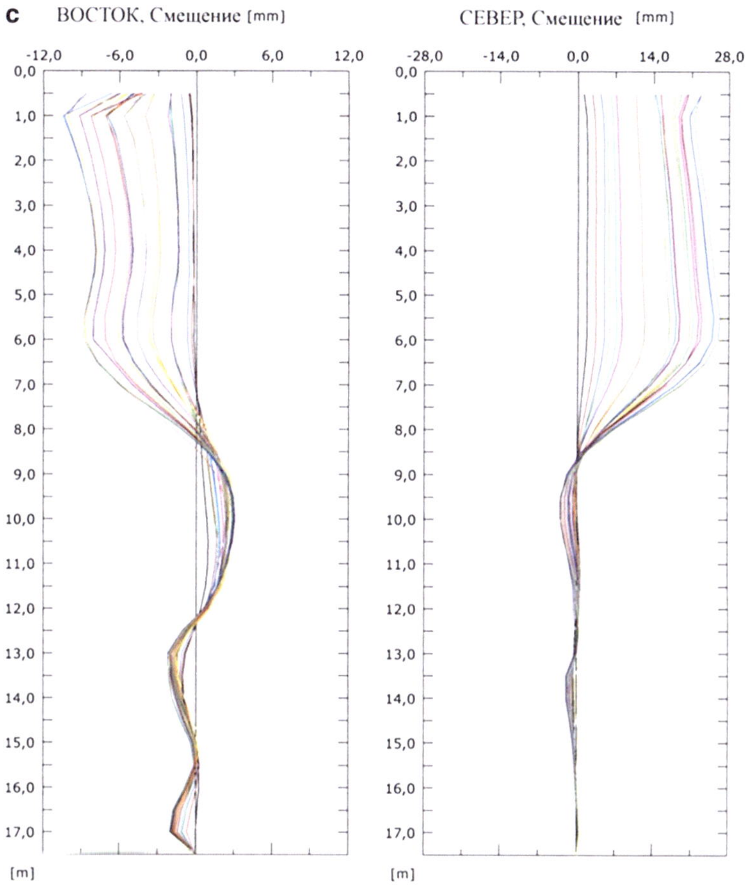

Fig. 4.16 (continued)

References

1. Kazeev, A. I., Postoev, G. P., & Fedotova, K. Y. (2014). Landslide hazard criteria for transportation safety of the 2014 Olympics in Sochi. In *14th Geoconference on science and technologies in geology, exploration and mining. Conference proceedings*. Volume II. (International Multidisciplinary scientific GeoConference; 17–26 June, 2014, Bulgaria) (pp. 567–572).
2. Nikolaev, A. V., Bashilov, I. P., Shou, K.-J., Svalova, V. B., Manukin, A. B., Zubko, Y. N., Behterev, S. V., Kazantseva, O. S., & Rebrov, V. I. (2011). Some directions of works on maintenance of geological safety of engineering constructions. In: *Proceedings of ENGEOPRO*, Moscow, p. 7.
3. Postoev, G. P., Erysh, I. F., Salomatin, V. N., et al. (1989). *Artificial activation of landslides* (p. 134). Moscow: Nedra.
4. Postoev, G. P., & Svalova, V. B. (2005). Landslides risk reduction and monitoring for urban territories in Russia. In *Proceedings of the first general assembly of ICL (International*

Consortium on Landslides), "Landslides: risk analysis and sustainable disaster management" (pp. 297–303). Washington: Springer.
5. Svalova, V. B. (2009). Mechanical-mathematical modeling for sedimentary movement and landslide processes. In *CD Proceedings of the international association for mathematical geosciences meeting (IAMG 2009)*, Stanford, p. 15.
6. Svalova, V. B. (2011). Mechanical-mathematical modeling and monitoring for landslide processes. *Journal of Environmental Science and Engineering, 5*(10), 1282–1287.
7. Svalova, V. (2011). Monitoring and modeling of landslide processes. *Monitoring Science and Technology, 2*(7), 19–27 (in Russian).
8. Svalova, V. B. (2011). Landslide process simulation and monitoring. In *Proceedings of ENGEOPRO*, Moscow, p. 7.
9. Svalova, V. B. (2012). Mechanical-mathematical modeling and monitoring for landslides. In *Proceedings of IPL (International Program on Landslides) symposium* (pp. 63–68). Paris: UNESCO.
10. Svalova, V. B. (2013). Risk reduction for landslide hazards. In *Modeling and monitoring, Proceedings of the international conference natural risks: analysis, estimation, mapping* (pp. 157–163). Moscow: MSU. (In Russian).
11. Svalova, V. B. (2014). Modeling and monitoring for landslide processes. In K. Linwood (Ed.), *Natural disasters – typhoons and landslides – risk prediction, crisis management and environmental impacts* (pp. 177–198). New York: Nova Science Publishers.
12. Svalova, V. B. (2014). Mechanical-mathematical modeling and monitoring for landslide processes. In *IPL 163 Project. Proceedings of the world landslide forum 3*. Volume 4. Beijing. (pp. 24–27).
13. Svalova, V. B. (2014). Modeling and monitoring for landslide processes: Case study of Moscow and Taiwan. In *Proceedings of the world landslide forum 3*. Beijing (Vol. 4, pp. 628–632)
14. Svalova, V.B. (2015). Mechanical modeling and geophysical monitoring for landslide processes. In *Proceedings of IAEG XII Congress "Engineering geology for society and territory"*, Torino-2014, (Vol. 2, pp. 345–348). Italy: Springer.
15. Svalova, V. (2017). *Landslide risk: Assessment, management and reduction* (p. 253). New York: Nova Science Publishers.
16. Svalova, V., & Postoev, G. (2008). Landslide process activization on sites of cultural heritage in Moscow, Russia. In *Proceedings of the first world landslide forum 2008*, Tokyo, p. 4

Chapter 5
Early-Warning Landslide Monitoring System

Alexander Ginzburg, Valentina Svalova, Alexey Nikolaev, Anatoliy Manukin, and Vladimir Savosin

Abstract Analysis of seismological phone changes will give possibility to elaborate early-warning system. Sergeev Institute of Environmental Geoscience RAS (IEG RAS) has developed and delivered the "System monitoring acceleration induced on the upper part of the offshore oil and gas platforms" for deposits Lunskoye-A (LUN-A) and Piltun-Astokhskoye-B (PA-B) for Sakhalin-2 project. The system is intended to ensure the safety of the operation of these facilities.

Experience of the system creation was used for real-time early-warning landslide monitoring system construction. This system was successfully used for landslides monitoring of coastal slope of the river Yenisei.

5.1 Introduction

Landslides, mudflows, debris flow, and avalanches are the most dangerous and destructive processes that cause significant damage to economic objects and often resulting in death.

There are numerous examples of negative consequences of influence of debris flows, landslides, mudflows, and avalanches in various regions of Russia. One of the main causes of such emergencies and consequences of the negative effect is the absence of a unified monitoring system for early warning of the dangerous state of controlled processes and the appearance of threats to economic objects [3–16].

In development of a method for automated monitoring of exogenous geological processes for the purpose of early warning about dangerous development, with identification of the main informative parameters of the processes and for their control, with development monitoring schemes (placement of equipment), and with development and testing of elements of systems for automated monitoring, these criteria for a hazardous condition are the main purposes of research [1–16].

A. Ginzburg · V. Svalova (✉) · A. Nikolaev · A. Manukin · V. Savosin
Sergeev Institute of Environmental Geoscience RAS (IEG RAS), Moscow, Russia
e-mail: inter@geoenv.ru

V. Svalova (ed.), *Natural Hazards and Risk Research in Russia*,
Innovation and Discovery in Russian Science and Engineering,
https://doi.org/10.1007/978-3-319-91833-4_5

The development of modern socioeconomic system provides for the establishment and operation of such grand and environmentally hazardous facilities like pipelines, nuclear power plants, chemical industry, etc.

Most ecologically dangerous objects or grandiose constructions are in seismic areas and tectonically active zone, in which there may be strong earthquakes, as well as landslides and mudflows. During operation it is necessary to ensure the safety of both the expensive facilities and safety of the environment.

Under these conditions, the timely detection of dangerous earthquakes and giving alarms and automatic shutdown of environmentally hazardous facilities is a paramount task.

IEG RAS for many years carries out all the work necessary to create the monitoring of environmental and seismic safety.

In 2005–2006, IEG RAS has developed and delivered the "System monitoring acceleration induced on the upper part of the offshore oil and gas platforms" for deposits Lunskoye-A (LUN-A) and Piltun-Astokhskoye-B (PA-B), located near Sakhalin for Sakhalin-2 project. It is intended to ensure the safety of the operation of these facilities.

A second example of environmental safety monitoring system is a system of monitoring of dangerous geological processes. The system is designed for monitoring of landslide coastal slope of the river Yenisei in real time. The system provides rapid collection of measurement data on the state of the observed landslides, processing and analysis of the distribution of monitoring results between users, and controls the security of shopping and entertainment complex, located on the coastal slope.

5.2 Topsides Induced Acceleration Monitoring System for Oil and Gas Offshore Platforms: TIAMS

According to the Sakhalin-2 Project, Sakhalin Energy Investment Company is building offshore oil and gas platforms PA-B and LUN-A at the Sakhalin Island shelf.

The platforms are situated within the seismically dangerous area where destructive earthquakes are likely to occur.

To reduce the risk of environmental accidents that can appear during oil and gas production, as a result of destructive earthquake, the client took a decision to provide platforms with Topsides Induced Acceleration Monitoring System (further referred as TIAMS).

Institute of Environmental Geoscience RAS has won the tender for design, development, and manufacturing of the TIAMS arranged by Sakhalin Energy Investment Company (further referred as SEIC). Basing on the technical assignment Information and Measuring Systems department, IEG RAS has designed and manufactured in 2005–2006 TIAMS packages for two offshore oil and gas platforms near Piltun-Astokhskoye-B (PA-B) and Lunskoye-A (LUN-A) fields [1, 2].

LUN-A and PA-B platforms are very complicated constructions. Each platform has three decks of the football ground size. The platforms are supported by four legs. Their diameters are from 16 to 24 m, height is approximately 60 m, and depth of the sea at the site is 30–35 m. The lower decks are placed at the height of ~27 m; the upper decks are at the height of 50–60 m above the sea surface.

Friction Pendulum Bearings are placed at the top of the legs to damp horizontal oscillations under seismic and load impacts to the platform supports.

The main function of the TIAMS is to determine dangerous earthquakes from other impacts induced to the platform (ice impacts, ship impacts, wave impacts, drill snatch, etc.) that can cause accelerations same to the dangerous earthquake accelerations at the topsides of the platforms. In case the destructive earthquake has been detected and its acceleration level exceed the threshold of 0.5 g (assigned by the client) in any key point of the platform, the TIAMS shall initiate the emergency shutdown signal (ESD).

In such a way, TIAMS shall provide safety of the oil and gas offshore platforms.

It is necessary to mention that there were no such systems in the world practice between earthquake detection systems that can detect earthquakes from other impacts that can cause the same accelerations as dangerous earthquakes.

IEG RAS has to solve the following tasks during the development and manufacturing that are due to the client's specifications and platform construction:

TIAMS shall work within severe ambient conditions, i.e., −40 °C–+50 °C in aggressive medium (salt fogs).

TIAMS shall work in the gas dangerous area, i.e., it shall comply with the requirements of intrinsically and explosion proof safety.

TIAMS shall comply with the requirements if the IEC 61508 "electrical, electronic, programmable electronic systems" are connected with functional safety.

TIAMS shall pass certification to obtain certificates for its elements and overall system.

IEG RAS shall simulate all input loads during Factory Acceptance Test.

IEG RAS shall determine requirements to the frequency and dynamic band of the measured accelerations.

IEG RAS shall determine platform responses to the dangerous loads.

IEG RAS shall select key points for the placement of sensors to minimize the configuration of the TIAMS.

The solution was based on the Unified automated equipment elements (UAE) designed in the seismological center by IEG RAS for forecasting polygons of seismic dangerous regions of Russia from 1996 to 1998 in the frames of the federal program "Development of the federal seismic monitoring system and forecasts of earthquakes in 1996–2000" (for detailed description of UAE, see IEG RAS website: www.geoenv.ru).

AMEC (affiliate of SEIC) has developed mathematical model of the platform in ABAQUS software and modeled seismic impact to the platform. IEG RAS specialists then formalized all external loads and simulated their impacts for ABAQUS platform models. Then IEG RAS specialists analyzed platform responses to the impact loads and earthquakes (more than 60,000 diagrams and schemes).

During the first stage, IEG RAS has done the following:

- Theoretical justification of the external non-seismic impacts to the platform, detection of their features: value and direction of the affecting forces and time dependences
- Modeled of 17 variant of impacts using ABAQUS platform model developed by AMEC
- Qualitative physical analyses of the topsides responses to the earthquakes and other impacts, detection of the main directions, and methods of the mathematical processing of the modeling results
- Developed the software program to process modeling results
- Analyses of modeling results from seismic and non-seismic impacts
- Determined the key point for the sensors and their numbers as 6
- Developed requirements for sensors installation
- Developed earthquake detection algorithm and algorithm of ESD signal initiation

An experience in development of such systems and its operation in the severe environment are very important for solving the same problems at other hazardous ecological objects, such as atomic power plants, chemical plants, high dams, and barrages. Such systems can also be used to provide safety of the mega polices.

5.3 Description of the Topsides Induced Acceleration Monitoring System (TIAMS)

TIAMS structural scheme is given at the Fig. 5.1.

TIAMS includes six Remote Measuring Units (RMU) and equipments placed in the Unit Control Panel (UCP).

RMU layout is given at the Figs. 5.2, 5.3, and 5.4. UCP layout is given at the Figs. 5.5 and 5.6.

Three-component accelerometers (TA, RMS, CCT-R) are constructively united into the Remote Measuring Unit (RMU), which is designed as the field device permitted to be used in the hazardous environment and severe severity conditions. RMU is located directly in the measuring point (Fig. 5.7).

The rest of the elements of the MCC are placed in the Unit Control Panel (UCP) which is placed in the Main Equipment Room (MER) of the oil and gas offshore platform.

Signals, coming from the RMU to the control panel, are transferred by the standard interface RS-485. Date rate through the COM port is 115,200 bit/s. Acquisition and preprocessing of the data received from the RMU are effected in the DSCPS. DSCPS-M carries out the temporary storage of data received from the

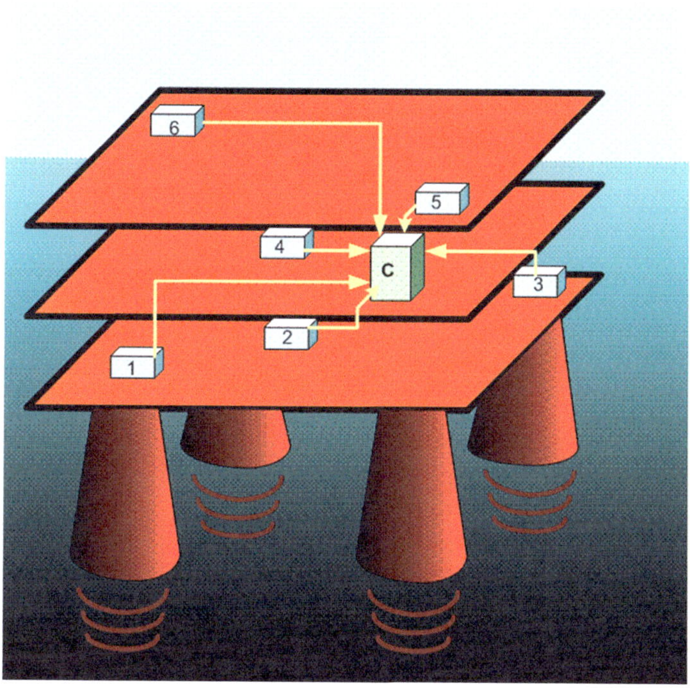

Fig. 5.1 TIAMS structural scheme

Fig. 5.2 RMU form

DSCPS, forms the data blocks in the defined format, processes the information according to the defined algorithm, and transfers it through the standard serial interface RS-485 to the controlling computer for registration, viewing, and current control of the equipment status. In case the DSCPS-M finds the hazardous seismic event, it initiates the ESD signal. Storage and processing of the received information and the viewing are done on the processing computer.

Fig. 5.3 RMU with open cover

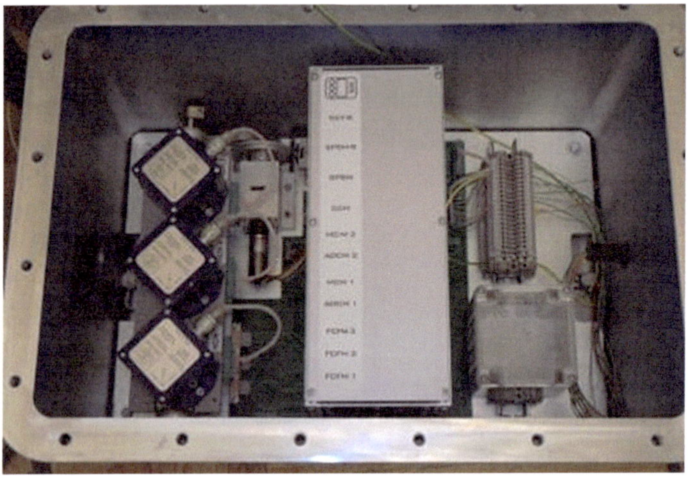

Fig. 5.4 RMU, view from above, open cover

TIAMS is designed for the automatic mode of operation, and it provides control and operation of all system elements automatically. TIAMS has the embed means of diagnostic of the hardware and provides the automatic finding of the emergency situations (at that TIAMS provides the "FAILURE" signal).

Fig. 5.5 UCP

The controlling computer and the processing computer provide the human – machine interface (HMI) within the operating system. The processing computer sets the operating modes for the TIAMS and makes changes the configuration. The current status of the system elements, results of the diagnostic, emergency reports, etc. are displayed at the monitor.

TIAMS Modes

TIAMS has the following operation modes: configuration mode (job assignment), operation mode, and testing mode.

Configuring mode allows determining the logic addresses of the RMU, calibration coefficients of the installed accelerometers, thresholds to track the dangerous seismic events, location and length of files with the recorded received data, etc.

After the power supply is switched on, the TIAMS proceeds to the operation mode, and the work begins from the selection of the job entered in the configuring mode. After the self-diagnostic and evaluation of the real configuration job adequacy, the TIAMS begins to monitor the accelerations in the selected points of the platform, then analyze the events, and provide the ESD signal initiation when it finds

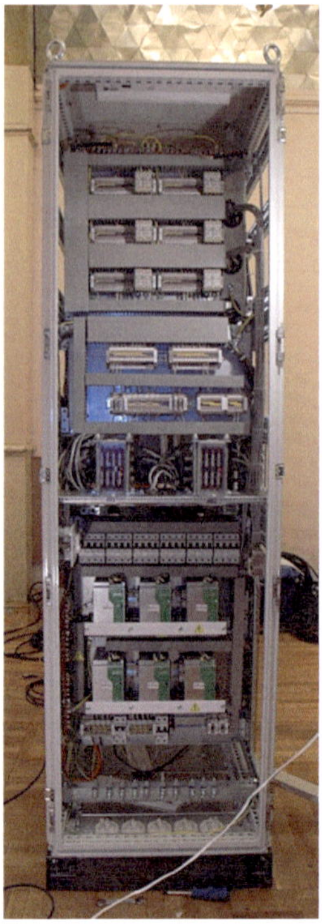

Fig. 5.6 UCP, view without back panel

Fig. 5.7 Three-component accelerometer (primary converters' module and frequency characteristic)

the dangerous earthquake according to the defined algorithm. It also provides the recurring calibration of the accelerometers.

The testing mode allows effecting the detailed examination of the modules capacity for work.

Switching of the System

The system switching is done in three stages as follows: self-testing, job of the configuration parameters, and system starting.

Self-Testing

On switching all programmable devices of the system (RMS, CCM.DSCPS, DSCPS-M, and controlling computer) are running the self-testing and inspection of the connection of the subordinate devices. The test functions of each device are described in the section "Efficiency monitoring".

After self-testing and detection of the efficiency of the connected programmable devices, the DCSPS is doing the initial calibration of the accelerometers. The results of self-testing are transmitted to the seignior devices upon requests. After self-testing of all modules have been finished, the results are displayed at the screen of the controlling computer and registered in the protocol. The results are as follows: devices' addresses three; self-testing results of each device, "no-fault" or list of the malfunctions; and calibration results from each accelerometer.

The DSCPS-M passes a "Failure" signal to the PCS in case a malfunction is found. If malfunctions are not resulted in a complete loss of system efficiency, then the system will go on operating with the following restrains: in case of failure of the calibration platform, the calibration shall not be running until it will be replaced; in case of failure of one of the accelerometers, the averaging out shall be done with two remained; in case of failure of the RMU or DSCPS, the system will go on operating, but the probability of false response will be increased; in case of failure of the active DSCPS-M, its reserved one takes up its functions; in case of failure of the port or connection line RS-485-1of the DSCPS-M, the connection shall be through the reserved port RS-485-2; in case of failure of the controlling computer, the doubling takes up its functions; and in case there are no fatal malfunctions, the system proceeds to the assigning of the parameters of configuration; otherwise the system is considered as inoperative until fault removal.

Assigning of the Configuration Parameters

After the self-testing has been completed, the system checks up the availability of the job assignment with system parameters and its configuration. Parameters are stored in the file with a fixed name at the controlling computer. If system configuration defined during self-testing concurs with the configuration from the job assignment file, then the automatic start-up of the system shall be done (if there is defined parameter of auto start).

In case there is no file with job assignment, or it is not correct, or it doesn't conform to the system configuration defined during the self-testing, then the system shall wait for the entering of the correct parameters which include the following: three of the serial numbers and addresses of the units, acceleration numeralization frequency, RMS channels in use, periodicity of testing and accelerometers calibration, factors of the analog-digital readout conversion into acceleration, threshold value of the acceleration, and earthquake detection algorithm parameters.

After the correct parameters have been entered, the specialist starts up the system.

System Start-Up

On start-up command, the following is carried out: time synchronization to the platform time system, transmission of the job assignment to the subordinate devices, and starting of all units.

Efficiency Monitoring

Each unit periodically does self-testing and informs the seignior device about results. Each unit monitors the subordinate devices. Each unit controls the connection with subordinate units. Two DSCPS-Ms check the operation of each other, and in case of failure of the main DSCPS-M, the other one takes up its functions. In case any of the DSCPS-M fails, then the "FAILURE" signal is initiated. The information of found malfunctions is displayed at the screen of the controlling computer and registered in the protocol. All diagnostics can be done in the automatic mode according to the program and under the commands from the operator. All processor units (RMS, CCM, DSCPS, DSCPS-M, and controlling computer) have watchdog timers to prevent the suspension of the program.

RMS Diagnostic Functions

RMS controller carries out the diagnostic of the following: RMS programs memory, RMS data memory, RMS EEPROM memory, efficiency of the analog-digital converter (ADC) of the RMS, conversion frequency error of the RMS ADC, correctness of the job assignment for the RMS ADC, and keeps statistic of the faults in the connection line to the DSC PS.

The diagnostic is done in the automatic mode and under the commands from the DSCPS. The results of diagnostic are transmitted to the DSCPS on demands.

CCM Diagnostic Functions

CCM microcontroller carries out the diagnostic of the following: CCM programs memory, CCM EEPROM memory, CCM nonvolatile backing memory, availability of the job for calibration of the accelerometers, efficiency of the engine, engine rotation speed, oscillation frequency of the of the calibration platform, position of the platform under operation mode, yields a "SELF TEST" signal for electronic test of the accelerometers, and keeps statistic of me faults in the connection line to the DSCPS. The diagnostic is done in the automatic mode and under the commands from the DSCPS. The results of diagnostic are transmitted to the DSCPS on demands.

DSCPS Diagnostic Functions

DSCPS microcontroller carries out the diagnostic of the following: DSCPS programs memory, DSCPS data memory, sends the commands for testing and calibration of the RMU (RMS and CCM), requests the results of the diagnostic from RMU, analyzes the data of accelerometers and estimates their efficiency and calibration parameters, analyzes the efficiency of connection line with RMU, and keeps the statistic of the faults in the connection line to the DSCPS-M. The diagnostic is done in the automatic mode and under the commands from the DSCPS-M. The results of diagnostic are transmitted to the DSCPS-M on demands.

DSCPS-M Diagnostic Functions

DSCPS-M microcontroller carries out the diagnostic of the following: DSCPS-M programs memory, DSCPS-M data memory, sends the commands for testing and calibration to all DSCPS, requests the results of the diagnostic from DSCPS, analyzes the efficiency of connection line with all DSCPS, analyzes the efficiency of the redundant DSCPS-M, keeps the statistic of the faults in the connection line to the redundant DSCPS-M, analyzes the efficiency of the main and redundant controlling computer, keeps the statistic of the faults in the connection line to the main and redundant controlling computer, analyzes the efficiency of the connection line to the main and redundant controlling computer, and controls the regularity of the secondary power supply sources. The diagnostic is done in the automatic mode and under the commands of the controlling computer. The results of diagnostic are transmitted to the controlling computer on demand. In case of failure, the DSCPS-M initiates the "FAILURE" signal to the PCS.

Controlling Computer (CC) Diagnostic Functions

The CC carries out the diagnostic as following: passes the job for caring out of the automatic diagnostic and calibration to the subordinate devices, passes the job for caring out of the automatic diagnostic and calibration to the subordinate device on command from operator, collects the results of diagnostic of all devices from DSCPS-M, displays at the screen the status of all units, registers in protocol the reports of all malfunctions and reestablishment of operation, and controls three levels of access to work with TIAMS by the systems of passwords.

All functions of diagnostic and data storage are carried out at the same time at the main and redundant controlling computer. The commands can be given from any of the controlling computers, laptop diagnostic functions. Laptop provides the possibility to do the autonomic diagnostic of the system modules (RMU, DSCPS, DSCPS-M) that are under repair or spared. The principle of the RMU operation is based on the RMU three-component accelerometer conversion of the induced linear accelerations into analog electrical signals with further analog-digital conversion. Three-component accelerometers do the conversion of the current values of the accelerations into output voltage through three mutually perpendicular axes. Each accelerometer has the self-testing and calibration devices. The self-testing is based on the sensing device inertial mass position changing upon the command "self-test" and under the impact of the electrostatic force that approximately equal to the 20% of the calibration measuring range (full scale). This is the way to check the efficiency of the overall mechanical structure and electric circuit of the accelerometer.

Electric motor with reducer, driving gear, calibration platform, platform position transducer, and calibration control module form the calibration unit that provides the calibration mode under which the position of the calibration platform should be changed according to the assigned principle. Control on the calibration and self-testing of the three-component accelerometer is done by the calibration control module that forms the "self-test" signal and drives the electric motor (sets in motion the calibration platform). The CP is intended for the placement, fastening, and mutual orientation of three Measuring Converter Modules. It is fixed at the bearings that installed at the platform supports. CP could change its position relatively to the long axis within the range of 0°–+45°. The moment that provides the oscillation movements of the platform revolving on its axis is passed by the crank-link mechanism that connects the drive shaft of the electric motor and CP. The oscillation frequency of the CP is equal to the value of the central frequency of the working range, which is 1, 0 Hz. The platform position transducer is used to provide the return of the platform to its starting (measuring) position. The signal of this transducer is used by CCM to control the operation of the electric motor.

The calibration of the accelerometer measuring channel is based on the periodical (with the frequency of 1 Hz) changing of the initial accelerometer converters position. These initial accelerometer converters are included into the Measuring Converter Module that is placed at the CP. This method shows the changing of the gravitational acceleration projection to the sensor sensitive axis. Variation amplitude of this projection is about 0.15 g (1.5 m/s2) and serves as the calibration constant. The value of the calibration constant is a reference value for the periodical calibration of the accelerometer measuring channels.

Two remote measuring systems do the conversion of the output analog signals from the three-component accelerometers into the digital code. Each RMS contains Analog Digital Conversion Module and Micro-Controlling Module, which control the analog-digital conversion process, self-testing, and receiving and transmitting of the data.

The changing of the acceleration amplitude, testing, and calibration are effected upon the commands from the UCP according to the operation software of TIAMS.

During the implementation of the contract, TIAMS has received the following certificates:

1. Permission for use from Rostekhnadzor
2. EMC certificate
3. Explosion proof certificate
4. Type approval certificate for system elements (RMS, DSCPS, CCT-C, CCT-R)
5. Type approval certificate for overall system is in process of issuing
6. Type approval certificate for three-component accelerometer is in process of issuing
7. IEG RAS has obtained Quality Management Certificate ISO 9001.

Oil and gas platforms for deposits Lunskoye-A (LUN-A) and Piltun-Astokhskoye-B (PA-B) for Sakhalin-2 project are represented in Figs. 5.8, 5.9, 5.10, 5.11, and 5.12.

5.4 Landslide Monitoring System for Coastal Slope of the River Yenisei

Geohazard monitoring system designed to monitor landslide coastal slope of the river Yenisei in real time. The system provides rapid collection of measurement data on the state of the observed landslides, processing and analysis of the distribution of monitoring results between users, and controls the security of "June" shopping and entertainment center, located on the coastal slope (Fig. 5.13).

The monitoring system includes eight mass displacements of ground points and two points of monitoring changes in the level of groundwater and automatic workplace of geologist and provides organizations with the process of collecting,

Fig. 5.8 Sakhalin Island and oil-gas platforms

processing, and distributing data and carrying out management of all its elements (Fig. 5.14).

In order to collect and exchange information using cellular communications channels, which included two GSM modems (primary and backup), each control point is a part of a GSM modem.

The system has two modes of user access to information. Access to each of the modes of operation is carried out by a password. The first mode of access – "Operator", which is only possible to view all parameters of the control points and

Fig. 5.9 Platform for PA-B deposit

Fig. 5.10 Commissioning works in South Korea

Fig. 5.11 Platform for Lunskoye-A (LUN-A) deposit

Fig. 5.12 Platform for Lunskoye-A (LUN-A) deposit

Fig. 5.13 Trade and amusement complex “JUNE”, located on the monitored coastal slope of the river Yenissei

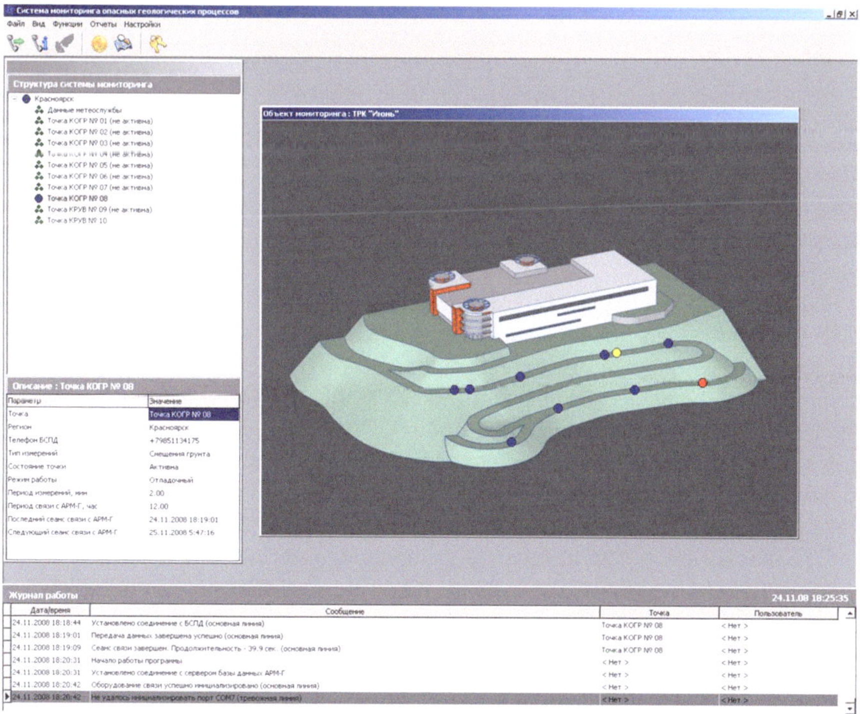

Fig. 5.14 The main window of the work program

Fig. 5.15 Equipment set deep frame

the region, measurement data and alarms, and change the system configuration or structure of any device in this mode cannot be. The second access mode, "Administrator", is possible to make a change of settings and system structure. To transmit alarm signals provided connectivity with unified duty-dispatch service of Krasnoyarsk by the Internet.

Equipment set deep frame is designed to measure linear displacement of soil that occurs when the landslide processes caused by natural and man-made causes, by its (linear movement) transformation into a digital code (Fig. 5.15). The kit has control equipment malfunctions and unauthorized access and allows you to quickly transfer the alarm information. Complete registration of groundwater level is designed for continuous automated measurement level, water temperature, and atmospheric pressure well and transfers the measurement results in digital form (Fig. 5.16). The kit has fault control equipment and unauthorized access to it.

The monitoring system has two operating modes: normal and abnormal. In any mode of functioning of the data from the hardware coastal slope, control points are processed in real time. If the ground speed displacement mass or velocity of groundwater level changes less than a predetermined threshold, the information is recorded and subsequently subjected to analysis and comparison with data obtained

Fig. 5.16 Complete registration of groundwater levels

previously. In another case, when the speed of the displacement of soil mass or rate of change of groundwater level with some – any control point exceeds a predetermined threshold, the equipment together with the data sends an alarm. Alarm is the basis for the transition to a freelance mode, in which decisions are made on a more detailed examination of the coastal slope and, if necessary, the evacuation of people from the building trade and entertainment complex "June" and further strengthening of the coastal slope (Figs. 5.17, 5.18, 5.19, 5.20, and 5.21).

5.5 Conclusion

The "System monitoring acceleration induced on the upper part of the offshore oil and gas platforms" for deposits Lunskoye-A (LUN-A) and Piltun-Astokhskoye-B (PA-B) for Sakhalin-2 project have been developed and constructed by Sergeev Institute of Environmental Geoscience RAS (IEG RAS). The system was successfully modified for landslides' alarm monitoring and used for coastal slope of the river Yenisei and for a number of large industrial objects and urban areas.

Fig. 5.17 Installation and commissioning of geohazards monitoring system

Fig. 5.18 Installation and commissioning of geohazards monitoring system

Fig. 5.19 Installation and commissioning of geohazards monitoring system

Fig. 5.20 Installation and commissioning of geohazards monitoring system

Fig. 5.21 After installation and commissioning of geohazards monitoring system

References

1. Ginzburg, A. A., Manukin, A. B., Mironov, O. K., & Novikova, A. V. (2009). The spatial and spectral characteristics caused by earthquakes and other effects of fluctuations in offshore oil and gas platforms. *Geoecology*, *N5*, 80–89.
2. Ginzburg, A. A., Itzko, A. S., Kondratenko, A. I., Manukin, A. B., Mironov, O. K., & Novikova, A. V. (2008). Simulation of different impacts on offshore oil and gas platforms for the detection of dangerous earthquakes systems. *Questions Engineering Seismology*, *N2*, *35*, 5–13.
3. Nikolaev, A. V., Bashilov, I. P., Shou, K.-J., Svalova, V. B. Manukin, A. B., Zubko, Y. N., Behterev, S. V., Kazantseva, O. S., & Rebrov, V. I. (2011). Some directions of works on maintenance of geological safety of engineering constructions. In *Proceedings of ENGEOPRO*, Moscow, p. 7.
4. Postoev, G. P., & Svalova, V. B. (2005). Landslides risk reduction and monitoring for urban territories in Russia. In *Proceedings of the first general assembly of ICL (International Consortium on Landslides), "Landslides: risk analysis and sustainable disaster management"* (pp 297–303), Washington, DC: Springer.
5. Svalova, V. B. (2009). Mechanical-mathematical modeling for sedimentary movement and landslide processes. In *CD proceedings of the international association for mathematical geosciences meeting (IAMG 2009)*, Stanford, p. 15.
6. Svalova, V. B. (2011). Mechanical-mathematical modeling and monitoring for landslide processes. *Journal of Environmental Science and Engineering, 5*(10), 1282–1287.
7. Svalova, V. (2011). Monitoring and modeling of landslide processes. *Monitoring Science and Technology, 2*(7), 19–27 (in Russian).
8. Svalova, V. B. (2011). Landslide process simulation and monitoring. In *Proceedings of ENGEOPRO*, Moscow, p. 7.

9. Svalova, V. B. (2012). Mechanical-mathematical modeling and monitoring for landslides. In *Proceedings of IPL (International Program on Landslides) symposium* (pp. 63–68). Paris: UNESCO.
10. Svalova, V. B. (2013). Risk reduction for landslide hazards. Modeling and monitoring. In *Proceedings of The International Conference Natural risks: analysis, estimation, mapping* (pp. 157–163), Moscow: MSU. (In Russian).
11. Svalova, V. B. (2014). Modeling and monitoring for landslide processes. In K. Linwood (Ed.), *Chapter in book: Natural disasters – typhoons and landslides – risk prediction, crisis management and environmental impacts* (pp. 177–198). New York: Nova Science Publishers.
12. Svalova, V. B. (2014). Mechanical-mathematical modeling and monitoring for landslide processes. In *IPL 163 Project. Proceedings of the world landslide forum 3* (Vol. 4, pp. 24–27). Beijing.
13. Svalova, V. B. (2014). Modeling and monitoring for landslide processes: Case study of Moscow and Taiwan. In *Proceedings of the world landslide forum 3* (Vol. 4, pp. 628–632). Beijing.
14. Svalova, V. B. (2015). Mechanical modeling and geophysical monitoring for landslide processes. In *Proceedings of IAEG XII congress "Engineering geology for society and territory" Torino-2014* (Vol. 2, pp. 345–348), Italy: Springer.
15. Svalova, V. (2017). *Landslide risk: Assessment, management and reduction* (p. 253). New York: Nova Science Publishers.
16. Svalova, V., & Postoev, G. (2008). Landslide process activization on sites of cultural heritage in Moscow, Russia. In *Proceedings of the first world landslide forum 2008*, Tokyo, p. 4.

Chapter 6
Problems in Forecasting Dangerous Processes and Phenomena on the Base of the Earth Remote Sensing Data

M. A. Bolsunovsky

Abstract The usage of the Earth remote sensing data (RS) optimally contributes to the solution of problems in forecasting dangerous processes and phenomena. The method of radar interferometry is especially effective for solving such problems. Radar imagery, in fact, does not depend on the weather nor on the lack of sunlight. It enables to monitor displacements and deformations of the Earth's surface at the highest level of accuracy and to determine the condition of buildings and constructions. The processed data of satellite radar imagery is used for creating maps fixing changes in the Earth's surface and structures as of each imagery date in vector and raster formats, being accompanied by technical reports. In addition, maps of vertical and horizontal shifts can be compiled. The analysis of the received materials allows to determine dynamics and direction of processes and to forecast the dangerous situation development objectively.

Lately, due to the increased number of satellites conducting regular satellite imagery in the visible and infrared bands with high and ultrahigh spatial resolution, it became possible to monitor changes in the state of the Earth's surface with a high degree of regularity. For example, the planet constellation of almost 200 small satellites provides imagery of every part of the Earth's surface daily with high resolution. Such combination of imagery frequency, resolution, and capture is a fundamentally new stage in remote sensing, and such kind of imagery can be applied successfully to a wide range of tasks, including monitoring and predicting hazardous processes.

Sovzond has developed a technology for regular aerospace monitoring of potentially hazardous regions (e.g., mining areas), with the integrated application of satellite and aerial (from unmanned space vehicles) imagery data. The proposed methodology enables to identify areas with adverse engineering-geological factors, to assess the impact of natural and anthropogenic factors on the development of unfavorable processes.

M. A. Bolsunovsky (✉)
Sovzond Company, Moscow, Russia
e-mail: sovzond@sovzond.ru

V. Svalova (ed.), *Natural Hazards and Risk Research in Russia*,
Innovation and Discovery in Russian Science and Engineering,
https://doi.org/10.1007/978-3-319-91833-4_6

The hazardous processes and phenomena monitoring is becoming an increasingly popular trend of information support in the modern world being complicated with every passing day.

The remote sensing data (RS) optimally contribute to solve the following problems of forecasting hazardous processes and phenomena:

- Creation of the most accurate, relevant, and visible electronic geospatial basis
- Quick search for the location of dangerous processes and phenomena and near-real-time monitoring of development at all stages
- Detection of local emergency situations – landslides, screes, avalanches, subsidence, swellings, and erosion, as well as the related ruptures of pipelines and railways and collapse of buildings on the base of optical imaging and radar interferometry
- Creation of some different-scale situational schemes (from the whole region (populated areas, roads, orography, hydrography) to a specific city (streets, large buildings, rivers, relief)) in the form of vector electronic maps for overlaying them on the images to analyze emergencies and further forecasting
- Accurate mapping of the emergency consequences with obtaining vector electronic layers of destroyed residential and public buildings, industrial constructions, structures, and communications
- Determination of all types of damage
- Creation of three-dimensional models of potentially hazardous objects and life support facilities for the population through satellite stereoscopic imaging

Let us designate separately the problems successfully solved through the methods and satellite monitoring technologies in areas of catastrophic natural disasters, such as earthquakes and volcanic eruptions:

- Identification of the changes that occurred in the natural and man-made landscape as a result of a devastating earthquake, including in automatic mode
- Creation of the digital terrain model (DTM) and research of the Earth's surface displacements caused by an earthquake
- High-precision detection of vertical and horizontal displacements of the Earth's surface on the basis of radar data
- Intelligence for rescue and recovery operations
- Rational use of human and technical resources for rescue and recovery operations
- Assessment and calculation of damages
- Detection of the eruption predecessors
- Assessment of the eruption scale
- Monitoring of the changes in the volcano craters' morphology
- Creation of the digital terrain model and research of the landscape changes caused by volcanic activity
- Assessment and estimation of damages

Types of emergencies on the suitability of satellite images for information support may be divided into the following classes:

In no way detected on the satellite images (prediction, search, monitoring, consequences):

- Crashes and accidents in the subway
- Emissions of pathogenic microorganisms
- Epidemics and poisoning
- Epizootics

The satellite images fix only consequences, rarely – the source:

- Accidents accompanied by the release of chemically hazardous substances
- Emissions of methane, carbon dioxide, etc.
- Accidents at nuclear power plants, reactors, and storage facilities
- Hazardous meteorological phenomena

Frequently occurring and small in size, clearly localized emergencies for which satellite monitoring is ineffective:

- Car accidents (except for major accidents)
- Fires in buildings and structures
- Accidents in utility networks

The satellite images enable to perform a full range of works (search, monitoring, damage assessment, forecasting)

- Crashes and accidents of freight trains and ships, aviation and space disasters, oil pipeline and drilling platform accidents, and emissions from oil and gas fields
- Sudden collapse of buildings and structures, accidents at power plants, power networks, and treatment plants, and hydrodynamic accidents
- Volcanic eruptions, earthquakes, landslides, landfalls, screes, karst processes, abrasion, erosion, kurums, avalanches, tsunamis, surges, ice formation, high water levels, low water levels, natural fires, damage to forests and crops by pests and diseases, and major terrorist acts

6.1 Satellite Radar Monitoring

The radar interferometry method is particularly effective for solving problems related to the dangerous processes and phenomena prediction. Radar imaging, in fact, does not depend on weather nor on the sunlight presence. It provides an opportunity to:

- Monitor the Earth's surface displacements and deformations with millimeter accuracy (differential interferometric processing), including detection of landslide
- Monitor the condition of buildings and structures
- Study earthquake consequences
- Carry out near-real-time monitoring of ice conditions and shipping

- Monitor the sea coastal water pollution by oil products as a result of fuel draining from ships
- Detect cuttings and windfalls
- Assess the damage from emergencies promptly

The result of the satellite radar imaging data processing are the maps that capture changes in the Earth's surface and structures as of each imaging date in vector and raster formats, accompanied by technical reports. In addition, maps of vertical and horizontal shifts can be compiled. The analysis of the received materials allows to define dynamics and a direction of processes and to make the objective forecast of the dangerous situation development.

Among all the technologies of interferometric processing of radar images, the most accurate measurement of displacements (with sub-centimeter accuracy) is achieved on the base of the interferometry of the radar signal constant scatterers, which, in the first stage of interferometric processing, enables to reveal the so-called constant scatterers (reflectors) of the radar signal, i.e., most stably reflecting the radar signal of a point or "reference points" for a satellite radar. After that only on these points the displacement and deformation are calculated.

Such approach enables to achieve sub-centimeter accuracy of displacement measurements for points – constant reflectors of the radar signal. At a site of 10×10 km, more than 2,000,000 permanent radar reflectors can be detected during radar image processing. That is, the average density of points with known displacements is about 20,000 reference points for radar points per 1 km^2 (local density of points varies greatly depending on the type of territory, reaching a maximum in built-up areas and a minimum on agricultural fields, in forests, and on water).

This technology enables to determine the displacements and deformations of the Earth's surface and structures with very high accuracy (up to several millimeters).

The main advantage of the method is regular independent remote estimation of displacements all over the area of the image. To calculate the displacements, an array of satellite data is used which is delivered from space vehicles on a periodic basis (up to eight times a month).

The main task to be solved is to regularly obtain information about the displacements and deformations of the Earth's surface and structures by remote (satellite) methods.

The Work Stages of Satellite Monitoring of Displacements and Deformations

The first stage – the planning and ordering of a target multi-visit radar satellite imaging – is conducted to obtain the initial array of radar data for further calculation of the displacement and deformation maps of the Earth's surface and structures.

The final results of the stage are 30 radar imageries of the territory of interest in 30 different dates. These 30 imageries can be held for a minimum period of

5–6 months (the optimal period, covering basically the snowless period of the year from April to October, that enables to monitor the intensive displacements up to 1 m per year) to several years (it is suitable for monitoring the displacements and deformations of the city structures, where the displacements are not too intense).

The second stage is the interferometric data processing of multi-visit radar satellite imaging – it is carried out to calculate the displacement and deformation maps of the Earth's surface and structures from the initial array of the 30-pass radar imaging data.

The final results of the stage are the displacement and deformation maps of the Earth's surface and structures as of each imagery date in vector and raster formats in electronic form, accompanied by technical reports. The displacement map is an array of points – the constant scatterers of the radar signal, in the attributes of which – the displacement in millimeters in the direction of the radar beam as of each imagery date relative to the date of the first imaging. A map of vertical displacements can be additionally calculated. In case of imaging in double geometry (30 imageries on the ascending coil of the orbit and 30 imageries on the downward coil of the orbit), one can obtain both vertical and horizontal (in the direction of west-east) displacement components. In the absence or a small amount of vegetation in the territory, it is possible to perform an additional area processing to the SBAS method, which yields bitmap displacement files (in pixels – the offset value in millimeters) and contour offsets.

Here are some examples of projects implemented by Sovzond, related to forecasting the hazardous process development based on radar satellite imaging.

Satellite Radar Monitoring of the Side Deformations of the Open-Cast Mines of JSC "Gaisky GOK"

The Project Purpose The displacement and deformation monitoring of the sides and ledges of the open-cast mines at the Gaisky copper ore deposit, buildings, and structures of the industrial site and tailing dumps, as well as the adjacent city of Guy, based on the satellite radar imaging data (Figs. 6.1 and 6.2).

Results

- A 30-pass satellite radar imaging of the Gay deposit territory from the COSMO-SkyMed-1-4 satellites has been planned and completed.
- Interferometric processing of 30 radar imageries using interferometric methods of pair constant reflectors (PSP-IfSAR) and also using the series of small baseline (SBAS) method has been performed.
- The customer has been provided with the displacement and deformation maps of structures with millimeter accuracy and of the Earth's surface with centimeter accuracy.
- Two hundred thousand points on the area of 100 km^2 (i.e., about 20,000 points per square kilometer) were identified, which are constant (stable) radar signal

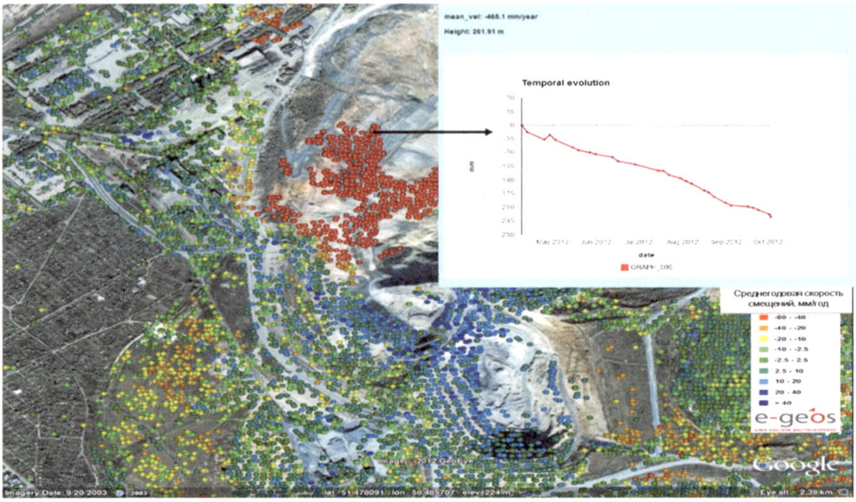

Fig. 6.1 Constant scatterers of the radar signal, revealed by the method of interferometry of permanent scatterers on the sides and ledges of open-cast mines of JSC "Gaisky GOK." From yellow to red, the growing subsidence, blue color, is rising. In the upper right corner – the dynamics of displacements for one of the most intensely settling points over the area of underground mining (the western side of quarry No. 1). The point is settling from April to October by about 25 cm. Background substrate – an optical snapshot from Google Earth

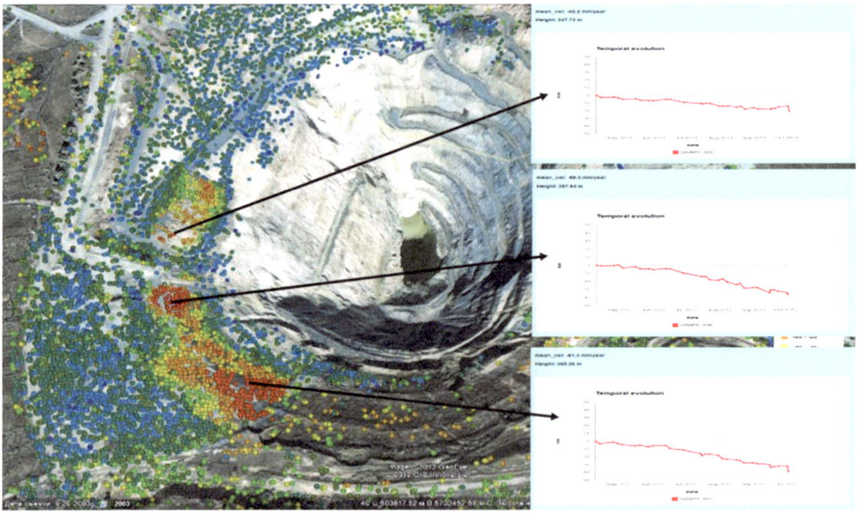

Fig. 6.2 Dynamics of displacement for April–October 2012 for points in the centers of deformation foci in the southwestern part of the pit No. 2 (the northern and southern foci are characterized by linear subsidence of 2 and 4 cm, respectively, and the focus between them – accelerating subsidence to 4 cm). Background is an optical image from Google Earth

reflectors, determined on the terrain during the interferometric processing of a series of radar images.
- For each of these reflectors, the displacements in millimeters are calculated as of the date of each of the 30 imageries made during the period from April to October 2012 (the offsets are measured relative to the date of the first imageries).
- The displacement measurements by constant reflectors are carried out with very high accuracy, but they are not evenly distributed in space. Thus, for example, the most intensive subsidence was expected on the northern side of quarry No. 1, where constant reflectors were not defined at all, since the points too much subsidenced cease to be permanent reflectors due to loss of coherence. Therefore, the same 30 radar imageries were additionally processed by the interferometry SBAS method, giving slightly less accurate but more spatially uniformly distributed displacement data.
- The correlated results of interferometry of pair constant scatterers and SBAS interferometry have been supplied to the customer.

Satellite Radar Monitoring of Displacements and Deformations of the Earth's Surface and Structures in the City of Moscow for the Period 2011–2015

The Project Purpose Reducing the risks of emergencies and reducing their possible consequences through the timely detection of displacements and deformations of the Earth's surface and structures (Figs. 6.3 and 6.4).

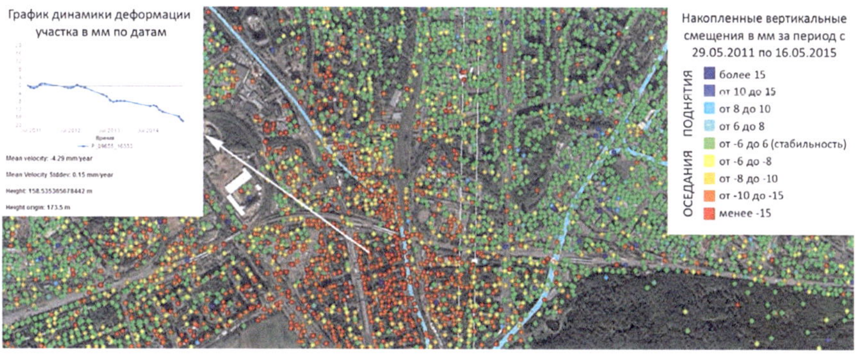

Fig. 6.3 Subsidence area in the north of Moscow. Crossing Lublinsko-Dmitrovskaya and Serpukhovsko-Timiryazevskaya subway lines

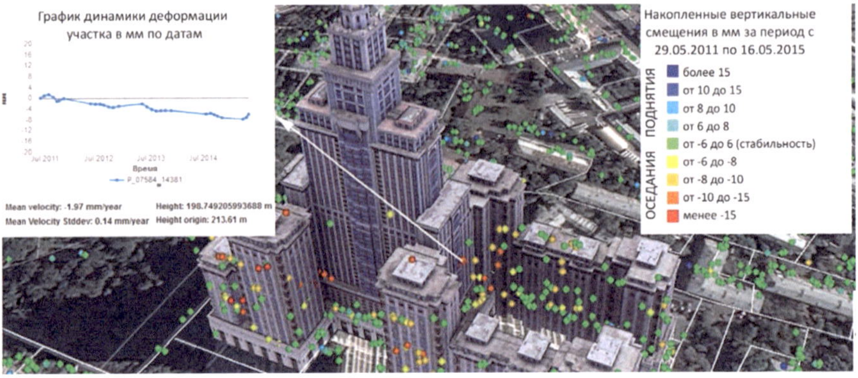

Fig. 6.4 Three-dimensional mapping of stable radar signal reflectors on the building of the residential complex "Triumph Palace"

Results

- The imagery from the COSMO-SkyMed-1-4 satellites (30-revisit satellite radar imaging obtained for the period May 2011–March 2015 in interferometric mode with 3-m spatial resolution) has been performed.
- Maps of the Earth's surface vertical displacements, buildings, and structures in the city of Moscow (within the MKAD) have been got in the form of a vector file of points, in attributes of every of them are vertical displacements as of each imagery date relative to the date of the first imaging.
- The obtained vertical displacement model contains 10,625,327 points – stable reflectors of the radar signal, i.e., averagely about 6600 points per 1 km^2. The point density in the sites with dense development reaches 20,000 points per 1 km^2.

The Main Result A vast area of sediment in the north of the city has been revealed. A section of about 5 × 3 km in size is characterized by spatially correlated subsidence developing from the site contour to its center. Subsidence at the site were nonlinear over time, intensified at the end of 2012, and reached 20 mm in the site center by May 2015, which is probably due to the new subway line construction.

6.2 Satellite Regular Monitoring Using the Data of Small EO Satellite Constellations

The recent growth in the number of satellites conducting regular satellite imaging in the visible and infrared ranges with high and ultrahigh spatial resolution facilitates to monitor changes in the state of the Earth's surface at a high degree of regularity. For example, the planet has a fleet of more than 200 small satellites providing imagery of every part of the Earth's surface daily with high resolution. Such combination of frequency, resolution, and coverage is a fundamentally new stage in Earth observation, and such imagery can be successfully applied to a wide range of tasks, including monitoring of hazardous processes and predicting them.

Possibilities to Prevent Flooding and to Analyze Its Impact with the Help of Planet Images, Implemented by Sovzond

The planet satellites' data allowed analyzing the flood development in the valley of the Ishim River in May 2017. Twelve images were selected for the area of interest (the vicinity of the Abatskoe village) for the period April 30–May 22. The minimum interval between the images was 1 day; the maximum, 3 days; and the average, 2 days.

The analysis was conducted on the base of the central part of the Abatsky District along the Ishim River on the area of 37,616 ha. A more detailed analysis was carried out on two sites: in the area of the Abatskoe village and the Likhachev village. The water level in the Ishim River did not change until May 1. On May 11, a dam was broken near the Ishim city, and the flood zone began to move downstream, and flooding covered most of the coastal areas of the Abatsky District (Figs. 6.5 and 6.6, green color).

On May 13, the water level in the Ishim River continued to rise, and the flood spread all the way down (yellow color). The flood peak in the Abatsky District fell on May 15–17, 2017 (additional flooded areas are shown in lilac color). At this time, the water level in the Ishim River has risen too high, and thus several houses, a horticultural partnership with low-rise buildings, and a section of the Abatskoe-Vikulovo road were flooded.

On May 19, the water began to recede from those settlements that were upstream, but continued to rise downstream, near the Abatskoe village. Since May 22, the water level in the Ishim River began to fall off gradually.

A more detailed analysis was carried out for the Likhachev village. To analyze the flood development, the PlanetScope images were used for the period May 11–22. The flooded areas were allocated in an automated mode, based on the algorithms designed in Sovzond.

The flooding peak in the Likhachev village of the Abatsky District took place on May 15, 2017. A total of 17 households were flooded (Fig. 6.7).

6.3 Aerial Survey

Aerial survey is one of the most effective methods of obtaining geospatial data for forecasting dangerous processes and phenomena due to high accuracy (up to several centimeters per pixel) and image quality. Aerial survey is carried out from airplanes, helicopters, and other aircraft using special cameras.

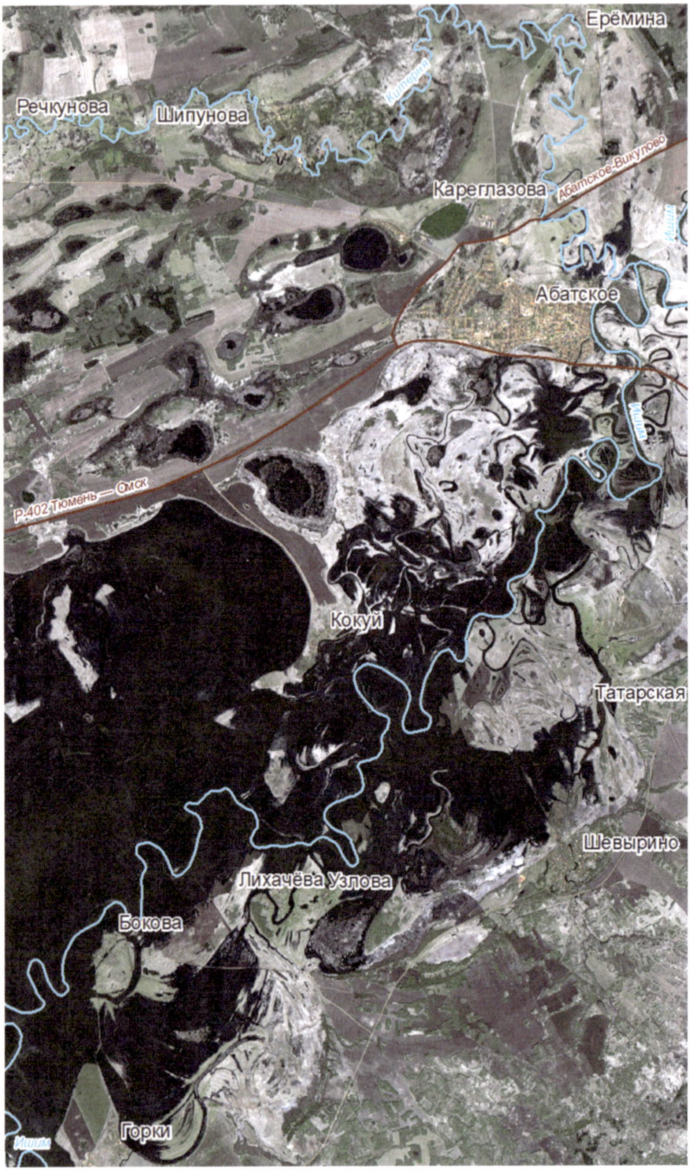

Fig. 6.5 The water level in the Ishim River on the planet image of May 11, 2017

The main advantages of aerial survey are the following:

- High visual information content.
- High measuring qualities can be provided by laser location.
- Use of quality digital aerial photographs as an information basis.

Fig. 6.6 The boundaries of the Abatsky District flooding by dates (green, May 11, 2017; yellow, May 13, 2017; pink, May 15–17, 2017; red, May 19, 2017)

Fig. 6.7 Automatically highlighted flooding boundaries of the Likhachev village on dates based on the planet image processing

- Use of inclined aerial photographs, as images, providing great recognition of the object.

At present, for example, survey from unmanned aerial vehicles (UAV) is becoming more popular to conduct operational aerial survey in man-made and natural disasters.

Aerial survey (including UAV-based survey) enables to obtain high-resolution digital images. Depending on the type of camera used, the following tasks can be performed:

- Automated obtaining of orthophoto mosaics with spatial resolution up to several cm
- Creation of photo-realistic digital 3D terrain models
- Monitoring of emergencies and its impact
- Control over emergency recovery operations, search for victims
- Analysis and evaluation of the terrain-change dynamics
- Industrial and environmental monitoring
- Thermal imaging

The air, ground, and mobile *laser scanning* technology is also used to map the territory and create 3D models of objects. Air laser scanning is used for high-precision mapping of linear and area objects in scales of 1:500–1:2000. The ground and mobile laser scanning technologies are used to obtain very detailed 3D models of objects in scale 1:50 and topographic plans in scale up to 1:500.

The main advantages of laser scanning technology are the following:

- The cost of imaging and modeling of objects is lower about three times as compared with classical technologies.
- The cumulative speed of imaging and data processing obtained through laser scanning is several times faster than conventional geodesy and aerial photography.

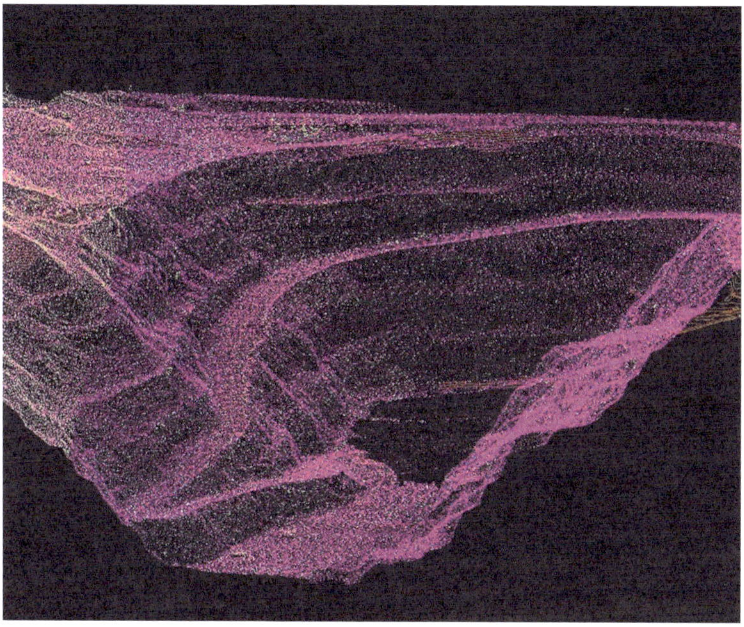

Fig. 6.8 Creation of an accurate digital elevation model basing on laser scanning results

- The accuracy of laser scanning is comparable to the accuracy of ground-based geodesy and much higher than the aerial photography accuracy.
- Vegetation, haze, and nighttime do not affect the performance of work.

Laser scanning for forecasting dangerous processes and phenomena is most effective for:

- 3D-modeling of objects (Fig. 6.8)
- High precision of structures surveying
- Monitoring of the Earth's surface displacements
- Rapid assessment and survey of the area

The Sovzond company has developed a technology for the regular aerospace monitoring of potentially hazardous areas (e.g., mining areas), comprehensively using satellite and aerial imaging data (aerial survey and airborne laser scanning). The proposed methodology allows to identify areas with unfavorable engineering and geological factors and to assess the impact of natural and man-made factors on the development of adverse processes.

In conclusion, we note that the examples given in the article do not exhaust all the satellite monitoring possibilities to solve the problems of forecasting dangerous processes and phenomena. In particular, the scope of the review did not include the satellite monitoring of forest fires – an actual and actively used direction of satellite and aerial survey.

Chapter 7
Assessment and Mapping Geological Risk for the Future Subsurface Linear Construction in Moscow

Olga Eremina, Irina Kozlyakova, Oleg Mironov, Nadezda Anisimova, and Irina Kozhevnikova

Abstract Geological risk as the qualitative or quantitative measure of geological hazard or a complex of hazards established for a certain object in the form of possible absolute or relative economic losses (damage) is the function of affecting hazard(s) value and the vulnerability of engineering structure. The georisk analysis in urban areas is usually aimed at the existing urban infrastructure, whereas assessment of geological risk for future construction appears to be a new approach in risk analysis. In urban areas, the risk of probable losses for a particular engineering structure in the course of its construction and operation may be comparatively analyzed for different types of engineering geological conditions distinguished in the area proceeding from the assessment of geological hazards that affect the engineering structures, since the vulnerability of engineering structure to these geological hazards is taken as constant in this case. Upon this approach, the qualitative characteristics of possible damage from geological hazards within the area with engineering geological conditions of a certain type serve as the risk index. According to the developed procedure, the map of geological risk upon the construction and operation of shallow (20 m deep) tunnels was compiled to a scale of 1:100,000 for the territory of Moscow. The possible damage was assessed proceeding from the analysis of such hazards as groundwater and quicksand outburst in the construction pit, suffusion, and karst-suffosion processes affecting the building structures. Very high geological risk arises upon the construction and running tunnels in water-saturated sandy ground, whereas the low risk is identified for the tunnels running in low-permeable Jurassic clay.

O. Eremina (✉) · I. Kozlyakova · O. Mironov · N. Anisimova · I. Kozhevnikova
Sergeev Institute of Environmental Geoscience RAS (IEG RAS), Moscow, Russia
e-mail: sci-council@geoenv.ru

V. Svalova (ed.), *Natural Hazards and Risk Research in Russia*,
Innovation and Discovery in Russian Science and Engineering,
https://doi.org/10.1007/978-3-319-91833-4_7

7.1 Introduction

Nowadays it is commonly recognized that in urban areas, the assessment and mapping geological hazards alone, without taking into consideration the risk of economic damage to urban engineering facilities, appear to be insufficient. In cities, the losses from geological processes are nearly equally controlled by engineering geological conditions as well as by the urban infrastructure load. The acute task of geological studies in urban areas involves the transition from the assessment and mapping of geohazards to the assessment and mapping of geological risk produced by the geohazards in question. This assessment will give us an adequate idea about the possible damage and economic loss. For the planned construction of subsurface engineering linear facilities, this assessment permits researchers to analyze the alternative project decisions and thus to minimize the economic losses during the construction and operation of engineering structures. The development of risk analysis technology caused by the hazardous geological processes is the most important task. Due to geodata uncertainties, this problem seems to have been solved on a quantitative level nowhere in the world [1–3].

Almost all researchers involved in risk assessment in urban areas agree that the combination of geohazard maps with the maps of urban environment vulnerability is the most promising approach to georisk assessment in cities [4–7]. It should be noted that the most efforts are usually aimed at the Georisk Analysis for the existing urban infrastructure, whereas mapping geological risk for future construction in cities appears to be a new approach in risk analysis. The present study was aimed at the development of the procedure of qualitative georisk assessment in urban areas for the future subsurface construction in Moscow megacity to be used at the preliminary stage of engineering survey.

7.2 Engineering Geological Conditions in Moscow

Engineering geological conditions of the Moscow region were comprehensively studied before by many researchers [8–14]. In terms of geology, Moscow is located within a vast syneclize in the center of the East European platform covered by a thick mantle of sedimentary deposits. Within the depth of technogenic impact, the sedimentary mantle in Moscow area consists of stratified Carboniferous, Jurassic, Cretaceous, and Quaternary deposits (Fig. 7.1). The bottom of engineering geological cross section is composed by marine terrigenous and carbonate deposits of Carboniferous system, with their upper limestone horizons being intensely karstified.

The Carboniferous massif is overlain by marine clayey-loamy-sandy deposits of Jurassic and Cretaceous age. The continuous strata of Jurassic clay form a regional aquiclude, protecting groundwater from contamination. These strata also show high physico-mechanical properties and serve as a reliable foundation and enclosing

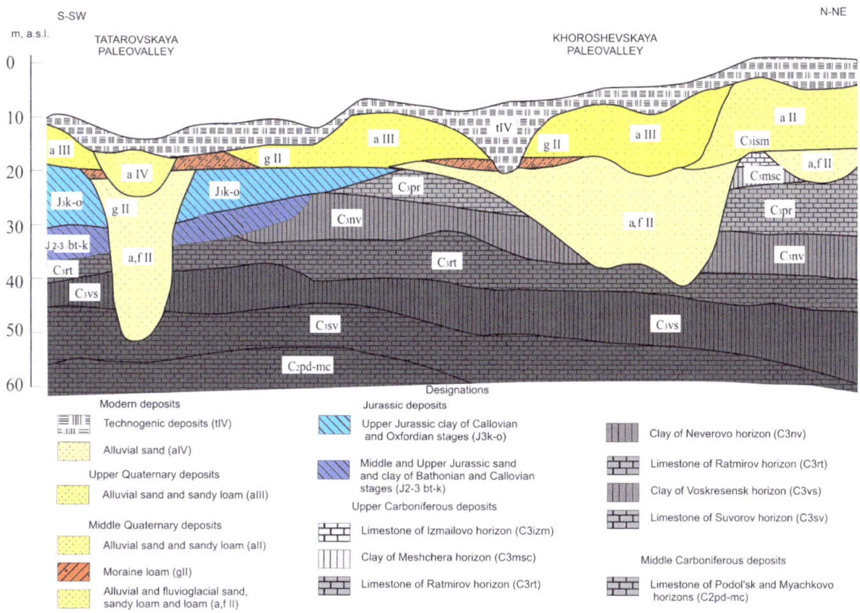

Fig. 7.1 Geological cross section through the center of Moscow

medium for both surface and subsurface engineering structures. Quaternary alluvial, fluvioglacial, lacustrine, and bog weak soils may manifest specific properties (e.g., high moisture, high organic content, low load-bearing capacity and strength, high compressibility, etc.) which occur at the top of the geological cross section. Another important feature of Moscow geology is the presence of three generations of buried river valleys and erosional cuttings formed during the long-lasting continental breaks by the Pra-Moscow River and its tributaries. These buried erosional channels disturb the hydrodynamic regime in the area and complicate significantly the ground conditions.

At the surface, the above mentioned specific features of Moscow engineering geological conditions are manifested in karst and suffusion phenomena, landslides, and waterlogging. Upon the subsurface development, the following problems arise: karstification and fracturing of limestone massif, decompaction and swelling of clay, liquefaction of sand and quicksand phenomena, as well as groundwater level rise, groundwater breakthrough in tunnels, and groundwater contamination.

7.3 Procedure of Mapping Geological Risk in Moscow for Planned Subsurface Facilities

Geological risk is defined as the qualitative or quantitative measure of geological hazard or a number of hazards determined for a particular object in the form of possible absolute or relative economic losses (damage) [15, 16]. This definition suggests that risk is a function of the hazard impact value and the engineering structure vulnerability. The comparative analysis of the risk of losses for a particular engineering structure at the stage of its construction and operation may be carried out only by assessing the impact of geohazards on this engineering structure, since the value of engineering structure vulnerability will be the same in this case, and assuming that this engineering structure may be built in different engineering geological conditions. The risk will be characterized by the possible damage caused by geohazards within the areas with certain type of engineering geological conditions. Upon this approach, the qualitative risk analysis in this case may be reduced to assessing the impact of geohazards on the given engineering structure by comparing the future expenditures for its construction and operation under different engineering ecological conditions.

Let us scrutinize the procedure of assessment and mapping risk for the construction of a metro tunnel in Moscow driven at a shallow depth of 20 m with a diameter of 6 m constructed by using an open-pit method and using the protective frame filling. The proposed procedure consists of the following steps:

1. Learning the geological structure and hydrogeological conditions in the foot and walls of the tunnel at the planned construction depth proceeding from the 3D model of geoenvironment
2. Typification of engineering geological conditions and the analysis of geohazard impact on the engineering structure for each type of engineering geological conditions
3. Determination of the possible losses (expenditures) during the construction and operation of the facility
4. Distinguishing and substantiation of the comparative qualitative categories of risk
5. Building the map of geological risk

In this analysis, the geological conditions at the construction depth are of particular importance. Figure 7.2a, b shows the geological structure of Moscow subsurface at the depths of 14 and 20 m, respectively, which correspond to the foot and top of the driven tunnel, obtained from the 3D geological model to a scale of 1:100,000.

We typified the possible ground conditions at the studied depths in three groups. The following types were distinguished:

- Meso-Cenozoic sand with interlayers and lenses of sandy loam and loam (Quaternary, Cretaceous, and Jurassic). For the most part, soils are water-saturated. Aquifers are confined and unconfined.
- Jurassic clay, weakly permeable.
- Carboniferous terrigenic and carbonate deposits: limestone, dolomite, marl, and clay. Unevenly watered with confined aquifers.

The combinations of these types in the walls and in the bottom of the tunnel result in seven types of possible engineering geological conditions along the tunnel route. The possible damage for the tunnel under construction was assessed proceeding from the analysis of the impact on it of such geohazards as groundwater and

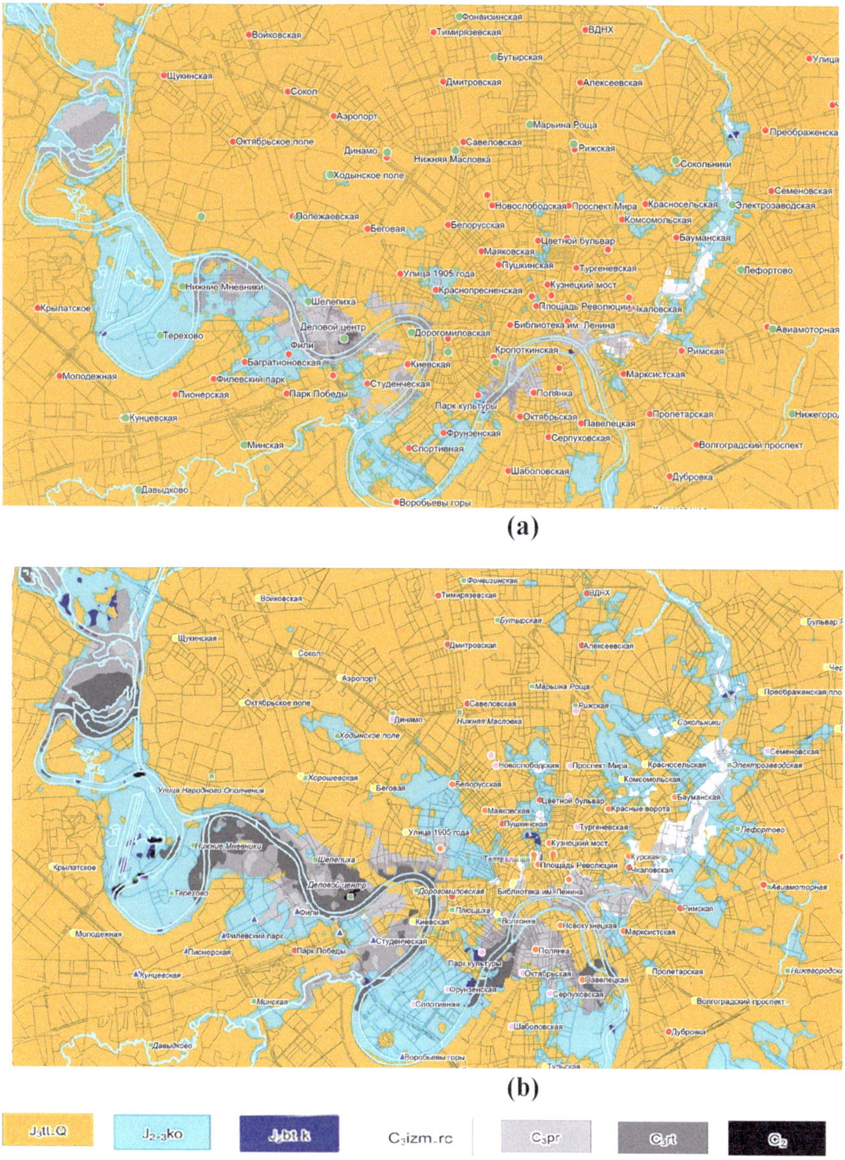

Fig. 7.2 Geological structure of Moscow subsurface at a depths of (**a**) 14 and (**b**) 20 m, scale 1:100,000

quicksand breakthrough to the construction pit, suffusion, and karst-suffosion processes. As a result, we distinguished and substantiated four risk categories: very high, high, moderate, and low (Table 7.1). Very high geological risk is recognized for the construction and operation of tunnels in Meso-Cenozoic water-

Table 7.1 Legend to the map of geological risk for the construction and operation of shallow tunnels (a depth of 20 m)

Deposits in the tunnel walls / Deposits in the tunnel foot	Meso-Cenozoic sand with interlayers and lenses of sandy loam and loam; mainly water-saturated; confined and unconfined aquifers	Weakly permeable Jurassic clay	Carboniferous terrigenic and carbonate deposits: limestone, dolomite, marl, and clay. Unevenly watered with confined aquifers .
Meso-Cenozoic sand with interlayers and lenses of sandy loam and loam; mainly water-saturated; confined and unconfined aquifers	Very high risk	Moderate risk	
Weakly permeable Jurassic clay	High risk	Low risk	
Carboniferous terrigenic and carbonate deposits: limestone, dolomite, marl, and clay. Unevenly watered with confined aquifers.	High risk	Moderate risk	Moderate risk

saturated sand, which is proved by many unfortunate cases of the shallow tunnel construction on Moscow. Meso-Cenozoic sand outcropping in the tunnel's walls gives rise to a high risk for an engineering structure irrespectively of the deposit types in the tunnel foot. Running a tunnel in the karstified Carboniferous deposits (limestone, dolomite, marl, and clay), due to their karstification and uneven watering with confined groundwater aquifers, produces a moderate risk to engineering structures. A low risk is usually identified for the tunnels drawn in the low-permeable Jurassic clay horizon.

The content of these risk categories is described in Table 7.2. It makes allowance for the distinguished type of engineering geological conditions, the possible manifestations of geological hazards, and the expected economic losses in the course of construction and operation of engineering facility.

7.4 The Map of Geological Risk in Moscow at a Depth of 14–20 m and Its Application for Subsurface Construction

According to the developed procedure, the map of geological risk for the construction and operation of shallow (a depth of 20 m) tunnels was compiled to a scale of 1:100,000. The territory of Moscow enclosed within the Moscow Automobile Ring

Table 7.2 The content of risk categories

Risk categories	Engineering geological conditions	Possible manifestations of geological hazards	Economic losses caused by geohazard manifestation in the course of construction and operation
Low	Low-permeable clay in the walls and floor of the tunnel	Destructive manifestations of geohazards are almost improbable upon using protective frame fillings	If construction technologies are followed adequately, the loss caused by geohazard manifestation is almost excluded
Moderate	Limestone containing a confined groundwater aquifer in the walls and in the floor of the tunnel	Groundwater breakthrough into the tunnel is possible upon the accidental disturbance of protective frame filling upon driving the tunnel	Additional expenditures for the elimination of accident consequences and the repair of protective frame filling in the period of construction
	Limestone containing a confined groundwater aquifer in the floor of the tunnel and low-permeable clay (or limestone and clay) in the tunnel walls		
	Water-saturated sand in the walls of the tunnel (with confined groundwater) and low-permeable clay in the walls of the tunnel		
High	Limestone in the floor of the tunnel and sand (or sand and limestone) in the walls. Water-saturated deposits. Confined and unconfined aquifers	Groundwater breakthrough into the tunnel is possible upon the acci dental disturbance of protective frame filling upon driving the tunnel. Inrush of quicksand or sandy-clayey filler of karstified zones in limestone is possible in the tunnel walls	Economic losses may be caused by surface subsidence and collapses above the decompacted zones near the tunnel, water leakage due to breaks in frame fillings
	Limestone in the floor of the tunnel; clay and sand in the walls. Limestone and clay contain confined and unconfined aquifers		
	Low-permeable clay in the floor of the tunnel. Water-saturated sand (or sand and clay) in the walls of the tunnel		Additional expenditures for the elimination of accident consequences and the repair of protective frame filling in the period of construction
Very high	Water-saturated sand in the tunnel floor. Water-saturated sand (or sand and clay) in the tunnel walls. Confined and unconfined groundwater	Groundwater and quicksand breakthrough into the tunnel walls and floor is possible upon the accidental disturbance of protective frame filling upon	Economic losses may be caused by surface subsidence and collapses above the decompacted zones near the tunnel, water leakage due to

(continued)

Table 7.2 (continued)

Risk categories	Engineering geological conditions	Possible manifestations of geological hazards	Economic losses caused by geohazard manifestation in the course of construction and operation
		driving the tunnel. Decompaction zones may develop in the ground massif around the tunnel	breaks in frame fillings. Deformation of protective lining above the decompaction zones cannot be excluded under the tunnel floor
			Additional expenditures for the elimination of accident consequences and the repair of protective frame filling in the period of construction

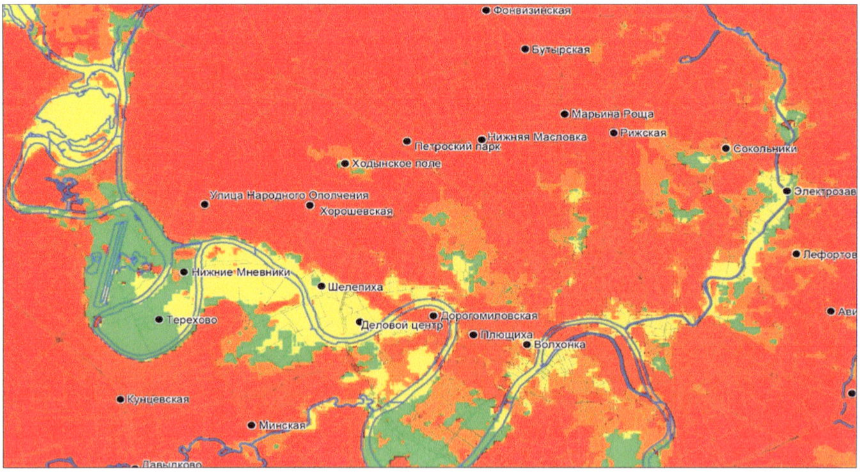

Fig. 7.3 Fragment of the map of geological risk for subsurface engineering structures at a shallow depth; the foot depth, 20 m; height, 6 m (see Table 7.1 for designations)

Road was taken as a representative case, for which the 3D model of geoenvironment has been created.

The map legend showing the different risk categories is based on the traffic light principle (Table 7.1). Figure 7.3 shows the fragment of the geological risk arising upon the construction of the subsurface linear facilities (tunnels) in the central part of Moscow cut at a depth of 20 m with a diameter of 14 m.

These maps may be compiled for different tunnel depths to come up with the 3D model of geological risk in Moscow. The task of further studies is to develop mapping technologies and to substantiate scientifically the compilation of the integral map of geological risk for the future subsurface construction by superposition and the analysis of the risk maps for two or three levels of subsurface development.

We should understand that the high risk in the territory does not mean that construction is prohibited in this area. In large cities, it appears improbable to decrease the geological risk by moving the planned engineering structure away from the hazardous regions. New metro facilities (as well as other linear transport structures) are planned in places where they are necessary for citizens. They also have to be linked to the already existing routes. Therefore, the subsurface facility may be moved to a very close distance only both in vertical and horizontal directions, and it is often impossible to avoid the zone of high geological risk. But it is possible to minimize risk or, in other words, to manage risk by scientifically grounded prediction of geohazard manifestation, optimizing estimates of possible losses, as well as the use of special construction technologies and engineering protection measures.

The principal aim of the proposed map is to reveal the zones at geological risk, within which a considerable damage may arise at any depth of laying engineering facility, and recommendations on special protective measures should be given. It will permit experts to compare the alternative project options in the viewpoint of their safety and economic efficiency at the early investment stage. It will permit, on one hand, to avoid undertaking the unnecessary protective measures raising the construction cost. On the other hand, the adequate assessment of geological risk for alternative options may permit avoiding huge amount of economic losses. The reckless choice of project decisions proceeding only from their relative cheapness may result in a very high damage.

7.5 Conclusion

Mapping geological risk is the principal method in risk analysis at the initial stages of projecting urban development. For a particular projected engineering structure, its risk is estimated as the value of probable expenses during its construction and operation. Mapping geological risk for future construction in cities appears to be a new approach in risk analysis. The principal aim of this map is to reveal the zones at geological risk, within which a considerable damage may arise at any depth of laying engineering facility, and recommendations on special protective measures should be given. Application of this approach and the analysis of the risk maps for two or three levels of subsurface development will permit experts to compare the alternative variants for driving linear engineering facilities (e.g., metro) in the viewpoint of their safety and economic efficiency at the early investment stage of implementation of large construction projects, e.g., upon laying of new metro routes in Moscow. The further progress in risk analysis and building a three-dimensional model of Moscow geoenvironment will permit urban planners to avoid substantial economic losses caused by geohazards.

References

1. Clayton, C. R. I. (2009). Urban site investigation. In M. G. Culshaw, H. J. Reeves, I. Jefferson, & T. W. Spink (Eds.), *Engineering geology for tomorrow's cities* (Vol. 22, pp. 15–141). London., Engineering Geology Special Publication: Geological Society.
2. Kalsnes, B., Nadim, F., & Lacasse, S. (2010). Managing geological risk. Geologically active. In Williams et al. (Eds.), *Proceedings of the 11th IAEG Congress*, Auckland, New Zealand, 5–10 September 2010 (pp. 111–126). London: Taylor & Francis Group.
3. Knill, J. (2003). Core values: The first Hans-Cloos lecture. *Bulletin of Engineering Geology and the Environment, 62*(1), 1–34.
4. Koff, G. L., Likhacheva, E. A., & Timofeev, D. A. (2006). *Geoecology of Moscow: Methodology and methods of assessing the urban environment state*. Moscow: Media-Press (in Russian).
5. Marchiori-Faria, D. G., Ferreira, C. J., et al. (2006). Hazard mapping as part of civil defense preventive and contingency actions: A case study from Diadema, Brazil. In *Engineering geology for tomorrow's cities. IAEG 2006*, 6–10 Sept. 2006, CD-rom, paper no. 4–154.
6. Mora, S. (2010). Disasters should not be protagonists of disaster risk. Geologically active. In Williams et al. (Eds.), *Proceedings of the 11th IAEG Congress*, Auckland, New Zealand, 5–10 September 2010 (pp. 89–110). London: Taylor & Francis Group.
7. Zhang, F., Yang Q., Jia, X., Liu, J., & Wang, B. (2006). Land-use optimization by geological hazard assessment in Nanjing City, China. In *Engineering geology for tomorrow's cities. IAEG 2006*, 6–10 Sept. 2006, CD-rom, paper no. 4–324.
8. Golodkovskaya, G. A., & Lebedeva, N. I. (1984). Inzhenerno-geologicheskoe raionirovanie Moskvy (Engineering geological zoning of Moscow). *Inzhenernaya Geologiya, 1984*(3), 87–102 (in Russian).
9. Kozlyakova, I., Eremina, O., Anisimova, N., & Kozhevnikova, I. (2016). Study of geology and carboniferous roof topography upon engineering geological mapping of Moscow territory. In M. J. Eggers, J. S. Griffiths, S. Parry, & M. G. Culshaw (Eds.), *Developments in engineering geology* (Vol. 27, pp. 45–53). London., Engineering Geology Special Publication: Geological Society. https://doi.org/10.1144/EGSP27.4.
10. Kozlyakova, I. V., Mironov, O. K., & Eremina, O. N. (2015). Engineering geological zoning of Moscow by the conditions for subsurface construction. In *Proceedings 12th IAEG Congress, Turin, Italy* (Vol. 5, pp. 923–926). Springer, 2015.
11. Kutepov, V. M., Anisimova, N. G., Eremina, O. N., Kozhevnikova, I. A., & Kozlyakova. (2011). The map of pre-quaternary deposits as a base for large-scale geological mapping of Moscow territory. *Geoekologiya (Environmental Geoscience), 5*, 399–411 (in Russian).
12. Osipov, V. I. (2014). Large-scale thematic geological mapping of Moscow area. In G. Lollino et al. (Eds.), *Engineering geology for society and territory* (Vol. 5, pp. 11–16). Switzerland: Springer International Publishing.
13. Osipov, V. I., & Medvedev, O. P. (Eds.). (1997). *Moscow. Geology and the city*. Moscow: Moskovskie uchebniki i kartolitografiya Publ (in Russian).
14. Osipov, V. I. (2008). *Geological conditions of Moscow urban development*. Moscow: ZAO Mir (in Russian).
15. Ragozin A. L., & Yolkin V. A. (2006). Geological risks, formation and assessment in urbanized areas in Russia. In *Engineering geology for tomorrow's cities*. IAEG 2006, 6–10 Sept. 2006, CD-rom, paper no. 4–282.
16. Ragozin, A. L. (Ed.). (2003). *Natural hazards of Russia. Assessment and management of natural risks. 2003. Topical vol* (p. 320). Moscow: KRUK (in Russian).

Chapter 8
Deep Extrusive Landslides; Peculiarities of Formation, Development, and Protection

Andrey Kazeev and German Postoev

Abstract The current study summarizes a vast body of knowledge encompassing an inventory of landslide cases, overview of mechanisms of landslide development, and slope stability assessments.

A new mechanism-based landslide classification is presented and discussed. The mechanism of formation of the limiting state in a soil mass during the preparation of the landslide is considered. The equations for assessment of a soil mass limited state prior landslide formation were obtained. Also the equation for calculation of a landslide slip surface depth was obtained, using slope morphological and soil strength parameters.

The article also contains scientific explanation of effective reinforcement of territories with deep block-type landslides. The suggested protection strategy bases on theoretical solutions of preparing of a destructive phase of landslide deformations, on taking into account mechanism peculiarities of deep block-type landslides, and also on investigation experience of such landslides in different engineering-geological conditions.

8.1 Introduction

First scientific publications dealing with landslide problem in Russia go back to the nineteenth century. However, major damage caused by large landslides has been historically documented even earlier. Landsliding for many times caused significant destruction of historical structures. For example, catastrophic development of a deep landslide in the slope of Volga River in 1597 resulted in a destruction of Voznesensky Pechersky monastery. This event was registered in the Chronicles. The signs of slope instability, such as deformations of a timber roadway, took place a week before

---

A. Kazeev (✉) · G. Postoev
Sergeev Institute of Environmental Geoscience of Russian Academy of Sciences (IEG RAS), Moscow, Russia
e-mail: opolzen@geoenv.ru

V. Svalova (ed.), *Natural Hazards and Risk Research in Russia*, Innovation and Discovery in Russian Science and Engineering,
https://doi.org/10.1007/978-3-319-91833-4_8

the main landslide movement, with the subsequent development of cracks over 1 km in length in the "mountain" that housed subterranean monastic caves and destruction of the buildings of the monastery.

Landslide research peaked after 1917. Theoretical and applied methods of slope stability assessment, as well as evaluations of efficiency of protective measures, have been a constant focus of specialized conferences, symposia, and workshops on landslides in Russia.

It is noteworthy that over 725 Russian cities are affected by landslides. Among them are such major cities as Moscow, Nizhny Novgorod, Kazan, Ulyanovsk, Volgograd, Cheboksary, Saratov, Saransk, Perm, Sochi, Rostov-on-Don, Tomsk, and Barnaul. Known landslides that involved deformations of Lower Cretaceous clays (i.e., observed depths of slip surfaces in excess of 100 m) occurred in Saratov on the slope of Sokolova Mountain in 1783, 1818, 1846, 1869, 1884, 1913, 1915, 1927, 1967, 1968, and 1986 and caused destructive deformations of the city area and bulging of Volga riverbed. The volume of sliding mass was estimated as about 10 million m^3 [42].

This article presents some results of our studies which deal with main aspects of the landslide hazard that might have a worldwide significance, specially landslide mechanics and categorization based on mechanism of their formation, monitoring of their development, and evaluations of stability of landslide massifs.

8.2 Landslide Mechanics and Mechanisms of Development. Definitions

Landslide researchers have always been concerned with understanding of physical and mechanical causes leading to formation and development of landslides.

Landsliding is a process of changes in stress-strain state of a slope groundmass leading to a mass separation and ground movement downslope while maintaining a continuous contact between sliding mass and underlying undisturbed ground. This definition is similar to that given by Emelyanova [8], Petrov [30], Ter-Martirosyan [40], and other authors. In accordance with this definition, the *mechanism* of a landslide is a systemic sequence of changes in stress-strain state of a slope groundmass under the influence of natural and anthropogenic factors. These factors lead to the formation and development of a landslide.

The formation of a landslide or other slope process is the result of a failure in the slope groundmass and loss of slope stability. The loss of soil strength and failure are preceded by a significant deformation. The initial stress-strain state of soil before the failure is determined by distribution of stresses in soil (gravity loading, pore water pressure, seismic forces, etc.). External forces may lead to the additional soil deformation that range from increased compactness of soil structural elements due to change of volume to a movement of soil mass caused by shear or mass wasting [31].

Soil mechanics consider such mechanisms of soil deformation as compression, shear, tension, and deformation of saturated soils ([4, 13, 21, 28, 40, 43], etc.).

At a present time, it is considered that one of the key objectives of a landslide investigation is an identification of landsliding mechanism. However, the definition of the mechanism of a landslide process given by different researches often differs. This may be partly due to the complexity of the landslide processes and a great variety of geological conditions that may lead to landsliding. Based on the definition above, the mechanism of the landslide process involves the following:

- Initiation stage (i.e., "preparatory stage" according to E.P. Emelyanova or a "phase of deep creep" according to G.I. Ter-Stepanyan) that involves physical separation of sliding masses under the influence of gravitational volumetric forces, pore water pressure, seismic forces, and factors associated with human activity
- The development (i.e., mostly "displacement") of a landslide after the separation under the influence of natural and anthropogenic factors

G.I. Ter-Stepanyan [41] emphasized that key elements of the mechanism are stress, strain, and time. Considering that a state of stress in the slope mass is realistically difficult to evaluate, G.I. Ter-Stepanyan recommends to focus on the kinematics of the process as the basis of understanding the mechanism, i.e., a relative movement of separate elements that constitute the landslide [41].

The majority of researchers reach the same conclusions, and as a result, majority of the landslide classifications are based on the mechanism of the landslide processes [3, 5, 9, 18–20, 22, 28, 30, 34, 35, 42, 47]. Furthermore, the mechanism of landslide movement is the basis for the classification in Russian guidance documents proposed for design and construction of buildings in landslide areas.

8.3 Mechanism-Based Landslide Classification

There are a great number of landslide classifications, based on certain characteristics: landsliding mechanism (i.e., the process), planar shape, magnitude of an area or volume, landslide age, depth of a slip surface, occurrence within a certain geologic formation, etc. Mechanism based classifications are the most common [5, 6, 9, 20, 30]. However, it is the mechanism of movement of a landslide mass (subsidence, uplift, sliding, rotation, translation, or flow) that is the most frequently considered, neglecting the mechanism of landslide formation (preparation, losing stability of a soil mass, and a separation of a landslide body). This often makes it difficult to correctly recognize a landslide type, because different landslide types may have a similar mechanism of *movement* in the intermediate stage of their development: i.e., movement of a landslide body along the slip surface.

For example, landslide classification based on a landslide mode of movement proposed by Cruden and Varnes [3] is quite popular. But we think that the mechanism of landslide formation (i.e., all factors, processes leading to physical movement or slide) is more important than the mechanism of displacement of the already separated part of the groundmass.

Accordingly, when classifying landslides, it is considered reasonable to take into account driving and resisting forces and a mode of deformation of the groundmass during initial (preparatory) stage of the landslide process, which would to a great degree determine the mechanism of formation and development of a landslide. Therefore, based on mechanism of formation, landslides may fall into three main categories [34, 35]:

- Deep-seated compression–extrusion type
- Shear-sliding type
- Liquefaction–flow type

The first category encompasses *relatively deep-seated block landslides of compression–extrusion* (Fig. 8.1).

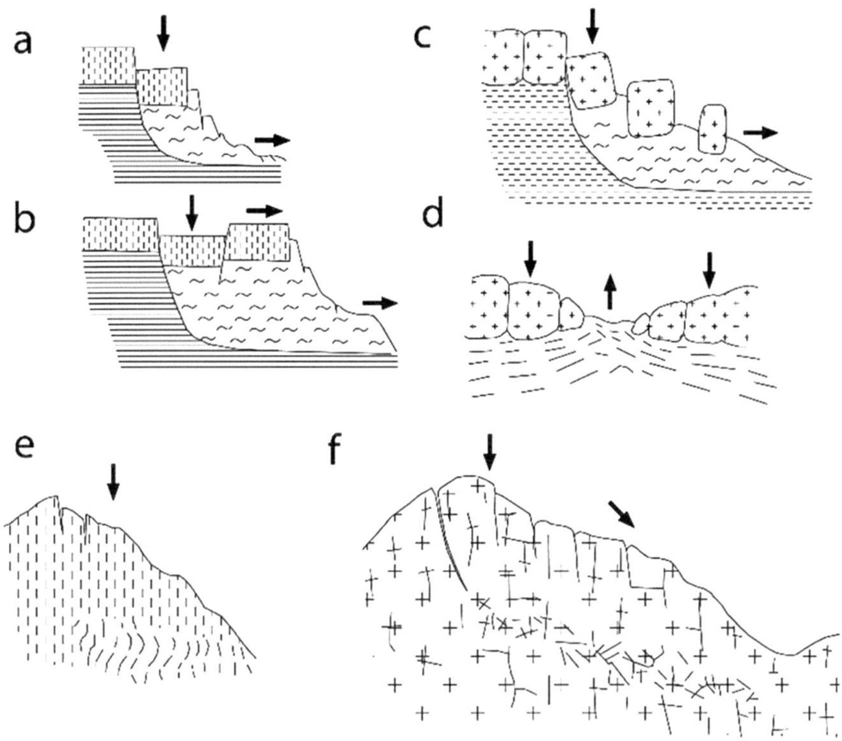

Fig. 8.1 Schemes of a landslide deformation with compression–extrusion mechanism: (**a**, **b**) the compression landslide in cohesive soils; (**c**) the subsidence and the spread of landslide blocks in rocks; (**d**) bulging of the valley bottom; (**e**) gravitational folds, deep creep with the S-shaped deformation of layers; (**f**) gravitational deformation of ridges

The block-type landslides of compression–extrusion mechanism are the most widespread and can be found in different engineering-geological conditions. For example, typical deep landslides can be found in Britain (Bromhead, Ibsen), Ukraine (Odessa, Mariupol), Russia (Moscow, Nizhny Novgorod, Kazan, Ulyanovsk, Volgograd, Cheboksary, Saratov, Saransk, Sochi et al.), etc.

The loss of massif's stability and its progressive deformation take place in accordance with the compression scheme. The horizon with soil strength σ_{str} below compressive stress caused by the weight of overlying strata undergoes deformation, which leads to first subsidence and depression in the overlying mass upslope, then increase in tensile stresses in a zone of subsidence, and then formation and downward propagation of a tension crack. Finally, a landslide block separates at this crack and moves down along a steep curved slip surface. The angle of the slip surface flattens downslope and may become nearly horizontal. The displacement of a new sliding block triggers a displacement of the previously separated blocks located further downslope.

Their slip surfaces are typically formed in clayey soils. They occur in coastal, river, and lake bluffs and may develop in slopes of excavations, embankments, and in pit walls. Landslides in fractured and/or weathered rock are less known. They occur in mountainous and foothill regions. They are characterized by slow development of deformations during the initial developmental stage that may take place for several hundreds of years.

However, very often this stage of the new block formation (with deformation of the mass based on the compression scheme) is not analyzed during investigation of a landslide mechanism. The subsidence of the new block along a steep curved slip surface is mistakenly thought of as a shear process, and a landslide is considered to have a shear-sliding mechanism.

Formation of the long front of these landslide blocks and, moreover, of landslide cirques is an important feature in recognition of this landslide type. As rule, the relief is stepped, reflective of a blocky structure of a landslide mass. This stepped geomorphology may be barely recognized in the relief at the end of the landslide cycle, at which point, as a result of prolonged repeated displacements, the upper blocks, as well as the blocks further downslope, reach the lowest position on the slope and form nearly horizontal landslide terrace.

The appearance of long continuous cracks is a feature of active displacement of landslide blocks. Crown cracks usually occur at the top of the slope on the contact of the landslide with undisturbed ground mass and also on the boundaries between landslide blocks. Cracks are also formed along the compressive bulges along the front of the blocks, as well as downslope at the toe. A scarp wall of the cirque is steep and curved and has a maximum height in the central part of the cirque. The slip surface formed during the separation of a landslide block from the original mass is steep and curved and has nearly circular profile. It flattens out downslope and meets the nearly horizontal slip surface of previously displaced landslide blocks constituting the landslide body in the existing landslide cirque.

Landslide blocks of the compression–extrusion type can be formed practically in any type of soil or rock, if the following condition takes place in the groundmass:

$$\sigma_{\text{str}} < \gamma h \tag{8.1}$$

σ_{str} – material strength of a layer in consideration
γ – unit weight of overburden material
h – depth of potentially deformable layer

Displacement of a landslide body is caused by the pressure of the upslope blocks; therefore, a compressional bulge or ridge is formed at the toe. This geomorphic feature is especially prominent during the main displacement stage, when a new landslide block separates from the plateau. In coastal areas when the sliding surface extends into the submerged part of the slope, this bulge often resembles an island. These characteristic bulges often occur in front of each landslide block that comprises the landslide body.

The second category of the *shear-sliding landslides* is characterized by a concentration of shear stresses in certain zones of a soil mass during initial stage of landslide formation and may occur when the slope is in quasi-equilibrium at a natural angle of repose or due to a surface creep of weathered near-surface slope sediments (shallow landslides). The movement is per infinite slope model or a shear along a zone of weakness determined by geological structure, such as a contact of weathered soil and bedrock or a weak layer between strong layers.

Slope deformation occurs as a progressive shear with a decrease of soil resistance with deformation, a reduction of soil strength from peak to residual strength, and a gradual formation of a slip surface [44–46]. On steep slopes, landsliding typically occurs along a curvilinear slip surface that extends into a toe of the slope or stops above it (Fig. 8.2a). Thus, a profile of equal-strength or equal-stability slope is being formed due to mass wasting (i.e., the displacement or fall) of weakened soil mass.

The slip surface may form along dipping boundaries of geological layers. This may involve significant geological strata (in accordance with Fig. 8.2b). The mechanism of displacement along polygonal subhorizontal slip surfaces is characteristic of sliding of eluvial-diluvial deposits over the bedrock (Fig. 8.2c). A frequent landslide case involves vegetated soil cover and is manifested by series of relatively small cracks (see Fig. 8.2d). A slow creep of near-surface soils can be observed on relatively stable slopes with steeply dipping bedrock (see Fig. 8.2e).

As was discussed above, the landslides in this category are caused by the shear force acting along an inclined boundary, i.e., slip surface [17]. They may have a well-defined landslide body and cirque, or a zone of slow soil creep extent which is difficult to define, or occur as a flow. They may be characterized by a hilly relief, or less often a stepped relief, documented in cases of the landslide body 10 m thick or more. Landslide cracks are typically discontinuous, crescent-shaped, and not extensive and may be better defined along the side borders. The scarp wall (or headwall) often has a gentle slope and exposed subhorizontal slicken-sided bedrock surface.

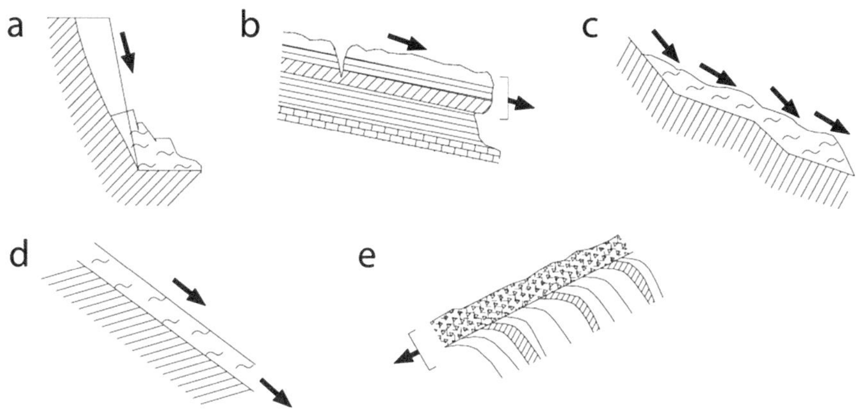

Fig. 8.2 Schemes of landslide deformation due to the *shear-sliding* mechanism: (**a**) shear-cutting, (**b**) shearing along layers, (**c**) shear-sliding of shallow surface cover, (**d**) shear of vegetated soil layer, and (**e**) bending of tops of steeply dipping layers

Both the landslide surface and the slip surface are typically inclined. They may be curvilinear in case of steep slopes. The displacing soil masses may form a bulge or ridge at the landslide toe.

The third landslide category represents *liquefaction–flow landslides*. A loss of slope stability is caused by liquefaction and the predominant force is groundwater. Soil liquefaction is a result of the increase in pore water pressure and thus a decrease of effective stress [1, 27, 38]. The pore water in saturated or partially saturated soils increases pressure on the mineral skeleton of soil in form of hydrostatic and filtration pressures caused by filtration volumetric forces. The magnitude and direction of these forces depend on the external factors, such as static and dynamic loads on the slope, filtration flow rate and changes in the groundwater level, the water-level fluctuations in reservoirs and surface streams, rainfall intensity, etc.

This mechanism of landslide formation is especially typical for fine-grained soils with a weak structural skeleton and low filtration capacity, such as recently deposited silts, young water-saturated clays and loams, and fine-grained water-saturated sands, topsoil, peat, and clays of different ages that have reduced strength due to weathering disintegration and hydration [26, 27].

Where groundwater exits onto ground surface of the slope, a landslide cirque with an hourglass shape and a narrow "neck" is usually formed (Fig. 8.3a). Liquefied soil masses (the product of destruction of the headwall and its sides) move from the "neck" to the slope in the form of a viscoplastic flow with formation of cones of deposition at the toe. Elevated groundwater levels due to heavy rains and abundant snowmelt and, as a result, upward seepage force can reduce internal friction in the soil to zero. The soil softening at low loads induced by surface layers may result in loss of cohesion between soil particles. In this case liquefaction of sandy and clayey soil may occur even at low slope gradients (1:10 or less) (see Fig. 8.3b). Local slope instability often occurs in areas of excessive moisture, resulting in soil deformation in the form of local flows (see Fig. 8.3c).

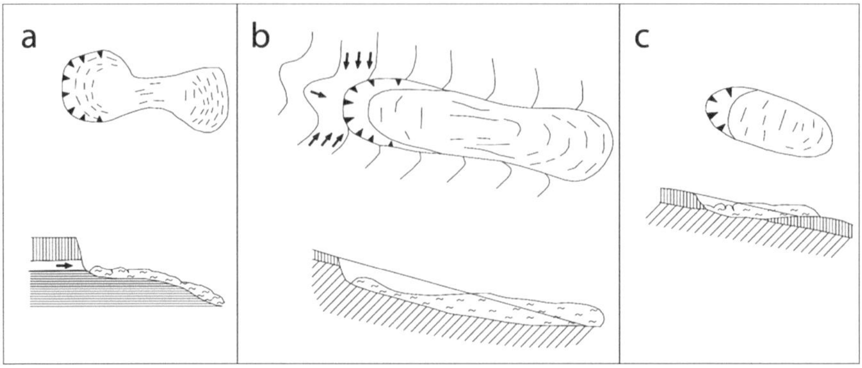

Fig. 8.3 Schemes of landslide deformation in accordance with liquefaction mechanism (plan and cross-section): (**a**) landslide cirque with a narrow neck formed at groundwater exit site, (**b**) landslide flow (earthflow), and (**c**) local flow

This type of landslide is often recognizable by the following indicative signs of the groundwater action: springs, low-relief depressions, temporary and permanent water courses, erosion undercuts, numerous cracks, low hilly relief, etc.

The subsidence of the soil mass that lies above the zone of deformation of the liquefied masses eventually leads to the formation of the front block-type landslides of compression-extrusion with a relatively deep deformation of the mass. There are known landslide cirques that start off with landslides of liquefaction-flow type forming bowl with a narrow neck. After reaching a certain size, they transform into block-type landslides of compression-extrusion. Such landslides occur for example along shoreline of Ovechka River near Cherkessk in the Caucasus, on the banks of the river Tom` near Tomsk, or on the banks of the Sura River [36] transition of landsliding from liquefaction-flow to compression-extrusion mechanism may be characterized qualitatively.

8.4 The Limit State of a Groundmass During the Formation of a Landslide

Reliable determination of the limit state of slope groundmass and landslide hazard assessment are essential during site exploration and development, as well as design safety of the adjacent buildings and structures. The landslide processes includes (1) the landslide initiation stage, (2) general downslope movement of the landslide body, and (3) distinctive phases of this movement during the landslide cycle, including the most important phase of catastrophic activation, when additional groundmass becomes separated from stable ground and joins with previously formed landslide body. Landslide movement becomes progressive and destructive during this phase [8, 14, 15, 19, 22, 28, 42].

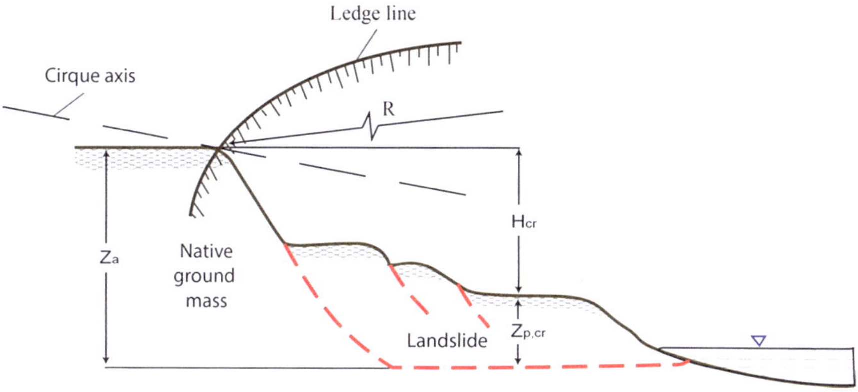

Fig. 8.4 The scheme for assessment of the stress-strain state of the soil at depth Z_a

The Institute of Environmental Geosciences of Russian Academy of Sciences (IEG RAS) has developed the criteria for the limit state of the soil mass and the methodology of the limit-state calculations [35]. These numerical solutions received three patents in the Russian Federation: No. 2340729 (V.I. Osipov, G.P. Postoev), No. 2412305 (G.P. Postoev, A.I. Kazeev), and No. 2413056 (G.P. Postoev, A.I. Kazeev).

The development of landsliding processes, such as deep block-type movements, dies out once a flat landslide terrace is formed, which represents the end of a landslide cycle. Near the axis of a landslide cirque, the height of a scarp wall reaches critical value H_{cr} (Fig. 8.4). At this moment there are temporarily no deformations on a landslide slope. The landslide body is motionless, but the native massif is in a limit state. Transition from this temporary quasi-equilibrium to a catastrophic activation of a landslide process (with formation of new sliding blocks and destructive deformations involving entire landslide cirque) may occur as a gradual change in a strain-stress state of a landslide-prone mass or as a sudden catastrophic event triggered by an external factor.

For the 2D case, considering landslide-prone slope mass with landslide deformation on i-th horizon [35] and assuming long, i.e., semi-infinite, slope with a curvature of slip plane approaching zero, utilizing the Mohr–Coulomb theory (for localized zone in a ground mass), the equation for limit state was developed as:

$$\gamma_i Z_{ai} - \sigma_{\mathrm{str},i} = \frac{\pi}{2}\gamma_i Z_{\mathrm{p,cr}} \tag{8.2}$$

γ_i – average unit weight of soils above the i-th horizon in a landslide-prone undisturbed mass (Fig. 8.5)

Z_{ai} – the depth of the i-th horizon in the undisturbed mass, $\sigma_{1a} = \gamma_i Z_a$

$\sigma_{\mathrm{str},i}$ – the soil structural strength of the i-th horizon of the undisturbed mass, $\sigma_{\mathrm{str}} = 2c \cdot tg\left(45 + \frac{\varphi}{2}\right)$

$Z_{\mathrm{p,cr}}$ – critical (calculated) depth to the i-th horizon of potential deformation with soil gravity loading considered as $\sigma_{1pi} = \gamma_i Z_{pi}$

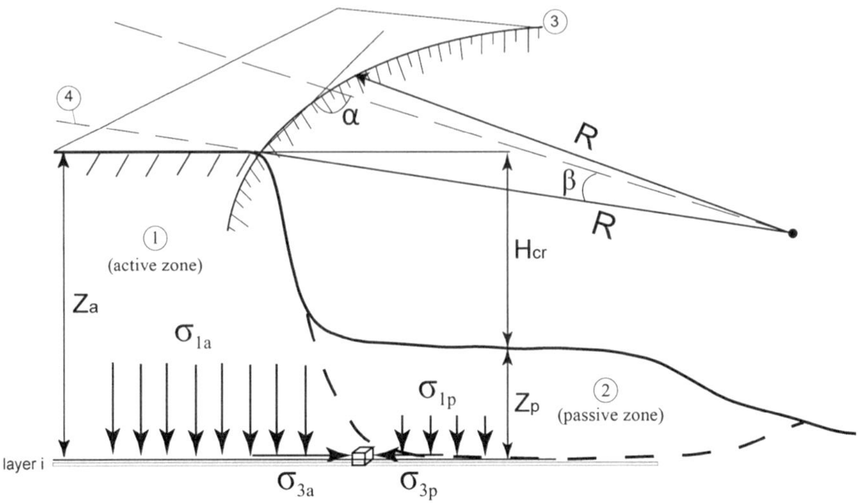

Fig. 8.5 The scheme for assessment for the stress-strain state of the soil at depth Z_a: 1, mass in consideration (active zone); 2, landslide mass deposited downslope (passive zone); 3, edge of cirque headwall; 4, central axis of the cirque; α, the angle between the front of active horizontal stress σ_{3a} (extent of the general outline of the cliff) and the direction to the considered section in the cirque; β, the angle between cross-section in consideration and central vertical section in the cirque; R, radius of the cirque

a and *p* indexes indicate active and passive horizontal earth pressures at the *i*-th horizon.

The results of the theoretical calculations were calibrated against a large number of real events where the limit state of landslide-prone soil masses and subsequent developments of massive landslide blocks were documented (Kazeev and Postoev 2017) (Fig. 8.6).

Considering the radius of the cirque R (Fig. 8.5), the limit state of the mass may be described by the following equation:

$$\gamma \cdot Z_{\mathrm{a}} - \sigma_{\mathrm{str}} = \frac{\pi}{2} \gamma Z_{\mathrm{pcr}} \frac{1 + \frac{1}{R}}{1 - \frac{1}{R}} \tag{8.3}$$

where $1/R$ is numerically equal to the curvature of the curved edge of active headwall cliff (1 and R in meters).

In general form, the equation of the limit state for potentially deformable horizon at the depth Z_a is as follows:

$$\sigma_{1\mathrm{a}} - \sigma_{\mathrm{str}} = \frac{\alpha}{1 - \cos \alpha} \sigma_{1\mathrm{p}} \frac{1 + \frac{1}{R}}{1 - \frac{1}{R}} \tag{8.4}$$

or

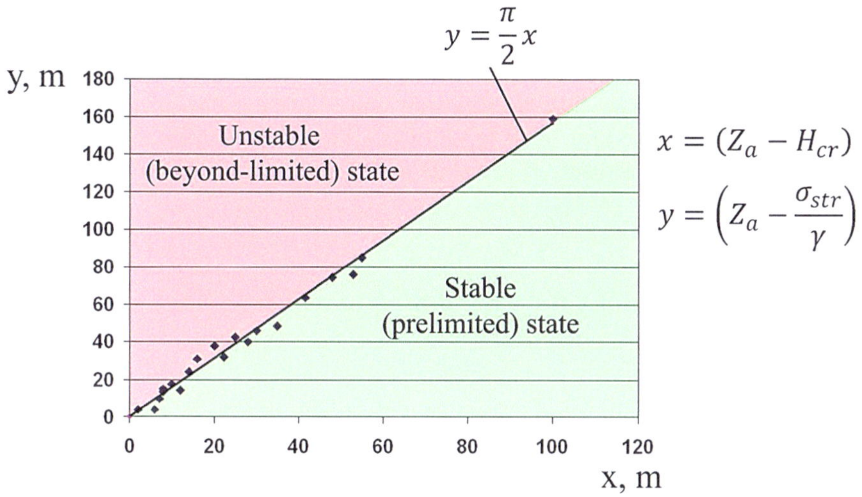

Fig. 8.6 Depth to the sliding surface vs. thickness of the landslide body for documented landslide cases, where the limit state was reached prior to catastrophic failure with the development of deep-seated block landslides of compression–extrusion type [35] (Refer to Fig. 8.4 for definitions)

$$\sigma_{1a} - \sigma_{str} = \frac{\frac{\pi}{2} - \beta}{1 - \sin\beta}\sigma_{1p}\frac{1 + \frac{1}{R}}{1 - \frac{1}{R}} \tag{8.5}$$

where $\beta = \pi/2 - \alpha$;

α – the angle between the front of active horizontal stress σ_{3a} (extent of the general outline of the cliff) and the direction to the considered section in the ledge

β – the angle between considered cross-section in the ledge and central vertical section in the cirque

The general Eqs. 8.4 and 8.5 define limit state in considered zone of landslide-prone soil mass at the horizon, where the principle vertical stresses in active zone σ_{1a} in undisturbed soil mass (exceeding the soil structural strength σ_{str} of this horizon) are counteracting with principle vertical stresses σ_{1p} in passive zone (zone of force disturbance with radius R).

The coefficient of stability, or factor of safety of the cliff (K_{st}), may be used to characterize how close is the slope to reaching the critical state:

$$K_{st} = {}^{\sigma_{1p_i,f}}/_{\sigma_{1p,cr}} \tag{8.6}$$

where $\sigma_{1pi,f}$ is the observed vertical pressure of the landslide mass and $\sigma_{1p,cr}$, the critical (minimal) vertical pressure of the landslide mass on the slope.

Thus, the equilibrium of horizontal stresses in the soil landslide-prone mass at depth Z_a at $\sigma_{1a} > \sigma_{str}$ is completely determined by the magnitude of horizontal stress σ_{3p} (resistance), which, in turn, is determined by the gravity load of slope deposits. When σ_{3p} reaches its critical value $\sigma_{3p,cr}$, the soil is in limit state. The decrease of resisting pressure ($\sigma_{3pi} < \sigma_{3p,cr}$) may cause development of vertical soil deformation and lateral spreading and possibly formation of a sliding surface.

It should be noted that a concept of a limit state that precedes landslide deformations as discussed above may be valid not only for landslides of compression–extrusion type but often for other types of landslides with similar mechanism of loss of slope stability and subsequent sliding or flow of the landslide mass.

8.5 New Technology for Calculation of a Slip Surface Depth

There are various methods for identification of a slip surface on landslide slopes. For example, a method is known of determining of sliding surfaces in a landslide mass through identification of weak zones and layers while drilling boreholes (Komarnitskii 1966). Also, different geophysical methods (seismic, electrical) are applied to identify the slip surface, by identifying horizons with anomalous values of geophysical parameters (Methods 1987). However, all these methods have significant disadvantages: they require drilling boreholes, laboratory soil testing, significant time costs, and often simply not applicable (in homogeneous clay mass).

Slope stability calculations are also widespread for assessment of possible landslide sizes by calculation of safety factor (K) (Grebnev 2008). The disadvantage of this method is that another mechanism of landslide formation is considered – shear-shifting scheme – when a prepared slip surface exists in a mass (e.g., defined by tilted lithological strata in the direction of the slope or weak layers).

The sliding surface can also be determined in the process of instrumental monitoring in boreholes (e.g., inclinometer observations), but this method requires some time (often significant) for fixing landslide displacements.

The developed method [37] allows to calculate and identify the depth of a slip surface, based on the data of geological survey on landslide areas.

The aim of the present invention is to eliminate disadvantages of the abovementioned methods of slip surface identification, to improve the accuracy of determining of a slip surface depth in a landslide mass at the stage of absence of landslide displacements along slip surfaces.

The method is carried out as follows. The landslide body is composed of blocks on the site with the existing landslide cirque (with ledge radius R) (see Fig. 8.5). This landslide blocks were separated from native groundmass situated upslope and displaced along the sliding surfaces at different times. These blocks are usually destroyed in various degrees. The height of the above-landslide slope mass should be measured in central part of the cirque (H_{cr}), as the excess of the ledge of the circus over the middle part of the landslide terrace. The radius (R) of the landslide circus ledge bending in the plan should be also defined. The depth of the slip surface is calculated as follows:

$$Z_a = \frac{2c_d \cdot tg\,(45 + \varphi_d/2)\left(1 - \frac{1}{R}\right) - \frac{\pi}{2}\gamma_{f,d} \cdot H_{cr} \cdot \left(1 + \frac{1}{R}\right)}{\gamma_d\left(1 - \frac{1}{R}\right) - \frac{\pi}{2}\gamma_{f,d} \cdot \left(1 + \frac{1}{R}\right)} \tag{8.7}$$

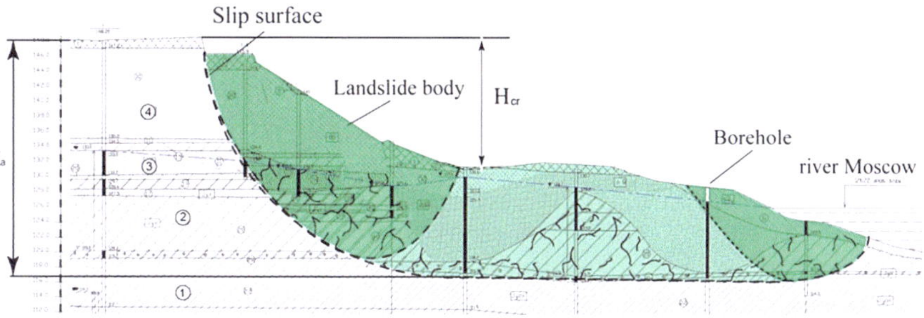

Fig. 8.7 Cross-section of the landslide slope thru central part of the cirque in Khoroshevo, Moscow. 1, 2, Jurassic Oxford and Tithonian clays, respectively; 3 and 4, cover deposits of loam and sand. Boreholes for inclinometric monitoring

where Z_a is the slip surface depth (relatively to cirque's ledge) /m/; c_d and φ_d, accordingly the cohesion (kN/m^2) and angle of internal friction (in degrees) of soil of potentially deformable horizon; R, the radius of a landslide cirque /m/; and γ_d and $\gamma_{f,d}$, accordingly the average unit weights of soils, overlaying potentially deformable horizon in native part of soil mass and on landslide terrace.

Example In 2006 a catastrophic activation of landslide process took place in a compression–extrusion landslide circus on the banks of Moscow river (Moscow, Khoroshevo district), with formation of a new landslide block and displacement of landslide masses over 350 m along the shore (Fig. 8.7). Landslide hazard appeared for residential homes and the Life-Giving Trinity Church (architectural monument of sixteenth century). Engineering-geological surveys and monitoring were carried out on the site of landslide deformations, including instrumental monitoring of deep deformation (inclinometric monitoring), which established the sliding surface at a depth of $Z_a = 31$ m.

The needed data for calculation of a slip surface depth were obtained in result of engineering-geological surveys and field observations: $c_d = 47.2$ kN/m^2; $\varphi_d = 22°$; $R = 180$ m; $H_{cr} = 16$ m; $\gamma_d = 20$ kN/m^3; and $\gamma_{f,d} = 20$ kN/m^3.

Setting these values into the formula (8.7) for determination of Z_a, we get $Z_d = 31.3$ kN/m^2, i.e., we receive almost full coincidence of results of calculation slip surface depth (Z_a) and actual inclinometric measurements in boreholes.

8.6 Peculiarities in the Development of Deep-Seated Landslides of the Compression–Extrusion Type and Protection Strategy

The following characteristic features in the development of landslides of this type were observed (Fig. 8.8):

I. *The degree of activity and frequency of displacements of a landslide in a cirque is directly correlated with a high position of landslide blocks adjacent to the*

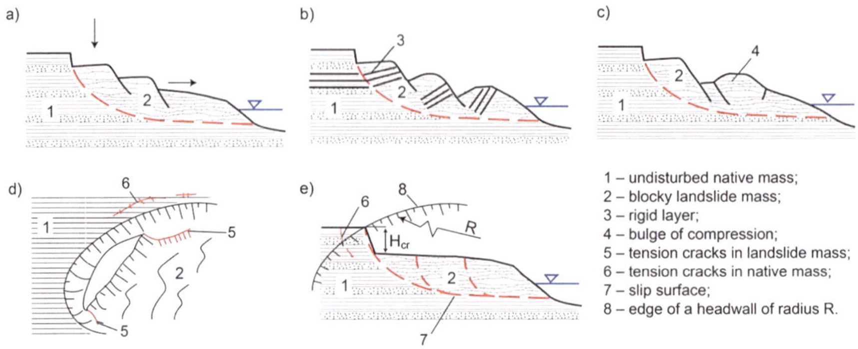

Fig. 8.8 Characteristic schemes of a landslide slope with deep-seated block-type landslide. (**a**) High position of the upper sliding block, regulating the displacement; (**b**) translation and rotary movements of blocks; (**c**) formation of a bulge of compression in a landslide body; (**d**) formation of a "cutting" failure at the bottom of a scarp wall and "main" (rear) failure in landslide native mass; (**e**) the limit state of the native mass at headwall, with preparation of a new landslide block

upper boundary (or head scarp) on a slope (Fig. 8.8a). As was shown above, the displacement of the landslide body is caused by block failure and stress increase in the remainder of the mass in the cirque. The magnitude of displacement depends on the size of sliding blocks, soil parameters, and other influencing factors. On high slopes in hard bedrock, this process can continue for many centuries. This is an important characteristic of deep displacements that needs to be taken into account when planning for construction in areas of known (or suspected) landslide hazard and adjacent water bodies, even when no visual deformations are apparent on the slope. This process leads to retreat of scarp wall and growth of landslide cirque.

II. *Translational and rotational movement of landslide blocks* is a specific feature of landslides of this type. It is caused to a great degree by the mechanism of formation of landslide blocks as a result of compression within the zone undergoing deformation and soil squeeze into the slope, along with failure and rotation of the blocks (Fig. 8.8b). Based on the landslide research along Odessa coast of the Black Sea, Naumenko [25] showed that the rotational movement of blocks occurs along nearly circular surfaces as they continue to move downslope. Inclination (i.e., angle of rotation) of a rigid marker layer (i.e., the 6–13-m-thick limestone layer of Pliocene, Pontic stage) increases with progressive translational movement of a landslide mass. Characteristic extrusion ridges form underneath the blocks, as soil mass continues to deform (Fig. 8.8b).

Therefore, it is necessary to take into account the complexity of movement of the blocks when considering building on any landslide-prone slope. Due to significant thickness and extent of landslide massif, significant continuous ground deformations need to be considered in structural design, as well as post-construction maintenance to provide building safety and functionality, which may be a difficult task.

III. *The formation of compressional ridge (bulge).* This element of relief can be found in landside local cirques, with landslide mass composed of a significant number of sliding blocks, which create a natural counterbanquet to any new block that has separated from a native groundmass (Fig. 8.8c). In this case, under pressure of a new block midpoint landslide compressional bulge may form in any part of a landslide slope [34]. Landslide deformations below this bulge are insignificant.

This peculiarity may influence the requirements for selecting the type and location of protective structures. The upper part of a slope may still have the potential for landslide development even after landsliding has occurred downslope. Moreover, a catastrophic activation of landslide deformations is possible, as will be shown below.

IV. *The self-sustaining effect* is a specific feature of deep-seated block-type landslides. During its movement, each new block pushes the entire remainder of the landslide mass away from the surface of rupture and a scarp wall, leading to a development of a deep tension crack along its margins (Fig. 8.8d). This landslide crack undercuts a potential landslide ground mass, forming "heading" tension cracks in the upper terrace which are telltales of the beginning of decrease of stability and initiation of a new landslide block.

This is a common occurrence, for instance, in landslide cirques that form long escarpments along coastal shorelines. Recorded instances of this phenomenon are repeated catastrophic activations in the same landslide amphitheater in Odessa. In 1964, a new block about 420 m in length formed along a coast. In 1965, the displacement of a new block with the length of 270 m occurred to the south of the 1964 location. The latter was attributed to the process of "self-development" and resulted in a significant magnitude of displacement [25].

It is often believed that side erosion (abrasion) determines the intensity of the development of deep landslide movements. But the effect of "self-development" gives evidence that once the process of formation of landslide blocks and the development of landslide deformations has begun, it may continue even without influence of external factors, such as undercutting or removal of the downslope areas and the toe. Thus, consideration of the mechanism of self-development is necessary for designing of appropriate mitigation and protection measures.

V. *The mechanism of a catastrophic activation.* As was illustrated above, the development of landsliding processes, such as deep block-type movements, dies out once a flat landslide terrace is formed (Fig. 8.8e), which represents the end of a landslide cycle. Near the axis of a landslide cirque, the height of a scarp wall reaches critical value H_{cr}. At this moment there are temporarily no deformations on a landslide slope. The landslide body is motionless, but the native massif is in a limit state. Transition from this temporary quasi-equilibrium to a catastrophic activation of a landslide process (with formation of new sliding blocks and destructive deformations involving entire landslide cirque) may occur as a gradual change in a strain-stress state of a landslide-prone mass or as a sudden catastrophic event triggered by an external factor.

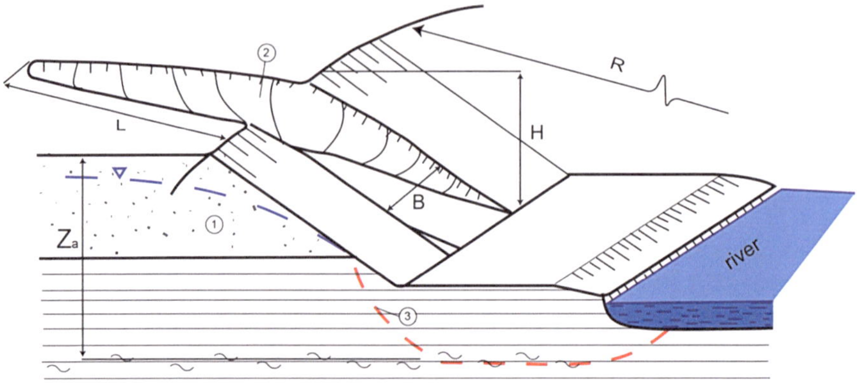

Fig. 8.9 Artificial "gully" cuts in a landslide-prone mass. 1, landslide-prone ground mass; 2, artificial local "gully" cut; 3, slip surface; *H*, height of a scarp wall; *L* and *B*, length and width of a "gully" cut; *R*, radius of the scarp wall; Z_a, depth to potentially deformable horizon

Thus, development of protective measures in areas with a potential for deep-seated block-type landsliding should include:

- Timely characterization of landslide-prone native ground masses, which are at near limit state
- Analysis of potential slip surfaces and lateral movements of blocks in a slope cross-section
- Development of effective engineered controls to increase the stability of the sites approaching a limit state to prevent formation of new landslide blocks and development of destructive deformations

Based on the above-presented methodology for limit-state analysis, a new practical method to increase slope resistance has been developed. It involves construction of artificial cuts similar to erosional downcuts of natural gully or ravine (patent for invention RF №2413056; authors, Postoev G.P., Kazeev A.I.). These cuts break long continuous sloped fronts of cliffs or escarpments that have high potential for landslide hazard (Fig. 8.9).

The proposed method would result in a decrease of active earth pressure and a reduction of driving force in accordance with the equation of equilibrium (8.3), an increased resisting force at the base of the slope, improved drainage at the cut sidewalls, and reduction of hydrostatic and hydrodynamic pressures. It should be noted that this leads to a general change of seepage path in cut sidewalls from original downslope direction (which results in increased hydrodynamic and static pressures in saturated landslide-prone zones) to a direction toward a thalweg of the artificial gully. Thus, the stability of sidewalls, and the overall slope stability, is increased. This effect is not considered in the calculations, but it is believed that the actual slope stability may be higher than the calculated values. Therefore, the limit state of the potentially landslide-prone slope with long continuous edge is altered by introducing artificial local "gully" cuts and creating local stable zones.

At each such site, the calculated resistance is increased *1.5 times*, as compared to an initial state of unmodified long slope that has a linear crown edge. The method is very effective in preventing huge massive landslides of compression–extrusion type, because the division of the groundmass into several discrete smaller bodies leads to elimination of one of the necessary conditions of their formation, i.e., a presence of the continuous slope front.

8.7 Landslide Monitoring

Monitoring of landslide process is an important part of any landslide investigation. The purpose of the monitoring is observation of landslide response to natural and anthropogenic factors as necessary for the assessment of a state of a landslide, predicting stability and development of its stabilization and building protection measures. Observations on natural landslide slopes have been performed by specialized industrial geological organizations for many years.

Notably, significant contributions to the development of the methodology and studies of mechanism and dynamics of landslides were made by Emelyanova E.P [7]. With the advancement of the landslide science, the methods of observation and monitoring were furthered by work of VSEGINGEO [18, 19, 24, 32]. The analysis of the accumulated information on the development of landslides in the different regions of the country leads to a deeper understanding of the landslide processes and further advancement of methods of landslide hazard assessment, prediction, and development of protective measures. Landslide monitoring is regularly presented in the reports of Russian scientists at numerous international forums ([10, 12], etc.).

In recent years, there has been a shift toward the use of modern remote control devices and automated systems of landslide monitoring [19, 32]. The techniques of controlling the sidewall stability and deformation of open-mining pits are being continuously developed and improved [9, 31]. In order to ensure the safety of the transportation systems, the methodological guidelines are prepared, and the monitoring of the state and deformations of the landslide slopes, as well as excavations, roadway cuts, and railway embankments, is carried out [2].

Geotechnical investigations on the landslide slopes include the monitoring of a landslide development in accordance with the regulatory documents [7, 19, 24, 32]. The main goal of these field investigations is the assessment of a landslide hazard, slope stability, and the prediction of possible landslide movements.

Review of monitoring data of landslide investigations has shown that periodic assessments of the slope stability based solely on geophysical parameters are not sufficient to adequately characterize critical changes in the development of a landslide and the beginning of its dangerous activation phase. Likewise, the monitoring of landslide deformations only by displacement rates cannot always detect a dangerous condition. In some cases, the rate of displacement of 75 mm per day was observed during the period of a full-scale experiment on artificial activation of a landslide, but did not lead to a catastrophic development [33]. In other cases, the lower rates have led to the destructive displacements in the open pit walls [29].

The Institute of Environmental Geoscience of Russian Academy of Sciences performed numerous studies leading to a development of effective technologies of automated monitoring. It was established that the most important parameters for characterization of the state of an active landslide and its dynamics (including the phase of a progressive hazardous activation) are the magnitude and the rate of soil deformation and displacement, the depth of a slip surface, and the areal distribution of active displacements on the monitored slope. Extensometer lines of control, remote wireless monitoring systems of displacement on the slope, and inclinometer measurements in boreholes are considered to constitute minimal requirements for the system of automated monitoring, based on the conducted research [29].

Influence of other factors, such as precipitation, changes of groundwater levels, and anthropogenic and other factors, is reflected in the deformational behavior of the landslide body and thus is also monitored using high-accuracy instrumentation.

For example, one of the most important transportation projects for the 2014 Winter Olympics in Sochi involved design and construction of a new transit corridor that connected Adler and Krasnaya Polyana and combined a highway with light rail. It passed along the valley of the river Mzymta, and landslide hazard potential had to be evaluated. Landslide slopes were heavily instrumented with primarily horizontal extensometers and inclinometers that provided the necessary monitoring data to develop landslide hazard criteria. Based on the rate of a displacement at the monitoring points (V), the following criteria and for landslide warning signals were developed and implemented: $V < 4$ mm/day, non-hazardous condition (green signal); $4 \leq V < 24$ mm/day, moderately hazardous condition (yellow signal); and $V \geq 24$ mm/day, very hazardous condition (red signal) [29].

8.8 Slope Stability Analysis

Quantitative characterization of static and dynamic factors that influence slope stability and landslide processes is required for reliable evaluations of landslide hazard and design of protective measures.

Slope stability analysis is typically required for any proposed development of slope, and thus it is the most widespread among the methods of local forecasting of a landslide process. A great variety of methods of slope stability analysis has been developed ([23, 42]; Ginzburg 1975). The existing analytical models are applicable for a variety of types of landslides. They include assumptions, such as soil mechanics models, failure criteria, simple slope geometry, or the shape of the slip surface, and come with limitations of limit-state analytical methods being used. The most popular are 2D models.

The experience shows that the reliability of numeric analysis is not as much determined by use of "correct" soil and analytical models (though they matter) but to a greater degree by taking into account numerical simulation of the deformation mechanism of a landslide, and ground behavior, such as realistic modeling of the soil shear resistance along sliding surface. Also, the actual influence of various natural and human factors that cause the change in the water saturation of the slope, the mass balance, and the soil properties on the stability of the existing slope needs to be correctly modeled [16].

The majority of Russian software for slope stability analysis includes calculation methods of K. Terzaghi–Fellenius, method of horizontal forces of Maslov–Berrer, and methods of G.M. Shahunyants [11, 16, 22, 39].

Often shallow landsliding by liquefaction–flow and shear-sliding mechanisms leads to a stress reduction at the toe of steep slopes. This factor is modeled by a removal of the downslope slices in a numerical simulation. The sequence of removal of vertical slices within a landslide body is also used as a basis of design decisions regarding construction sequence on the slope, such as excavations for various engineering structures.

In Russia, investigations of the stress-strain state of slope soil mass and the slope stability assessments are typically performed using finite element methods, limit-equilibrium numerical modeling, as well as old-fashioned pencil-and-paper engineering calculations methods. Both Russian software and international programs (such as Plaxis, Geostab, Slide, SlopeW, etc.) are utilized.

8.9 Conclusions

Numerous research and industrial organizations conducted landslide investigations on the vast territory of the former Soviet Union (one sixth of the world). Thousands of publications presented summaries of findings and/or focused on the certain important issues concerning the landslide distribution and development in different geological settings across the country. Russian scientists and investigators engaged in the landslide hazard evaluations made a great contribution to the development of state of knowledge regarding the landslide processes accumulated by the scientific community worldwide.

The landslide hazard assessment and effective prediction of possible landslide activation, coupled with the implementation of mitigative and protective measures, are impossible without characterization of a possible landslide based on the mechanism of its formation, as well as conditions and mechanics of the stress-strain state changes in the landslide mass prior to reaching their threshold limit values. Neither they are possible without taking into account the dynamics of the landslide processes and the criteria that mark transition of the landslide development into a catastrophic phase.

For the past several decades and into the present, the Russian scientists carried out a high-level research of landslides that included landslide classifications, specific issues in the landslide assessment, and the development of the necessary protective measures, often achieving distinctive results, some of which are presented in this article.

Acknowledgments This research is supported by Russian Scientific Foundation (Project № 16-17-00125).

References

1. Abelev, M. Y. (1983). *Building of industrial and civil constructions on weak saturated soils*. Moscow: Stroyizdat.
2. Ashpiz, E. S. (2002). *Monitoring of road beds during the exploitation of railroads* (p. 112). Moscow: Put'-Press (in Russian).
3. Cruden, D. M., & Varnes, D. J. (1996). Landslide types and processes. In A. K. Turner & R. L. Schuster (Eds.), *Landslides investigation and mitigation: Transportation research board* (pp. 36–75). Washington, DC: US National Research Council., Special Report 247.
4. Dalmatov, B. I. (1988). *Soil mechanics, bases and foundations (including special course of engineering geology)*. Moscow: Stroyizdat 415 p. (in Russian).
5. Demin, A. M. (1981). *Regularities of manifestations of deformations of slopes in open pits*. Moscow: Nauka 144 p. (in Russian).
6. Demin, A. M. (2009). *Landslides in quarries: Analysis and forecast*. Moscow: GEOS 79 p. (in Russian).
7. Emelyanova, E. P. (1956). *Methodological guidance on stationary study of landslides*. Moscow: Gosgeoltehizdat 256 p. (in Russian).
8. Emelyanova, E. P. (1972). *Main regularities of landslide process*. Moscow: Nedra 310 p. (in Russian).
9. Fisenko, G. L. (1965). *Stability of pit sides and blades*. Moscow: Nedra 378 p. (in Russian).
10. General reports of the VIII International Congress on soil mechanics and Foundation engineering. (1975). *Translation from English*. Moscow: Stroyizdat 190 p. (in Russian).
11. Ginzburg, L. K. (1986). *Recommendations on the choice of methods of calculation of stability factor and landslide pressure. Central Bureau of sci.-tech*. Moscow: Inform (in Russian).
12. Goldstein, M. N., Turovskaya, A. Y., & Babitskaya, S. S. (1969). *About the strength of water-saturated plastic clays* (pp. 5–12). Proceedings of the VII International Congress on soil mechanics and Foundation engineering. Stroyizdat, Moscow. (in Russian).
13. Goldstein, M. N. (1979). *Mechanical properties of soils (stress-strain and strength characteristics)*. Moscow: Stroyizdat 304 p. (in Russian).
14. Gulakyan, K. A., Kuntsel, V. V., & Postoev, G. P. (1970). The mechanism and dynamics of the landslide process on the example of landslide in Zerafshan valley. In *Materials of sci.-tech. meeting on methods of study and forecast of debris flows, rockfalls and landslides* (pp. 59–65). Dushanbe. (in Russian).
15. Gulakyan, K. A., Kuntsel, V. V., & Postoev, G. P. (1977). *Forecasting of landslide processes*. Moscow: Nedra 135 p. (in Russian).
16. Kazeev, A. I., Lapochkin, B. K., & Postoev, G. P. (2009). Peculiarities of computer modelling of stability of landslide slopes by the program AKNARK. In *Proceedings of the conference «Sergeev's readings»* (pp. 296–300), Moscow. (in Russian).
17. Kostomarov, V. M. (1964). Designing of landslide protection structures. In *Proceedings of the conference on the study of landslides and protective* (pp. 44–53). Kiev. (in Russian).
18. Kuntsel, V. V. (1980). *Regularities of the landslide process in the European part of the USSR and its regional forecast*. Moscow: Nedra 213 p. (in Russian).
19. Landslides and mudflows. (1984). *Under the editorship of E.A. Kozlovskiy* (Vol. 1). Moscow: Centre Int. Projects GKNT 352 p. (in Russian).
20. Lomtadze, V. D. (1977). *Engineering geology. Engineering geodynamics*. Leningrad: Nedra 475 p. (in Russian).

21. Malyshev, M. V. (1980). *Soil strength and stability of foundations of structures*. Moscow: Stroyizdat 136 p. (in Russian).
22. Maslov, N. N. (1977). *Soil mechanics in construction practice (landslides and their control)*. Moscow: Stroyizdat 318 p. (in Russian).
23. Matsiy, S. I., & Bezuglova, E. V. (2010). *Management of landslide risk*. Krasnodar: Alvi design 240 p. (in Russian).
24. Sheko, A. I., & Grechischev, S. E. (Eds.). (1988). *Methodology of studying and forecasting of exogenous geological processes*. Moscow: Nedra 216 p. (in Russian).
25. Naumenko, P. N. (1977). *Conditions of formation and geotechnical characteristics of landslides on the Black sea along the coast of Odessa. Landslides on the Black sea coast of Ukraine* (pp. 57–100). Moscow: Nedra (in Russian).
26. Niyazov, R. A. (1974). *Landslides in the loess rocks of the South-Eastern part of Central Asia*. Tashkent: Fan 148 p.
27. Osipov, V. I. (1988). Dynamic liquefaction of saturated soils: Nature and determine factors (scientific review). *Engineering Geology, 2*, 3–31. (in Russian).
28. Osipov, V. I., Kutepov, V. M., Zverev, V. P., et al. (1999). *Dangerous exogenous processes. Under the editorship of V.I. Osipov*. Moscow: GEOS 290 p. (in Russian).
29. Osipov, V. I., Postoev, G. P., & Kazeev, A. I. (2015). Justification of the choice of criteria for landslide hazard in the system of automated monitoring of the landslide process on the coastal slopes of the river Mzymta. *Geoecology, 2*, 133–146 (in Russian).
30. Petrov, N. F. (1987). *Landslide system. Simple landslides (aspects of classification)*. Kishinev: Shtiintsa 164 p. (in Russian).
31. Pevzner, M. E. (1992). *Deformation of rocks in quarries*. Moscow: Nedra 335 p. (in Russian).
32. Postoev, G. P., Erysh, I. F., Kruglov, A. V., et al. (1982). *Studying of the regime of landslide processes*. Moscow: Nedra 255p. (in Russian).
33. Postoev, G. P., Erysh, I. F., Salomatin, V. N., et al. (1989). *Artificial activation of landslides*. Moscow: Nedra 134 p. (in Russian).
34. Postoev, G. P. (2010). Main characteristics and features of landslides. *Geoecology, 2*, 140–148 (in Russian).
35. Postoev, G. P. (2013). *Limit state and deformation of soils in the mass (landslides, karst holes, settlings of ground foundations)*. Moscow: Nestor-istoriya 100 p. (in Russian).
36. Postoev, G. P., Lapochkin, B. K., & Kazeev, A. I. (2015). The assessment of the extent of a landslide hazard by forcing the activization of a landslide and conducting of inclinometric monitoring. *Engineering Protection, 1*(Special issue), 128–135. (in Russian).
37. Postoev, G. P., & Kazeev, A. I. (2011). Method of determining in a landslide prone mass of potentially deformed horizon depth. Patent RF № 2412305. 20.02.2011. Bull. №5. (in Russian).
38. Shadunts, K. S. (1983). *Landslides-flows*. Moscow: Nedra 120 p. (in Russian).
39. Shahunyants, G. M. (1953). *The roadbed of railways*. Moscow: Transzheldorizdat 827 p. (in Russian).
40. Ter-Martirosyan, Z. G. (1986). *Prediction of mechanical processes in masses of multi-phase soils*. Moscow: Nedra 292 p. (in Russian).
41. Ter-Stepanyan, G. I. (1978). *New methods of landslide investigations*. Yerevan: Publishing House of AAS 152 p. (in Russian).
42. Tihvinsky, I. O. (1988). *Evaluation and prediction of stability of landslide slopes*. Moscow: Nauka 144 p. (in Russian).
43. Tsytovich, N. A., & Ter-Martirosyan, Z. G. (1981). *Fundamentals of applied geomechanics in building*. Moscow: Higher School 317 p. (in Russian).
44. Turovskaya, A. Y. (1979). *Regularities of development of landslide processes depending on strength and deformation characteristics of clayey soils*. Abstract of Dissertation. Moscow: Publishing House of Moscow State University. (p. 39) (in Russian).
45. Vyalov, S. S. (1978). *Rheological bases of soil mechanics*. Moscow: Vysshaya shkola 447 p. (in Russian).

46. Zhihovich, B. V. (2007). About the presence of extrusion of landslide slopes of Odessa coast. *The Bases, Foundations and Soil Mechanics, 5*, 7–11 (in Russian).
47. Zolotarev, G. S. (1983). *Engineering geodynamics*. Moscow: Publishing House of MSU 328 p. (in Russian).

Chapter 9
The Research of Surface Runoff in Engineering Geological Zoning

Ekaterina Karfidova and Gleb Batrak

Abstract In the continuation of the Moscow Project of geological mapping, the following works are carried out: at the mid-scale level – calculation of the network and the zone of influence of surface runoff, analysis of the stability of slopes, and the construction of an integrated map of engineering geological zoning taking into account the influence of surface runoff – and on a large scale, analysis of the distribution of tributaries and the growth of accumulation zones for the modern and buried (the roof of quaternary sediments) relief.

9.1 Introduction

In recent years (2011–2017), the development of the Moscow Project for geological and geoenvironmental mapping [1, 2] is associated with the use of advanced morphometric methods of analysis of digital elevation models at two-scale levels of study: at the mid-scale level (M 1:50,000) from radar surveying and at a large-scale level (M 1:10,000) using terrestrial topographic surveying at local area of Tarakanovka River.

As the main tasks, the morphometric analysis by the digital elevation models for surface runoff study was directed:

1. To study and to determine the factors of flow accumulation zones of surface runoff for real (modern) and buried relief at the mid-scale level [3]
2. To determine the main characteristics of flow accumulation zones of surface runoff for real (modern) and buried relief at the large-scale level
3. To compare large-scale calculated characteristics of flow accumulation zones for real and buried relief
4. To compare the calculated characteristics of surface runoff between the mid-scale level and large scale of study
5. To compose the zone of influence of surface runoff

E. Karfidova (✉) · G. Batrak
IEG RAS, Moscow, Russia

V. Svalova (ed.), *Natural Hazards and Risk Research in Russia*,
Innovation and Discovery in Russian Science and Engineering,
https://doi.org/10.1007/978-3-319-91833-4_9

6. To determine zones of varying stability of the slopes
7. To merge engineering geological zoning map, the zone of influence of surface runoff and zones of varying stability of the slopes

9.2 Data Sources and Preprocessing

The initial data of the research of surface runoff included:

1. The cartographic materials about historic data of the river network: historical data of the museum's collection of Shchusev State Museum of Architecture, encyclopedia, and atlas of Moscow [4, 5].
2. The NASA Shuttle Radar Topographic Mission (SRTM) has provided digital elevation data that are available in $5 \times 5°$ tiles, in geographic decimal degrees (latitude and longitude) projection, with WGS84 horizontal datum and Earth gravitational model (EGM96) vertical datum by the CGIAR-CSI GeoPortal http://srtm.csi.cgiar.org/ [5]; SRTM 90 m v.4.1 data were used to generate an elevation model for Moscow city area. Since 2000, the SRTM DEM has been widely used for mapping and classifying environments in natural system (Amazonia and other [6–8]) and more rarely for urban ecology.
3. The terrestrial large-scale topographic data from the project mainline tunnel construction in the northwest district of Moscow (2014–2015). The local study area is 15 km^2 and point density 260/1 km^2. Digital surfer model used GRID (cell −10 m) [9].
4. The engineering geological zoning map from the project of Moscow geological mapping, 2010 [2] in scale M 1: 10,000.
5. The technogenic deposit map from the project of Moscow geological mapping in the scale 1:10,000.

We calculated DEM SRTM for Moscow city area inside Moscow ring road (cell 100 m) in the scale 1:50,000, and results were compared with ground topographic data in local study area at Tarakanovka River.

9.3 Methods of Modeling and Comparison

The geostatistical analysis made it possible to obtain a wide range of estimates of the digital elevation models: elevation, slope, aspect, and curvature; estimates included some statistic parameters (average, minimum, maximum, standard deviation, and moda). The comparison of estimates allows us to better understand the nature of these processes of accumulation of surface runoff and slope stability.

1. At the first stage in 2016, the model of surface runoff of urban territory was developed on the base DEM, including filling digital elevation model (DEM_{Fl})

and GRID local lowland [9]. Then GRID local lowland ($GRID_{locll}$) was calculated using Raster calculator:

$$GRID_{locll} = DEM - DEM_{fl}. \tag{9.1}$$

The closed local lowlands with a minimum depth of 1 m were calculated; the total area of local lowland is a significant value of 15 km^2 (1.8% of city area inside Moscow ring road) [10]. These areas are of particular importance if it placed in riverbed with the middle part of the watershed, as it takes place in the valley of the Tarakanovka River.

In morphometric hydrological studies, several GRIDs were calculated (flow direction, flow length, flow accumulation, and watersheds). The runoff net calculations were carried for mid-scale level and for the local study area, and then it was compared.

2. Some researchers have noted [6, 7]: the city terrain model and its features derived from DEM are correlated with hydrologic environments, valleys of real and buried rivers, and thickness of technogenic deposits. In our study, we checked this assertion for the local area. We calculated buried relief (DEM_{br}) using the formula of Raster calculator:

$$DEM_{br} = DEM_t - Th \tag{9.2}$$

where DEM_t – calculated DEM from ground topographic data and Th is the thickness of technogenic deposits.

Data on man-made deposit distribution and thickness allow to find anthropogenic disturbances of relief. One of the main violations is creation of closed depressions on the drain net. Their area and the impact on the urban environment can be considerable.

In addition to the traditional methods for analyzing the formation of surface runoff [11], the innovative method was proposed: to locate the sample of the accumulation zone grid in increasing order and calculate the percent growth as a percentage of the total amount of runoff accumulation and to compose graph flow accumulation (Y) vs X – Ln (S) and S square of watershed (Figs. 9.1, 9.2, and 9.3).

The most difficult in the study of surface runoff in the Tarakanovka River valley is the sharply differing land use conditions: the middle part of the river is channeled, and bottom part of the river is an open watercourse. Along with quarters of dense buildings, with asphalt-covered lands, there are quarters of a protected landscape with a rare low-rise building, as well as quarters of a protected natural complex.

The results of digital comparison of erosion networks of real and buried relief allow to note:

- The ratio of the total area of the catchment basin of the modern relief to the buried Ks = 1, 13
- The ratio of the maximum value of accumulation of drains of the modern relief to the buried Ka = 1.05

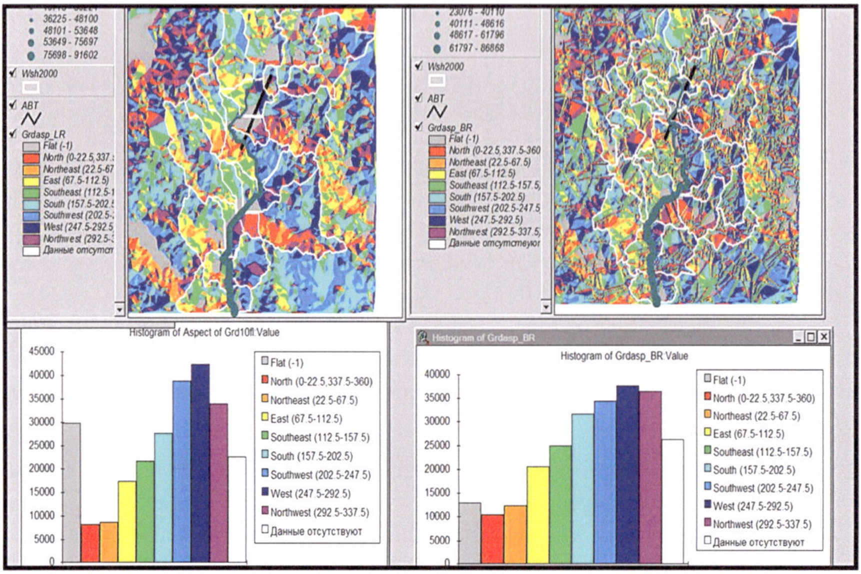

Fig. 9.1 Comparative analysis aspect of DEM (10 m) of real (left) and buried relief (right) in local study region. At bottom, histograms of distribution aspect slopes by eight rumbas and flat surface

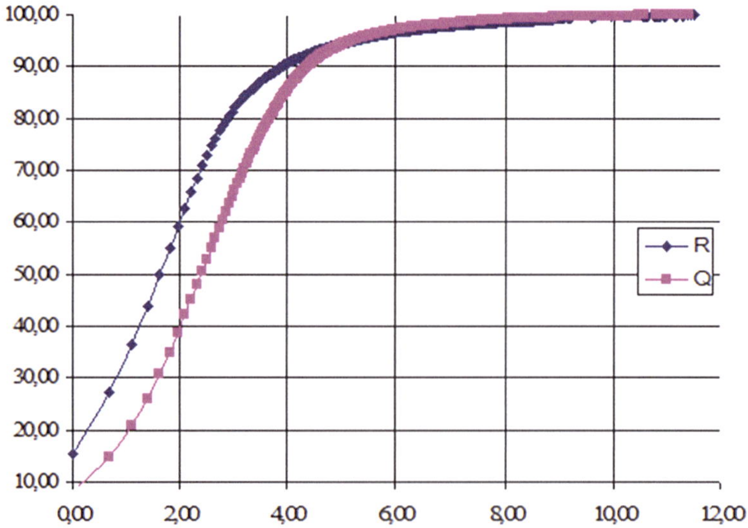

Fig. 9.2 Graph flow accumulation (Y) vs X – Ln (S), S square of watershed, R – modern relief, Q – buried horizon

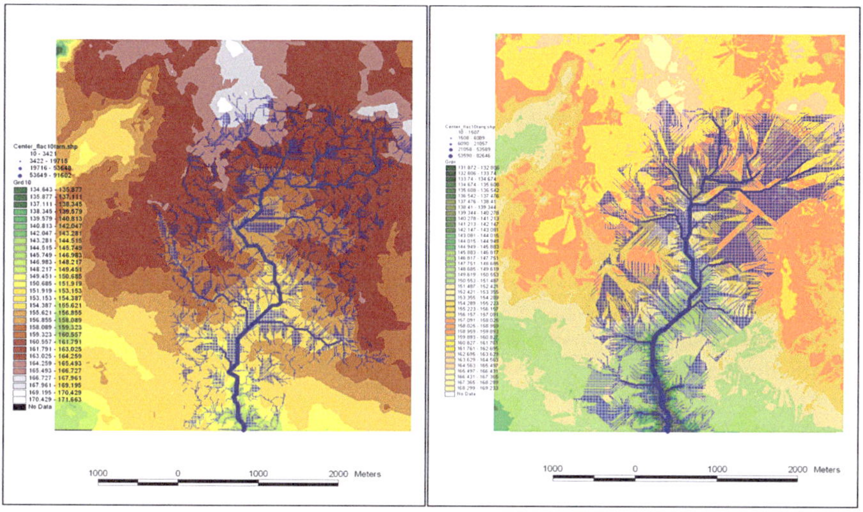

Fig. 9.3 Large-scale level of study. At the left, flow accumulation of surface relief by DEMz (cell 10 m), and at the right, flow accumulation of buried relief by DEMbr. Square of watershed: real watershed – SWSHz = 9.2 km^2, buried relief – SWSHbr = 8.2. The depth of erosion of the river valley (in brackets – value for buried relief): at the top, 10.5м (6,6); in the middle part, 7.6 m (9,3); at the bottom, 4.9 (6,6)

Table 9.1 Statistics of calculated classes by SMORPH method

N	Statistics	Value in cells	
		Class caution	Class unstable
1	Count	173	12
2	Sum	327.0	25.1
3	Mean	1.9	2.1
4	Min	0.6	0.7
5	Max	10.4	6.9
6	Standard deviation	1.8	1.8

- The ratio of the average slope of the modern relief to the buried relief Ku = 1.40

3. The SMORPH method was used to determine zones of varying stability of the slopes. This model, known as the slope morphology model (SMORPH), was developed in 1995 by Shaw and Johnson to run on the ESRI software platform [8]. The SMORPH model evaluates the potential for shallow, rapidly moving landslides based on slope curvature and slope aspect in urban areas (Table 9.1).

Calculation method of landform classes:

Curvature Class	Slope Class S1	S2	S3	S4	S5
C1	1	2	3	3	3
C2	1	1	1	2	3
C3	1	1	1	1	2

Output Classes: 1 = Green, Stable, 2 = Yellow, Caution, 3 = Red, Unstable, where: Slope classes:

Slope Class 1: V <= 2% [Relatively Flat], Slope Class 2: 2% < CV <= 5% [Low Steepness]
Slope Class 3: 5%< CV <= 10% [Moderate Steepness], Slope Class 4: 10% < CV <= 13% [Very Steep]
Slope Class 5: SV > 13% [Extremely Steep].
CURVATURE CLASSES: Predefined Set CUR_01
Curvature Class 1: CV < −0.01 [Concave surface]
Curvature Class 2: −0.01 <= CV <= 0.01 [Planar surface]
Curvature Class 3: CV > 0.01 [Convex surface]

4. The mapping model was proposed in order to compose a zone of influence of surface runoff. ModelBuilder (ArcGIS/ArcView) was used (Fig. 9.4).

We have proposed to allocate the total contour flow accumulation of the value of more than 1,200,000 м^2 or 1.2 km^2.

5. At the final stage of the study, an engineering geological zoning map was used. On the map of engineering geological zoning, zones of low (green color), medium (yellow color), and high hazard (pink color) and engineering geological massifs with an index of a dangerous process (number of brackets) are allocated.

The hazard processes have index: (1) flooding, (2) shallow landslides, (3) potential karst-suffusion, (4) karst-suffusion, (5) deep landslides, (6) technogenic soils, and (7) weak soils.

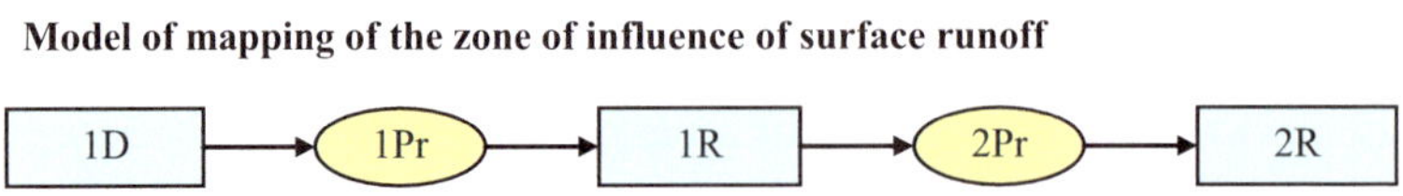

Fig. 9.4 Model of mapping of the zone of influence of surface runoff. Rectangle, data; Oval, process. где: 1D, GRID zone of flow accumulation; 1Pr, calculation zone of summer flow accumulation in radius 250 м; 1R, GRID summer flow accumulation; 2Pr, calculation izolines by interval 200,000 м^2 or 0.2 км^2; 2R, izolines of summer flow accumulation

The additional layers of integrated map are:

- Zones of local lowland
- Zones of stability of the slopes with class caution and class unstable
- Zones of influence of surface runoff

The cartographic result (fragment in local area of Tarakanovka River) of integrated map is presented at Fig. 9.5.

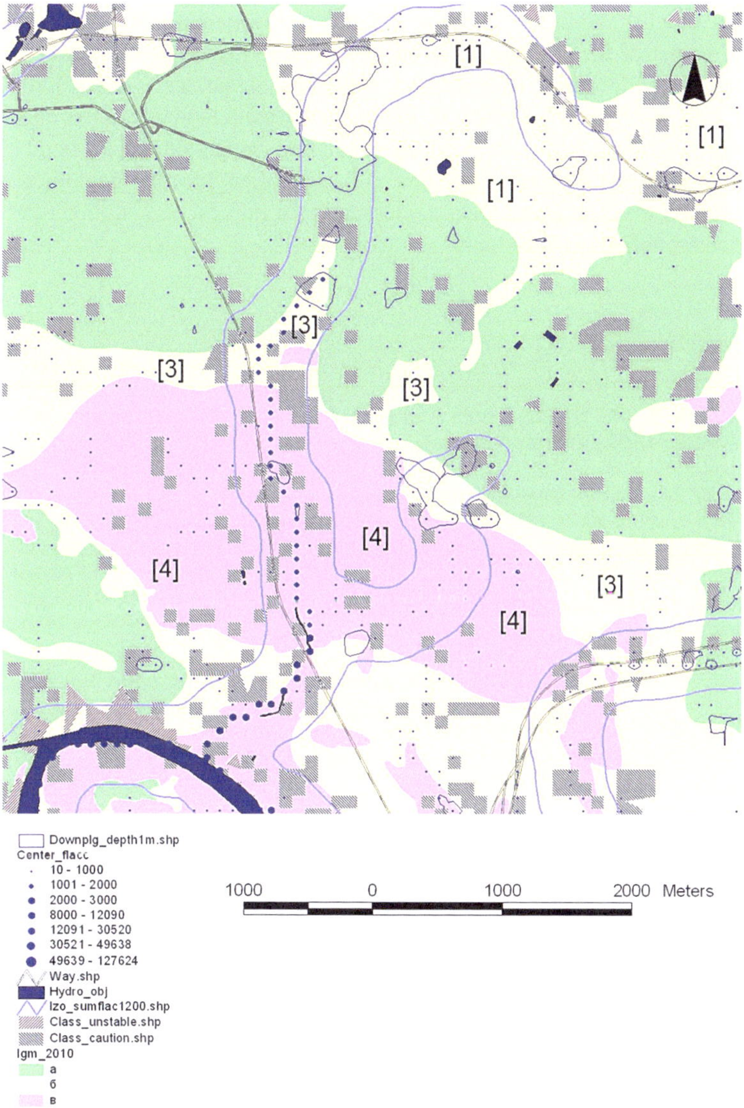

Fig. 9.5 Cartographic result (fragment in local area of Tarakanovka River) of overlay engineering geological zoning map and additional calculated layer zone of influence of surface runoff and zones of landform zoning (class caution and class unstable)

9.4 Conclusion

1. The studies showed that the radar remote sensing data and ground surveying can be used for adequate modeling of the runoff and drain network in urban areas.
2. The comparison of the distribution of accumulation zones of modern and buried terrain allows one to find anthropogenic disturbances of relief and understand the transformation of the relief.
3. The proposed methods make it possible to determine the zone of influence of surface runoff and to identify zones of different slope stability.
4. The merger of the engineering geological zoning map and calculation layers of zones (local lowland (depressions), influence of surface runoff, and various stabilities of the slopes) makes it possible to create a new integrated map.
5. At the next stage of the project, based on a new integrated map, it is possible to determine new values of the degree of hazard of engineering geological conditions, taking into account local depressions, the influence of surface runoff, and various stabilities of the slopes

References

1. Osipov, V. I. (2011). Large scale geological mapping of Moscow territory. *Geoecology, Engineering Geology, Hydrogeology, Geocryology 3*(3), 195–197. (on the Russian).
2. Osipov, V. I., Burova, V. N., Zaikanov, V. G., Molodykh, I. I., Pyrchenko, V. A., & Savis'ko, I. S. (2012). A map of large-scale (detail) engineering geological zoning of Moscow territory. *Water Resources, 39*(7), 737–746 © Pleiades Publishing, Ltd., ISSN 009778078.
3. Tarboton, D. G., Bras, R. L., & Rodriguez–Iturbe, I. (1991). On the extraction of channel networks from digital elevation data. *Hydrological Processes*. Special Issue: Digital Terrain Modeling in Hydrology, *5*(1), 81–100.
4. Moscow encyclopedia//Nauchnoe izdatelstvo "Bolshaia rossiskaia enciklopedia, Moskva, 1998 (on the Russian).
5. Pozdnyak G. (2013). *Bolshoi atlas moskvy*. Moskva: Feoria. ISSN 978-5-91796-032-6. (on the Russian).
6. CGIAR-CSI. (2015). The CGIAR consortium for spatial information, appling geospatial science for a sustainable future. http://srtm.csi.cgiar.org/SELECTION/inputCoord.asp.
7. Vaugeois, L. M., Jenness, J., & Rosenberg, L. I. (2007). SMORPH slope stability software, ArcView 3.x implementation of Washington Department of Natural Resources ARC/INFO SMORPH program: Public domain software developed by Laura Vaugeois, L.M. and adapted by Jenness Enterprises for the San Luis Obispo Planning County Department.
8. Shaw, S. C., & Vaugeois, L. M. *Comparison of GIS-based models of shallow landsliding for application to watershed management, Final report*. Olympia: Washington Department of Natural Resources Forest Practices Division.
9. Shaw, S. C., & Jonson, D. H. (1995). Slope morphology model derived from digital elevation model. http://www.krisweb.com/biblio/gen_xxxx_shawetal_1995_slopemorph.pdf.
10. Jenson, S. K., & Domingue, J. O. (1988). Extracting topographic structure from digital elevation data for geographic information system analysis. *Photogrammetric Engineering and Remote Sensing*., 0099–1112/88/5411–1593$02.25/0, *54*(11), 1593–1600.
11. Osipov, V., Karfidova, E., & Batrak, G. (2016). The drain net modeling on the base terrestrial and radar data. Comparative analysis. In *5th annual international conference on geological and earth sciences (GEOS 2016)*. https://doi.org/10.5176/2251-3353_GEOS16.35. ISSN 2251-3353.

Chapter 10
Assessment of Seismic Hazards for Extractive Installations of the Oil and Gas Complex on the Caspian Sea Shelf

I. G. Mindel, M. D. Kaurkin, V. V. Nesynov, and D. O. Andreev

Abstract At present time exploration and development of oil and gas fields on the shelf of the Caspian Sea is actively conducted. In turn, the water area of the Caspian Sea is earthquake-prone, and in the construction of oil and gas facilities it is necessary to assess the seismic hazards. In the article the calculation method used in seismic microzoning is considered in detail. The methodology of seismic influence calculation from possible centers of earthquakes is visually shown. The result of the work was a set of accelerograms, which can be used to calculate the seismic loads on the object.

10.1 Introduction

In connection with the active exploitation of oil and gas condensate on the Caspian shelf, there appeared a need to assess the natural hazards and risks in the area. One of the urgent natural hazards of the southern part of Russia is its high seismicity. At the time of the earthquake the stability of the ground base of the drilling platform could be violated, which could lead to the damage of some components of the borehole construction and the oil spill into the waters of the Caspian Sea.

The seismic microzoning of some objects of the oil and gas complex located on the Caspian Sea shelf was carried out by the Laboratory of Engineering Geophysics and Seismic Microzoning of the Federal State Budget Institution of the Institute of Geoecology named by E.M. Sergeyev of the Russian Academy of Sciences.

The assessment of seismic hazards on the shelf is characterized by a number of problems not peculiar to similar overland work, including limited number of records of strong motions on the surface of the floor, difficulties in organizing observations on the shelf, presence of an unsustainable surface water-saturated layer in the geological section, susceptible to liquefaction in seismic effects, etc.

I. G. Mindel (Deceased) (✉) · M. D. Kaurkin · V. V. Nesynov · D. O. Andreev
IEG RAS, Moscow, Russia

V. Svalova (ed.), *Natural Hazards and Risk Research in Russia*,
Innovation and Discovery in Russian Science and Engineering,
https://doi.org/10.1007/978-3-319-91833-4_10

The studies for each individual object can be conventionally divided into two major phases: refinement of the source (background) seismicity of the area and seismic microzoning.

The source seismicity was refined on the basis of detailed maps of general seismic zoning developed at the Institute of Earth Physics named by O.Y. Schmidt and the results of previous studies. The result of the work is the refined maximum intensity of seismic concussions with the precision right up to the tenth fraction of the required repeat event.

Seismic microzoning was carried out in a series of methods, including methods of engineering and geological analogies and of seismic rigidity and the calculation method. The result of this phase is seismic microzoning maps with different periods of repeated concussions and a set of synthetic accelerograms (three oscillation components of X, Y, Z), taking into account the impact of the most dangerous areas of earthquake occurrence. In this article we will review the calculation method used in seismic microzoning.

The development of hydrocarbon deposits at different stages requires the building of constructions of various types and designations. For example, we consider the seismic hazard assessment methodology for the site of the jack-up rig “Neptune” during the well drilling in the “Titonskaja” structure. The results can then be used to assess seismic hazards of already-available production platforms and related infrastructure.

10.2 Seismic Hazard Assessment for the Location of the Jack-Up Rig “Neptune”

Figure 10.1 shows the layout of the “Titonskaja” structure. The purpose of the research was the seismic microzoning of the jack-up rig 200×200 m, with the quantification of predicted seismic effects from possible earthquakes.

Refining the Original Seismicity of the Study Area

In accordance with the regulations in force at that time (GSZ-97 and DC 14.13330.2014), in integral points the territory of the nearest large locality (Makhachkala) refers to the 8-point area on a map A (repetition of concussions 1 time at 500 years), on a map B to the 9-point area (repeatability 1 time at 1000 years), and on a map C to the 10-point area (repeatability 1 time in 5000 years).

On the refined map GSZ 97-C, built in the Institute of Earth Physics for the same technology as the maps in integer points but more detailed (step between isoseismal lines equals 0.1 points), the “Titonskaja-1” site is located near the isoseismal line of

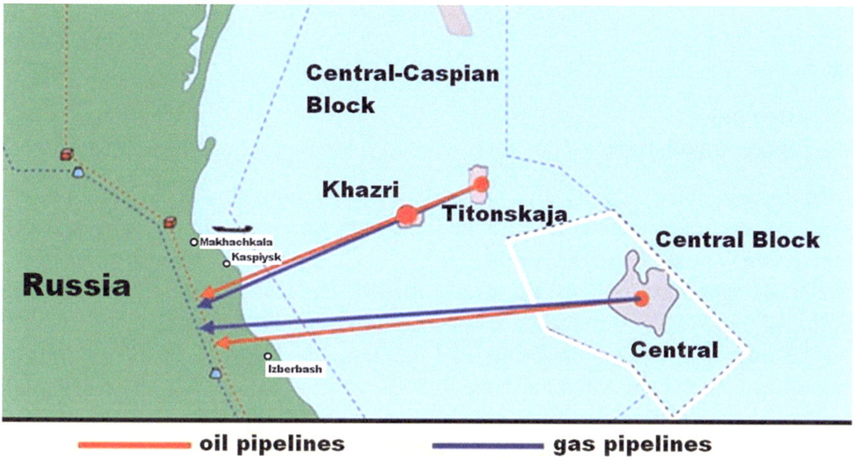

Fig. 10.1 A fragment of the Caspian Sea mining scheme, including the location of the "Titonskaja" structure

Table 10.1 Seismic impact parameters dangerous for "Titonskaja-1" site

Seismogenerating area	Magnitude, M_S	Depth of the center, km	Epicentral distance to the site, km	Intensity *I*, point
Makhachkalinskaja area	7.0	10–15	45–50	8.9
Local earthquake from D0153 domain	5.5	8–12	0	8.1
Krasnovodskaja area	8.0	20–50	450	6.0

9.0 points. Thus, for soil category II seismic properties on the "Titonskaja-1" site, seismic intensity should be taken close to 9 points (with repeated concussions 1 time in 5000 years).

Initial Parameters of Seismic Effects

According to the original seismicity and based on the analysis of the seismic situation in the site area, it was decided to calculate the seismic hazard from the following areas of possible earthquake center appearance (AEC): a close Makhachkalinskaja area with $M = 7.0$, a local area within the D0153 domain with $M = 5.5$, and a center at about 10 km depth, in the most unfavorable case, located directly beneath the site, as well as a remote Krasnovodskaja area with $M = 8.0$. Table 10.1 shows the main parameters of the three scenario-specific earthquakes that are dangerous for the "Titonskaja-1" site.

For local earthquakes a known formula of macroseismic field with the refined coefficient for Dagestan was applied to assessing the concussion strength in points:

$$I_0 = 1.5M - 3.6\lg(R) + 3.1,$$

where

M – magnitude
R – hypocentral distance

In the case of the close Makhachkalinskaja area, seismicity was adopted in points (with minor adjustments) on a refined map of isoseismal lines, with repeated concussions one time in 5000 years.

For a remote Krasnovodskaja area, a map of isoseismal lines of Krasnovodsk earthquake in 1895 [7] was taken into account in estimating the points.

The intensities shown in Table 10.1 refer to the category II soil of seismic properties, which is repeated one time in 5000 years.

The seismicity in points mentioned in maps of GSZ-97 [6] refers to the category II soil of seismic properties. However, it is not worth forgetting the diversity of the category II lithological and physico-mechanical properties. Thus, the quantitative characteristics of seismic impacts – in points, accelerations (velocities or relocations) mentioned in the maps – are essentially classified as abstract "middle" soil of the category II. In each case, it is necessary to take into account the geological structure of the soils, the size of the layers in the section, the relief, and other local factors, which are reflected not only at the amplitude level of the fluctuations from the earthquakes but also at the frequency content and duration [2].

Based on Table 10.1, the formulas below gave us the possibility to calculate the original parameters of seismic effects, depending on the magnitude M, hypocentral distance R, and seismic intensity I.

The maximum acceleration amplitudes for the "medium" soil of the category II, depending on seismic intensity I, were obtained according to a known formula of V.I. Ulomov [5]:

$$\lg\left(A\left(\mathrm{cm/c^2}\right)\right) = 0.301 * I - 0.107,$$

within the current scale of MSK-64.

For the local and near areas, the peak acceleration values were assessed according to a known dependency of F.F. Aptikaev [1, 3]:

$$\lg(\mathrm{PGA}) = C * (0.33 * M_S - 0.61 - \lg(R)) + 2.23 \pm 0.16;$$

where the C parameter depends on the type of movement in the center (if the shear is predominant with $C = 0.633$, in the case of a drop-shift with $C = 0.717$), the near area does not have any significant influence of the ground conditions.

For the Krasnovodskaja area:

$$\lg(\mathrm{PGA}) = 0.63 * M_s - 1.92 * \lg(R) + 1.076 \pm 0.18$$

[1, 3].

Table 10.2 The resulting seismic impact parameters

Name of area	Category of soil	Max. acceleration cm/s^2	Prevailing period T, s	Duration d, s	Seismic intensity I, point
Makhachkalinskaja	II	385	0.36	17	8.9
	I	191	0.31	13	7.9
D0153 domain	II	220	0.20	4	8.1
	I	110	0.17	3	7.1
Krasnovodskaja	II	50	1.1	45	6.0
	I	25	1.0	35	5.0

In this case, for soils of category I, the amplitude level is lowered by 0.15 logarithmic unit.

The prevailing periods for accelerations were assessed by the dependence of F.F. Aptikaev [1, 3], which was in annex 4 of the Gosatomnadzor Russian manual, SG-006-98:

$$\lg(T(A)) = 0.15^* M_S + 0.25^* \lg(R) - 1.9 \pm 0.16.$$

The effective duration of the vibrations was assessed for acceleration also according to the known dependency of F.F. Aptikaev [1, 3]:

$$\lg(d(A)) = 0.2^* M_S + 0.51^* \lg(R) + C_1 + C_2 \pm 0.3,$$

where

$C_1 = 0.0$ for medium soils
$C_1 = -0.15$ for rock soils
$C_2 = -1.3$ for accelerations

Table 10.2 shows the resulting seismic impact parameters for the location of the jack-up rig from the three most dangerous areas of earthquakes on the abstract soil of the category II and on the surface of the conventional rock.

Calculation of Seismic Effects

At the Institute of Ocean Sciences named by P.P. Shirshov of the Russian Academy of Sciences, there was a research of seismic oscillations on the Black Sea shelf using bottom stations. As a result of this research, it has been found that, in the depths of the sea, up to 50–80 m, the pattern of the response of bottom sediments to earthquake variations is almost identical to that of similar soils on land. On the basis of this work, it was assumed that the seismic risk assessment methodology could be applied to sites located on the Caspian Sea shelf.

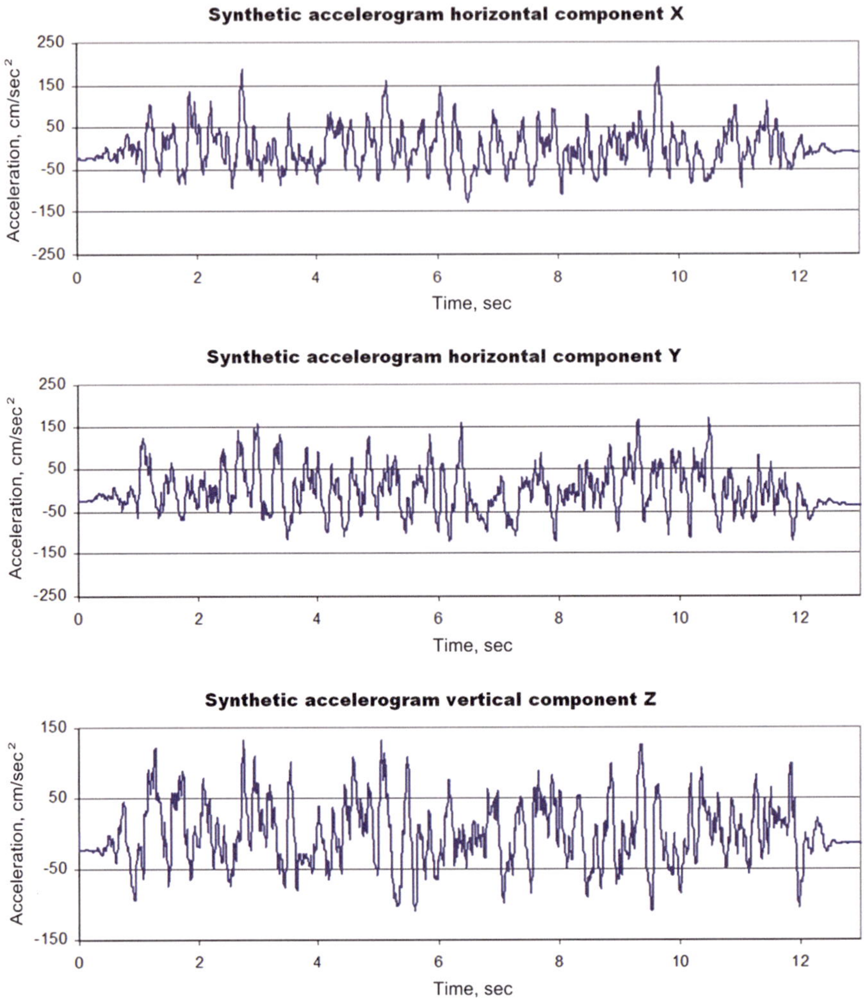

Fig. 10.2 Synthetic accelerograms (horizontal components X and Y and vertical Z) on the surface of a conventional rock from the Makhachkalinskaja area (AEC)

When simulating the propagation of seismic waves through a geological environment, an initial signal is required in the form of accelerogram at the level of the so-called elastic half-space, for which the roof of the rocks of category I is commonly accepted. The original accelerograms at the conventional rock level were generated (according to Table 10.2) in the "PSEQGN" [9] program. As an example in Fig. 10.2, are synthesized accelerograms (horizontal components of X and Y and vertical Z oscillations) on the surface of the conventional rock from the most dangerous site of the "Titonskaja-1" of Makhachkalinskaja area (AEC).

When synthesizing the accelerograms, the ratio of acceleration amplitudes between the two horizontal components of the vibrations on the basis of work [1] was adopted as follows: the more intense of the two horizontal components exceeds the randomly oriented component of 1.133 times. The ratio between the acceleration amplitudes on the vertical and horizontal components, depending on the acceleration value on the horizontal component in the range from 250 to 600 cm/sec^2, varies from 0.5 to 0.8 [1]. In most cases, in practice, acceleration on the vertical component relative to acceleration on a more intensive horizontal component is taken with a factor of 0.7, which is used later. The prevailing period of vibrations on the vertical component is adopted at −0.1 logarithm less than horizontal. The resulting synthetic accelerograms on the surface of the conventional rock will be used as input signals in the calculation of the accelerograms on the bottom of the "Titonskaja-1" site, taking into account the specific seismogeological section.

Seismic effects, from close, local, and remote earthquakes, were calculated in the NERA (*N*onlinear *E*arthquake Site *R*esponse *A*nalyses of Layered Soil Deposits), which was developed by J. P. Bardet and T. Tobita [8]. The NERA program is implemented as an add-in for Microsoft Excel, which makes it much easier to work and process the results of calculations. In the program, calculations are based on a model that takes into account the mechanisms for nonlinear behavior of soils with strong seismic effects. A layered ground pattern reaction is calculated by simulating the propagation process through the seismic wave that is defined with the corresponding parameters at the level of elastic half-space that underlies the mass. The wave spreads in a vertical direction and is given by the digitized accelerogram. The speed section (seismogeological model) is set by the system of layers, for each of which is given the power, the transverse velocity, the specific gravity (in kN/m^3), and the soil type – clay (1), sand (2), or semi-rock (0). According to the soil type, the program assigns the shear module and damping (absorption) constraints of the soil, reflecting the nonlinear nature of the relationship between the voltage and the deformation. The geoseismic model of a ground base in the NERA program format is shown in Table 10.3.

The result of calculations is a set of accelerograms for the three vibration components (X, Y, Z) and each AEC area. As an example in Fig. 10.3 as a result of the calculation, there are shown accelerograms (horizontal components of X and Y and vertical Z) on the surface of the bottom at the point of jack-up rig at the "Titonskaja-1" site from the earthquake in Makhachkalinskaja area.

As was to be expected, despite the damping properties of the bottom soil of about 90 m, represented mainly by the saturated sands (about 70% thick) and to a lesser extent clay (about 30%), the most intense fluctuations in the surface of the bottom are expected from the earthquakes from the Makhachkalinskaja area – the maximum acceleration on component X is 182 cm/sec^2 (Fig. 10.3). From a local earthquake from the D0153 domain, the most disadvantaged of the location of the center directly beneath the site at a depth of about 10 km, the maximum acceleration is assessed according to the calculations of the same intensive component X in 126 cm/sec^2. In the case of earthquakes from the remote Krasnovodskaja area, the maximum acceleration on the horizontal component X is estimated at 54 cm/sec^2, which is equivalent to 6 points in terms of point scoring.

The calculated accelerograms at the point of jack-up rig from the most dangerous Makhachkalinskaja area and from the D0153 domain for all three vibration

Table 10.3 Geoseismic models of the earth cover at the jack-up rig site are used to calculate seismic impacts

	Layer number	Soil material type	Thickness of layer (m)	Maximum shear modulus G_{max} (MPa)	Total unit weight (kN/m^3)	Shear wave velocity (m/s)	Depth at top of layer (m)	Vertical effective stress (kPa)
Surface	1	2	3.0	78.70	19.30	200	0.0	0.00
	2	2	5.5	124.24	19.50	250	3.0	28.47
	3	2	5.0	134.37	19.50	260	8.5	81.76
	4	2	4.0	163.11	19.70	285	13.5	130.21
	5	1	2.5	77.06	18.90	200	17.5	169.77
	6	1	5.5	196.41	20.05	310	20.0	192.50
	7	1	1.5	111.56	19.00	240	25.5	248.82
	8	2	4.0	229.79	19.50	340	27.0	262.60
	9	1	1.5	114.50	19.50	240	31.0	301.36
	10	2	6.0	243.50	19.50	350	32.5	315.90
	11	2	12.0	264.22	20.00	360	38.5	374.04
	12	1	5.0	111.56	19.00	240	50.5	496.32
	13	2	17.5	294.39	20.00	380	55.5	542.27
	14	1	7.0	324.57	19.90	400	73.0	720.59
	15	1	12.0	489.30	19.20	500	80.0	791.22
Bedrock	16	0		882.90	20.50	650	92.0	903.90

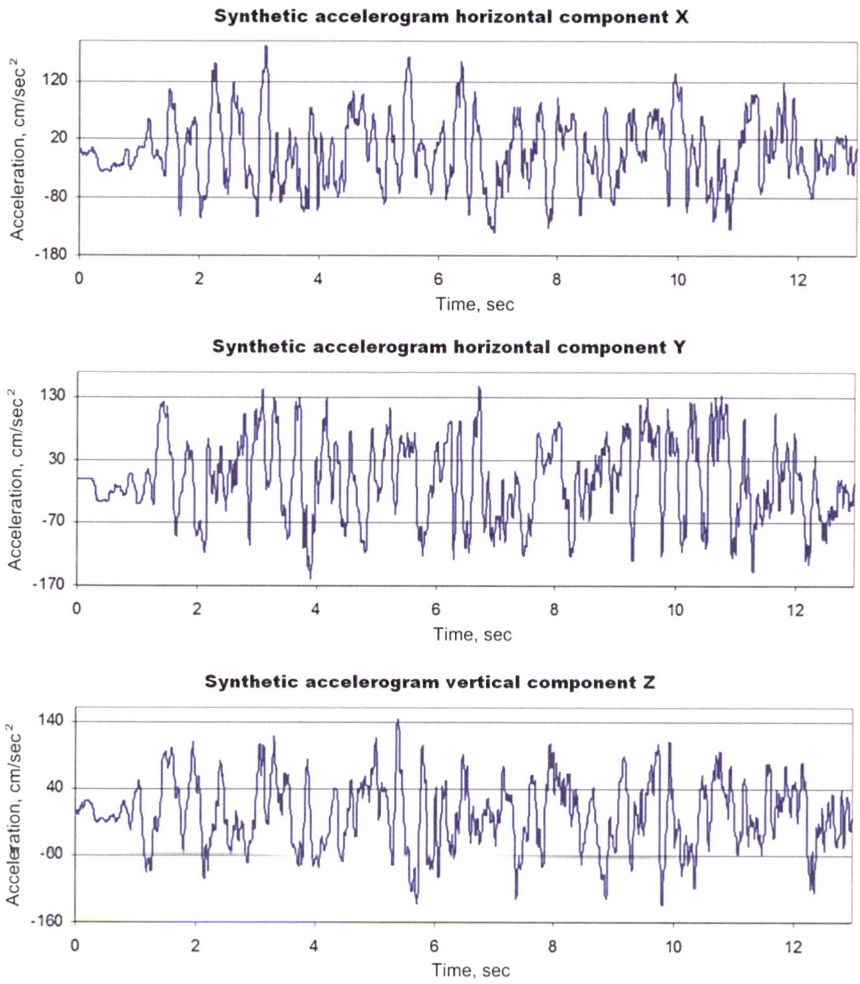

Fig. 10.3 Synthetic accelerograms on the surface of the bottom at the point of jack-up rig on the "Titonskaja-1" site from the earthquake in Makhachkalinskaja area with $Ms = 7.0$

components can be used to calculate seismic loads in the temporary area on structures at the maximum calculated earthquake (MCE) in the recurrence of earthquakes one time every 5000 years. If necessary to take into account relatively small frequency fluctuations of slightly more than half a minute, it is possible to apply accelerograms received for the Krasnovodskaja area.

Since the jack-up rig design implies a significant cover of supports in the bottom of the sediments, a study of the change in accelerations with depth for the events from the AEC areas under review was carried out to assess the seismic hazard. Figure 10.4 shows a timetable for changing accelerations with depth for the most intensive components of X oscillations in an earthquake from the

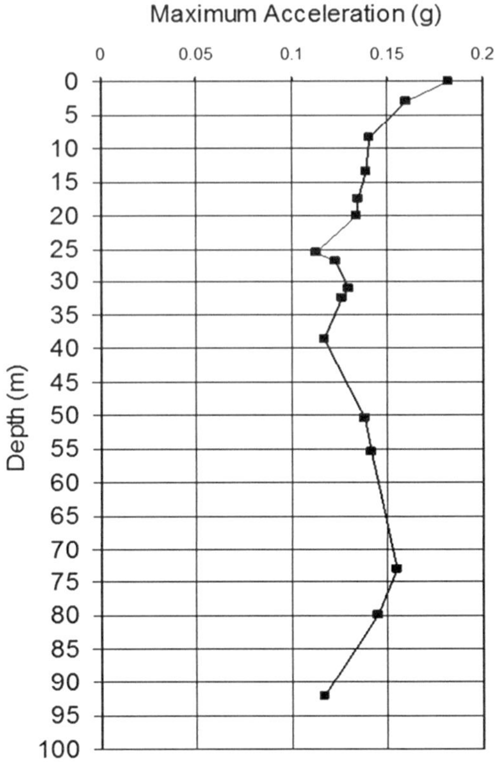

Fig. 10.4 The dependence of magnitude of the acceleration from depth (from the surface of the bottom) in an earthquake from the Makhachkalinskaja area at the point of jack-up rig (horizontal component of oscillations X)

Makhachkalinskaja area, the most dangerous for the amplitude of the jack-up rig point. In the case of Fig. 10.4, it can be seen that in the interior of the earth cover, between 5 and 60 m from the surface of the bottom, the acceleration values range from 110 to 150 cm/sec^2. This is a positive factor and can be stocked in the calculation of seismic loads for the structure.

The resulting set of accelerograms enables calculations of seismic loads on an object, but it is useful to get an accelerogram that would unite all possible variants of earthquakes from the most dangerous near and local seismogenic zones. This, in turn, would reduce the computation of seismic loads on the structure and take into account the full range of possible periods of fluctuation from $0.15 \div 0.20$ to 0.30 with $\div 0.80$ s. This approach does not contravene the provisions of DC 14.13330.2014 [4] concerning the form values of the dynamic β_1 coefficient for soils of categories III and IV by seismic properties.

In order to accomplish this task, the first phase was to build a generalized spectrum response for each component of vibrations from the most dangerous earthquakes in the Makhachkalinskaja area and the D0153 domain, based on the respective spectra. For example, Fig. 10.5 shows the spectra reactions from the most dangerous earthquakes and the generalized spectrum for the most intense vibration components of X.

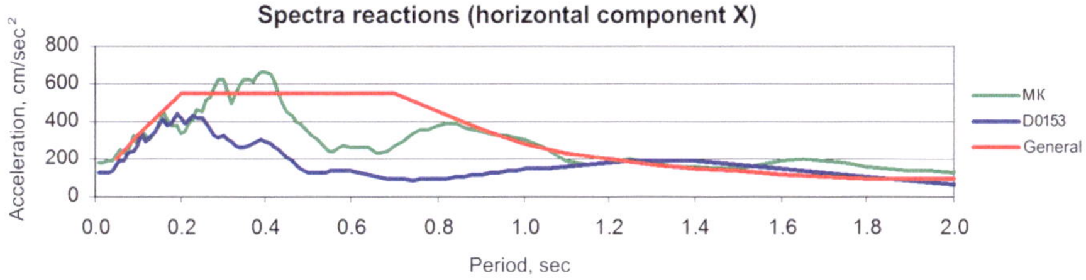

Fig. 10.5 Spectra reactions from calculated accelerograms on the surface of the bottom at the point of jack-up rig from the near earthquake from the Makhachkalinskaja area and from local earthquakes from the D0153 domain

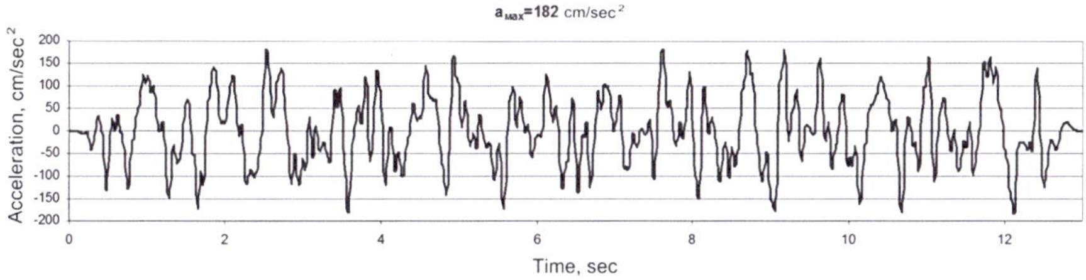

Fig. 10.6 Synthesized accelerogram, calculated in a generic response spectrum (horizontal component X)

The next phase was to calculate the accelerograms of the corresponding generalized spectrum reaction. Figure 10.6 shows the calculated accelerogram on the surface of the jack-up rig, obtained from the generalized spectrum on the oscillation component X. In turn, Fig. 10.7 shows the specified generic spectrum of reaction and derived from the calculated accelerogram (with normative critical decay of 5%). As you can see in Fig. 10.7, the specified and calculated spectra reactions have no significant differences; on this basis it is possible to judge the correctness of the calculated accelerogram.

The accelerogram set obtained by this method contains entries for the three vibration components and allows the seismic hazard to be assessed from the most dangerous Makhachkalinskaja area and from the D0153 domain.

It is worth recalling that the calculated seismic microzoning method described in the work is one of the set of methods. In the study seismic hazard assessments were carried out by other methods: method of engineering and geological analogies and method of seismic rigidities. On comparing the results of the various methods received, it can be noted that the calculation values of accelerations on the research site are due to explicit nonlinear phenomena in the thicker of the saturated sand and clay bottom soils of category III of the power about 90 m (due to the damping effect) corresponds to 7-8 points, rather than 9-10 points on the MSK-64 scale, as measured by the engineering-geological indicators and the comparison of seismic rigidities of soils, which do not take into account nonlinear phenomena in relatively weak soils with severe earthquakes.

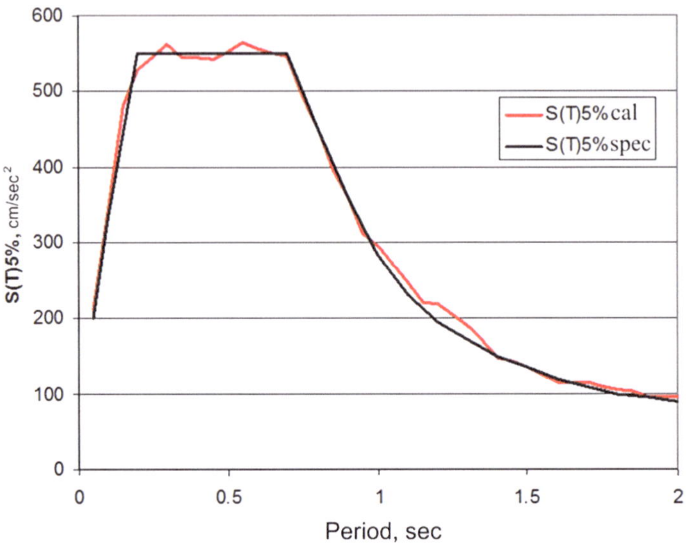

Fig. 10.7 Generalized and calculated on the basis of the accelerogram spectra reactions for soils in the research section (horizontal component X)

10.3 Conclusion

The data obtained from the seismic hazard assessment provide important characteristics for the design, such as the maximum possible acceleration value, the period, and the spectral composition of the surface variations. The results of the work can be summarized in more detail by the following items:

1. The area of the "Titonskaja-1" site in the middle of the Caspian belongs to one of the most seismically active regions of the Caucasus. Clarification of the original seismic area was carried out with the latest data on the seismic zoning map with a fractional step of 0.1 points (Institute of Earth Physics of the Russian Academy of Science, Ulomov V.I.). The "Titonskaja-1" site on the detailed SRF map-97 from the area (repeatability of earthquakes one time in 5000 years) is assigned to the section with isoseismic lines of 8.9–9.0 points on the MSK-64 scale (on soil II of the seismic category). Seismic hazards for the site are conditioned by possible earthquakes from three earthquake areas (AEC): from near centers in Dagestan of the Makhachkalinskaja area with magnitude $M = 7.0$ at epicenter distance of 45–50 km, from local centers from the D0153 domain with $M = 5.5$ at a depth of about 10 km (in the most unfavorable case with the center directly beneath the site), and from the centers of remote catastrophic earthquakes from the Krasnovodskaja area with $M = 8.0$ at an epicenter distance of 450 km.
2. As a result of the calculation of the seismic microzoning at the point of jack-up rig, computed accelerograms (three components of X, Y, Z) were obtained separately for the nearby (Makhachkalinskaja) and local (D0153 domain) areas

of AEC. A set of synthetic accelerograms was also obtained, combining the effects from the AEC's near and local areas.

3. The maximum acceleration on the most intensive X component at the jack-up rig point is estimated at 182 cm/sec^2. This in turn confirms the phenomenon of a significant decrease in the acceleration on the surface of sediments at the expense of nonlinear deformation.
4. A set of synthetic accelerograms (three components of X, Y, Z), taking into account the impact of the dangerously close and local areas of AEC, both separately and together (general), and, where necessary, a set of earthquakes from the Krasnovodskaja area with much lower speeds, but with low frequencies and long fluctuations, may be applied to calculate seismic effects for the jack-up rig.

References

1. Aptikaev, F. F. (2012). *Seismic intensity tool scale*. Moscow: Science and Education.
2. Engineering Research for Construction. (1986). Seismic microzoning. Standards for the production of works. RCN 60-86. The Russian Federation-13. The order of the RSFSR, Moscow.
3. Determination of the initial seismic effects for the design framework. (1998). Gosatomnadzor of Russia. SG-006-98, Moscow.
4. Construction in Seismic Areas. (2014). DC 14.13330.2014. Ministry of construction of the RF, Moscow.
5. Ulomov, V. I. (2005). Probability-deformation assessment of seismic impacts based on SRF-97 and total earthquake maps. Earthquake construction. Safety of installations. No.4 60-69.
6. Ulomov, V. N., & Shumilina, L. S. (1999). *Set of maps of general seismic zoning in the territory of the RF. SRF-97*. Moscow: UIEP RAS.
7. Shebalin, N. V. (1974). *Centers of strong earthquakes on the territory of the USSR*. Moscow: Science.
8. Bardet, J. P., & Tobita, T. (2001). *NERA: A computer program for nonlinear earthquake site response analyses of the layered soil deposits*. Los Angeles: University of Southern California.
9. Ruiz, & Penzien, J. (1969). *PSEQGN probabilistic study of the behavior of structures during earthquake, Earthquake engineering Earthquake Center, Report No. UCB/EERC-69/3*. Berkley: University of California.

Chapter 11
Geodynamic Stability of Territories with Specially Important Engineering Objects

Natalia Makarova, Vladimir Makeev, Tatyana Sukhanova, and Irina Korobova

Abstract The presented research most fully characterizes the geological structure of the territories chosen for construction of important engineering facilities. The research consists of three blocks. The structural and compositional block characterizes a structure of the basement and the sedimentary cover, structural unconformities, and faults, which can be activated in new geodynamic conditions. The structural-geomorphological (neotectonic) block is aimed at defining the type, amplitudes, and development rates of the newest regional and local structures. The structural and geodynamic block includes the definition of geodynamic conditions and trends of the modern development of structures that could cause negative processes. The concept of geodynamic safety of the territories with environmentally hazardous objects, based on this research, has been developed.

11.1 Introduction

When constructing especially important engineering facilities (nuclear power plants, hydroelectric power plants, etc.), it is necessary to take into account natural and man-made factors that determine the safe conditions for the construction and operation of these facilities. The main ones are the geological structure of the territories and the development of new, mainly, modern tectonic deformations that can cause negative processes and disrupt the geodynamic stability of the territories.

Geodynamic stability is the ability of the territory of engineering facilities to maintain their physical state, including strength and equilibrium within the norms for all types of external and internal influences [1].

The research was carried out in different regions of the Eastern European and Siberian Platforms – on the territories of the location of nuclear power stations, oil

N. Makarova (✉) · T. Sukhanova
Moscow State University (MSU), Geological Faculty, Moscow, Russia

V. Makeev · I. Korobova
IEG RAS, Moscow, Russia

V. Svalova (ed.), *Natural Hazards and Risk Research in Russia*,
Innovation and Discovery in Russian Science and Engineering,
https://doi.org/10.1007/978-3-319-91833-4_11

and gas pipelines, the creation of storage facilities for hazardous waste products, the development of mineral deposits, etc. In the process of research, a methodology was developed that allows the most complete and comprehensive study and to characterize the geological environment of different territories, to envisage a possible manifestation of negative endogenous and exogenous processes. It includes a set of studies that can be divided into three blocks: (1) geological (material-structural), (2) structural-geomorphological (neotectonic), and (3) structural-geodynamic. Studies are usually conducted on two scales: relatively reduced scale covering an area of 30–50 km radius, in order to take into account the regional structures within which the construction site is located, and on a detailed scale directly on the site of its deployment. In this case, published and stock materials are used: cartographic, geological, tectonic, geophysical, geochemical, aerospace, drilling data, etc., as well as GIS technologies. In addition to laboratory work, special field studies are of great importance. Below are briefly shown some examples of the use of this technique in the study of the territories of some nuclear power plants on the Eastern European Platform (hereinafter referred to as "EEP"). These are the areas of Ostrovets (Belorusskaya NPP), west of the EEP; the Monakovskiy (Nizhny Novgorod NPP), in its central part; and the Kurchatov (Kursk NPP-2) (Fig. 11.1), in the southern part of the EEP.

11.2 Geological Research

The Geological Research Block

The geological research block includes the study of ancient material-structural heterogeneities of the basement and sedimentary cover (excluding the Paleogene-Neogene and Quaternary deposits) of the location of the objects. These are rocks and ancient structures, fault-folded zones, flexures, structural noses, bays, and sharp lithologic-facies boundaries. They were formed in the past epochs of tectonic-magmatic activity. In new modern geodynamic conditions, these ancient heterogeneities remain passive or are activated with the development of additional stresses, secondary deformations (zones of increased fracturing, lineaments, etc.), geodynamical active zones, and various negative processes. All this can lead to a disturbance of the stability of rocks, which serve as the basis for engineering structures.

In tectonic structure Ostrovetsky region is located within the Baltic monocline, representing the adjoining slope of the Belarusian anteclise and the Baltic syneclise. The crystalline basement here lies at a depth of 220–380 m with a slope of 002–003. The Monakovo region is located on the northern slope of the Tokmov arch, where the foundation is immersed on the NW up to – 1000 m with a gradient of 002. The Kurchatov district is located within the Shchigrovsky arch of the Voronezh anteclise. The depth of the foundation is rapidly increasing to the south from + (100–200) m to – 200 m with a slope of 0.005–0.01.

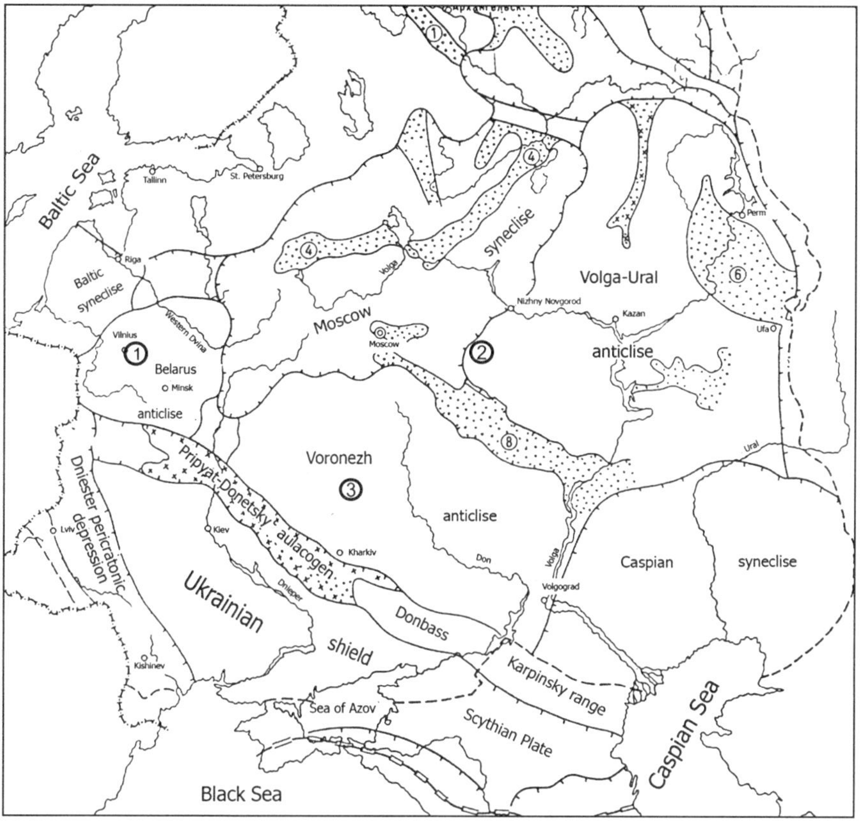

Fig. 11.1 Areas of research on the tectonic regionalization scheme of the Eastern European Platform (EEP) [2]. Large circles show areas: (1) Ostrovetsky, (2) Monakovskiy, (3) Kurchatov. The other circles with numbers – names of different structures of the platform: (8) Pachelmsky avlakogen, (6) Kotlas avlakogen, (4) Srednerussky avlakogen

In the structure of the sedimentary cover, there are breaks in sedimentation and disagreement and changes in the thicknesses of horizons and areas of their development. This reflects changes in structural plans in different epochs of tectonic-magmatic activation. In connection with this, a method was developed for estimating the structural azimuth and angular misalignment of the support horizons of the cover to identify areas of increased tension [3]. These zones were identified in all three regions.

Great importance in the study of ancient structures is given to the foundation faults and cover. Particularly fractured are the foundation in Ostrovets and Kurchatov areas, and some faults are located or are expected in the immediate vicinity of the construction sites of the nuclear power plant. The fractured foundation is less disturbed in the Monakovo district. Here Pavlovskiy fault, partially located in the near area of the nuclear power plant, in the late Proterozoic (Vendian) and in the subsequent time did not show its activity. From the faults of the Ostrovets region, it is most reflected in the bearing horizons of the cover, including the quaternary buried surface, the Oshmyansky fault. In the Kurchatov region, the Nameless and Western faults are partially reflected in the reference Cretaceous horizons. The most famous of them is the Zheleznogorsk fault in the surface of the crystalline basement.

The Structural-Geomorphological (Neotectonic) Block

The structural-geomorphological (neotectonic) block of studies includes the identification of the main heterogeneous neotectonic structures based on relief and geological data, often hidden under the powerful Quaternary cover, their position in the regional neotectonic structure, and a complete description. It includes different tasks: the definition of the morphology of structures and their types (folded, blocky), age, stages of development; the degree of inheritance of ancient structures; study of the newest including Quaternary sediments; analysis of deformations of marking levels (erosion-denudation surfaces, terraces, soles of the newest deposits); cartography of lineaments and their comparison with buried fractures of the basement and fracturing of rocks; the determination of the amplitudes of neotectonic (total) and modern (Holocene) deformations and calculation of the stepwise velocities of motions; and the study of various indicators of the manifestation of modern movements. Part of these studies is carried out in a detailed scale directly on the sites of objects under construction or already existing. To solve these problems, a large set of desk work is carried out. In addition to deciphering various-scale cartographic, aerospace materials, including digital elevation models, an important part of the work is the construction of a network of geological and geomorphological profiles that allow determining a number of the above characteristics [4]. As a result of desk and field studies on deformations of the pre-Neogene or pre-Quaternary surface, recent uplifts and depressions in the above areas were identified. Basically, these are gently sloping positive and negative bends of the surface of the foundation and cover, complicated by lineaments. The latter reflects zones of tectonic or planetary fissuring of rocks and some faults in the basement. Uplifts and depressions are often asymmetric. In the Monakovo and Kurchatov regions, they form a latitudinal zoning, within which individual local structures are isometric or have a diagonal orientation. They do not always inherit the structure of the foundation and cover, being displaced from them in one direction or another (Monakovo region), or do not agree with them (Kurchatov region). In the Ostrovets area, the structural pattern is represented by a gently sloping development of the uplifts and depressions of the NW and NE strike and their complicating steps, identified by the base of the Quaternary sediments. Structures of the NW strike, developed mainly in the eastern part of the region, are consistent with the Hercynian structures and the structures of the NE strike with the Caledonian structures. Thus, the formation of neotectonic structures is influenced by the ancient material-structural inhomogeneities of the platform cover. The age of the uplifts in the Kurchatov region is Oligocene-Miocene, the Early Pliocene in the

Ostrovets and Pliocene-Early Pleistocene in Monakovo. Accordingly, the total amplitudes of the uplifts are 250–260, 50–55 m, and 120–130 m. The small amplitudes in the Ostrovets region are probably due to the significant pressure effect of at least three glacial coverings. The amplitudes of the staged Quaternary uplifts range from 12 to 45 m. During the Quaternary time, the gradual increase in the rate of uplift occurs, and in the Holocene they are 3, 0.6, and 2.2 mm/year, respectively. Some ancient faults, activated at the newest stage, are expressed in modern relief by ledges, development of fracture zones used by the erosion network in the form of lineaments, geochemical anomalies, etc. In Ostrovetsky region, such an activated fault is Oshmyansky, in which the zone of origin of earthquake foci (OEF) of the same name is designated in connection with the confinement to her epicenter of the earthquake of 1908–1909 (6–7 points on the MSK-64 scale). Maybe that activation of the Oshmyansky fault is connected with the glacioisostatic factor. Modern movements, in addition to special geodesic measurements, can be judged by the various signs of their manifestation. These include the change in the facies and thickness of the Holocene sediments; the springing of the surface runoff of rivers and streams, leading to swamping of sites; the changing of the shape of the channel and the pattern of the hydroset; the appearance of the erosion floodplain; the increase in the depth of the gully network; and the activation of landslide and karst processes. On the basis of these features, modern local uplifts and subsidence have been revealed in the valley of the river Seimas. Signs of the right-shearing stresses are shown in the figure of lineaments and the pull-apart type of the structure of the floodplain areas at the same valley.

The Structural-Geodynamic Block

The structural-geodynamic block is aimed at determining the geodynamic conditions for the formation of neotectonic structures and their development trends. For this, the data obtained during the research of the first two blocks are used. They make it possible to identify specific sources of tectonic stresses and forces, in the field of action of which the investigated areas and engineering objects are located and structures are formed. These include the latest geodynamic systems of global, regional, and local [2, 5–7]: the orogenes of the Caucasus, the Carpathians and the Urals, the North Atlantic rift zone, as well as the basins of the Black and Caspian seas, the intraplatform arches of the Voronezh, the Tokmov and other uplifts, the Oka-Don, the Dnieper-Donets basins, and others. The morphology of the structures and the data of the kinematic of jointing and lineaments analysis allow us to determine the main and secondary sources of stresses. In general, there are usually several stress sources that weaken or enhance each other's actions. As a result, complex sometimes contradictory parageneses of the conjugate structures of compression, extension, and shearing are created [8]. Geodynamic conditions for the development of structures in the regions studied were determined on the basis of the obtained geological and structural-geomorphological data. So the tectonic structure of the Monakovo region is a young

(Late Pliocene-Early Pleistocene) uplift. It is part of a system of sublatitudinal neotectonic zoning, caused by the imposition of several types of stresses. First of all, it is a submeridional compression, proceeding from the south from the developing Tokmov new arch. To the north, the structures experience extension due to the influence of the developing and advancing Mari-Chuvash trough, which is part of the dead Moscow syneclise. From the west, the structures are cut by the Nizhneoka depression, which represents the geodynamical active zone [9], which develops under conditions of shearing stresses of compression. Complex geodynamic conditions for the development of the structures of the Monakovo region cause the active development of erosion, landslide, and karst processes. The recent and most modern geodynamic environment of the Ostrovets region is characterized by lateral compressive and tension stresses. The former are directed from the Carpathian-Caucasian neotectonic region and the latter from the emerging young (Middle Quaternary) East Baltic rift zone. Extension structures are marked by wide flat bogging plains, the valley of the river Viliya, chains of lakes of NW strike, predominance of faulting deformations in quaternary fluvioglacial deposits. The extended lineaments of the northwest strike, some of which reflect faults in the foundation, within the region under study correspond to extension zones with a dextral shear component and the northeast to compression zones with a sinistral shear component. Orthogonal lineaments develop under compression. The recent structure of the Kurchatov district is also formed under the influence of several sources of stress [10]. The northern part, including the main Kursk uplift, is under pressure from the north-northeast by the Voronezh uplift. A consequence of this is the widespread development of submeridional lineaments, which are extension zones. Some of them are associated with radon anomalies. In the formation of the structure of the southern part of the region, tectono-gravitational processes may possibly take part – the movement of the sedimentary mantle toward the south of the Dnieper-Donets basin. The basis for this is the relatively steep slope of the foundation surface and the development of clay deposits at the base of the sedimentary cover. The latitudinal Seim trough separating the southern and northern parts of the Kurchatov district is a geodynamically active zone. It is characterized by a setting of right-lateral compressive stresses oriented in the southeast direction, under conditions of additional extension.

11.3 Conclusion

Thus, complex studies of ancient structural-material inhomogeneities and structural-geomorphological (neotectonic) and geodynamic conditions for the formation of new structures make it possible to comprehensively characterize the geological environment of the areas of location of especially dangerous engineering facilities. In many ways, the state of this environment depends on the influence of different sources of tectonic forces and stresses. They determine the activity of manifestation at the present stage of ancient structures, the formation of various types of new structures, and the development of dangerous endogenous and exogenous processes.

An important task is an objective determination of the degree of influence of each of the sources on the geological environment of the study area. The results of the studies are the basis for assessing the geodynamic stability of each territory. The wide development of extension structures in all three regions is the reason for the increased water-fluid permeability of rocks. This is especially manifested and expressed in Kurchatov region by changing engineering and geological properties of soils. The conditions of extension in the Monakovo region cause a wide development of suffosion-karst processes. In the Ostrovets region, the influence of glaciotectonics and, possibly, seismicity is significant. The presented complex of researches, applied in the territories of especially important engineering objects, allowed to create the concept of geodynamic safety of territories of environmentally hazardous production facilities [11] and recommend it when studying other engineering facilities.

References

1. Lomtadze, V. D. (1999). Dictionary of engineering geology. SPb.
2. Deep structure and geodynamic processes in the lithosphere of the East European platform (2003). Ekaterinburg: UrO RAS, 299 p.
3. Makeev, V. M., Karfidova, E. A., Miklyaev, P. S., Avkhimovich, A. N., & Dorozhko, A. L. (2013). A method for studying the azimuthal disagreement of the different-age and sub-deep horizons of the sedimentary cover under geodynamic studies // Problems in the Region. geology and sex exploration. anthropology. Materials of the VII University Geological. readings. Minsk. Ed. Center of BSU, P. 18–20.
4. Kostenko, N. P., Makarova, N. V., & Korchuganova, N. I. (1999). *Expression in the relief of folded and discontinuous deformations. Structural-geomorphological interpretation of aerial photographs, space images and topographic maps: Textbook*. Moscow: Publishing House of Moscow State University 120 p.
5. Kopp, M. L. (2004). *Mobilistic neotectonics platforms of South-Eastern Europe*. Moscow: Nauka 340 p. (Tr. GIN RAS, issue 552).
6. Makarov, V. I., Dorozhko, A. L., Makarova, N. V., & Makeev, V. M. (2007). Geodynamically active zones of platforms. *Geoecology*,(2), 990–110.
7. Makarova, N. V., Makeev, V. M., Dorozhko, A. L., Sukhanova, T. V., & Korobova, I. V. (2016. 91). Geodynamic systems and geodynamically active zones of the east European platform. *Bulletin Moscow of the Testers of Nature Otdel Geology, 4–5*, 9–26.
8. Kopp, M. L. (2011). *Imposition of differently oriented Cenozoic stresses in the center of the Russian plate. Modern. Problems of geodynamics and geoecology of intraplate orogens. Materials V intern. Simpos* (Vol. T.2, pp. 37–41). Bishkek: Scientific. Station of the Russian Academy of Sciences.
9. Makarova, N. V., Makeev, V. M., Sukhanova T. V., & Miklyaev P. S., et al. (2012). The newest tectonics and geodynamics of the Nizhnekoksky district, Vestn. Moscow. un-ta. Ser. 4. *Geology*. № 4.. 3–11. Vestn. Moscow. un-ta. Ser. 4. Geology. 2012. № 4. P. 3–11.
10. Makarova, N. V., Makeev, V. M., Sukhanova, T. V., Dorozhko, A. L., & Korobova, I. V. (2015). The newest tectonics and geodynamics of the southwestern wing of the Voronezh anteclise // Vestn. Moscow. un-ta. Ser. 4. *Geology*. . № 1. P. 10–19.
11. Makeev, V. M., Makarova, N. V., Ledenev, V. N., Dorozhko, A. L., Sukhanova, T. V., Karfidova, E. A., & Korobova, I. V. (2015). Bases of the concept of geodynamic safety of ecologically dangerous constructions. *Geoecology*. (2), 99–110.

Chapter 12
Combined Measures on Providing Safety of Technogenic Objects Within the Mined Territories of Potassium Deposits in Perm Region, Russia

Yu. A. Mamaev, S. V. Kozlovskiy, and A. A. Yastrebov

Abstract The paper covers a complex of protective measures providing safety during the construction and operation of potassium salt mines within the Solikamsk-Berezniki urban and industrial agglomeration in Perm region, Russia. At the municipal level, the protective measures imply the adoption of planning and administrative managerial decisions. At the local level, the programs of providing safety of mining works should be developed, the measures of engineering protection of engineering structures should be worked out, the monitoring network schemes should be substantiated, and the sets of special geological research should be performed. These works are aimed at minimizing risks of accidents and emergencies both at operating and projected objects.

Important objects of researches, assessment and prediction of natural (geological) dangers and risks are territories of the developed mineral deposits. Irrespective of ways of development (mining, quarry, surface, polygon, etc.), these objects include constructions, different in the functional purpose, networks, and systems which have significant effect on changes of a condition of the geological environment and activation of natural and techno-natural processes. The economic categories characterizing conditionally realized damage from these processes in social, economic, technological, and ecological spheres of human activity and expressed in a natural or monetary form are also indicators of natural risks.

A qualitative or quantitative assessment of natural risks, on any of the above called objects, complex geological researches, and monitoring of the environmental environment for definition of the main conditions and factors of emergence and (or) development of geological dangers precede.

Y. A. Mamaev (✉) · S. V. Kozlovskiy · A. A. Yastrebov
Sergeev Institute of Environmental Geoscience, Russian Academy of Sciences, Moscow, Russia

V. Svalova (ed.), *Natural Hazards and Risk Research in Russia*,
Innovation and Discovery in Russian Science and Engineering,
https://doi.org/10.1007/978-3-319-91833-4_12

Such researches in the constant mode are carried out in the territory of the Verkhnekamsky field of potassium salts (VFPC) in Perm region of the Russian Federation and, in particular, in the territory of the city Berezniki.

Perm region is one of the most industrially developed regions of the Russian Federation which is characterized by a larger variety and complexity of an environment, scale and complexity of development of geological processes, and intensity of technogenic transformations of the environment [1, 2].

The largest urban agglomeration in the territory of the region is the Solikamsk-Berezniki agglomeration, located within the Cis-Urals edge of the trough, in the center of Russia's largest potassium salt deposit. Within the boundaries of this agglomeration, the largest industrial area is the city of Berezniki with a population of about 150 thousand people, considerable areas of which are located within the boundaries of the depleted rock massifs.

The industrial zone of the city and part of its residential development are located in the wide left bank bend of the river. Berezniki is on the vast surface of the first above-floodplain terrace of the Kama River, in the rear part of which fragments of the second and third above-floodplain terraces were preserved (Fig. 12.1). The territory includes estuaries of the valleys of small rivers – the left tributaries of the Kama, the width of valleys of which reaches 0.7–1.3 km. The well-developed terraced river valleys have ancient riverine depressions up to 20 m. Within the industrial zone, parts of the channels of these rivers are covered; others are slightly flooded with the creation of reservoirs and ponds.

In the eastern part of the city territory, alluvial terraces adjoin the aboriginal slope of the ancient valley of the river. Kama, which passes into the structures of the Berezniki uplift, is composed of Paleozoic rocks of lower Permian (P_1usl_1) and overlapped from the surface by eluvial, deluvial, and technogenic formations of low power. In the geological section of the territory of the city of Berezniki, the basic breeds of the lower Permian age of the Ufa stage (P_1usl_1), lying from the surface and overlapping layers of salts, are of primary importance. According to the predominant petrographic composition of the rocks, several strata are distinguished from above:

- A terrigenous-carbonate stratum (TCS), with a thickness of more than 140 m, represented by marls, clay-plastered limestones with interlayers of sandstones, siltstones, and carbonate clays
- Salt-marl layer (SML), up to 70 m thick, composed of marls with interlayers of limestones, calcareous clays, and gypsum, including up to seven rock salt beds, up to 3 m thick
- Salt stratum (SS) with a total thickness of up to 600 m, consisting of powerful rock salt horizons, containing the layers of potassium salts of sylvinite and carnallite from 0.5 to 5.5 m thick, which are being developed by the mine method (Fig. 12.2)

In the top of the salt layer, there is a cover rock salt bed up to 20 m thick. The deepest layer of sylvinite (Red II), 5.5 m thick, is also underlain by a thick (up to 400 m) rock salt.

The territory has a complex structural-tectonic plan with numerous disparate disparities of different orders, which form a diversified fault-block structure of massifs of bedrock. Throughout the considered territory, there is a marked

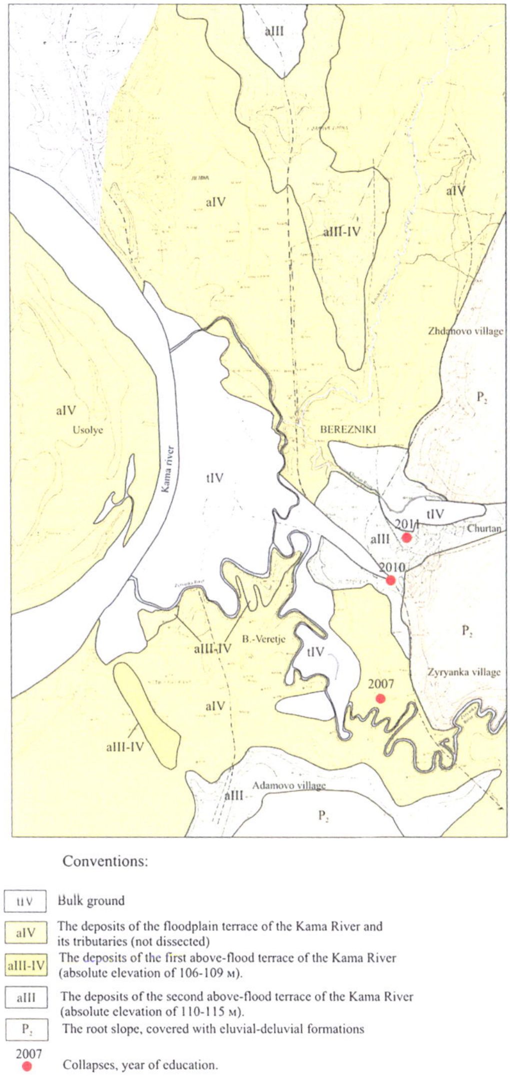

Fig. 12.1 Geomorphological scheme of the territory of the district Berezniki (based on the materials of the "Azot" branch of OJSC Uralkim)

overlapping of the marking horizons even within adjacent territories (blocks). Several systems of tectonic disturbances are distinguished here: sublatitudinal, submeridional, and northwestern. There are other systems of cracks of higher orders, which largely determine the nature and regimes of groundwater movement, as well as the conditions for the stability of rock massifs, first of all, in the underworked territories.

On the studied territory, there are two main hydrogeological groundwater complexes. These are a complex of Quaternary alluvial groundwater and a complex of

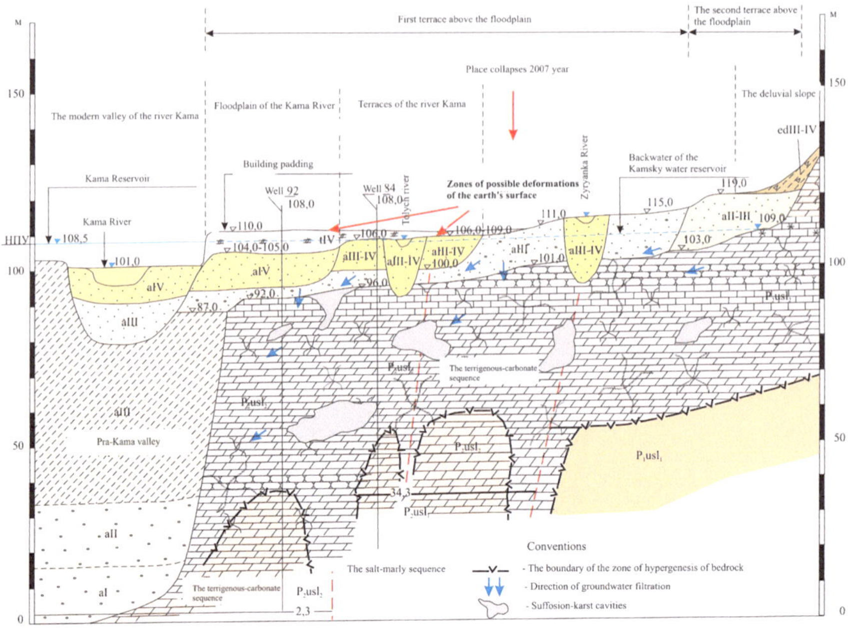

Fig. 12.2 Active processes in the hypergenesis zone

groundwater in the complex of built-up sedimentary rocks of the Permian age. The complex of alluvial groundwater is very widely developed in the given territory within the various and all-age alluvial terraces. Mostly it is stratum-pore nonpressure horizons with close hydraulic connection, including the level of the Kama Reservoir. Due to intense infiltration and overpressure in the Kama reservoir, the water level varies cyclically by 8–9 m, allowing water of the Kama River to penetrate far into the territory of a bypass Quaternary deposits, and into eluviated and fissured zones of massifs of bedrock. This leads to waterlogging and flooding of certain areas of the territory.

The groundwater bedrock complex of the Perm age depending on the filtration conditions are allocated in fractured ground, stratum-fractured, and the fractured karst types of groundwater. The mineralization and salt composition of the waters are very different, from fresh to brine with salinity up to 300 g/dm^3. Infiltration recharge is due to overflow from Quaternary aquifers to weathered fractured zones.

Unloading of groundwater occurs, in part, in the ancient over-deep valley of the Kama River with a depth of more than 100 m.

From dangerous geological processes and events in the city of Berezniki observed: natural seismicity (5–6 points), areal subsidence of the earth's surface over the open area, large-scale failures of the earth's surface over a diameter of 100 m and a depth of 80–100 m (Figs. 12.3 and 12.4), the active development of suffusion-karst processes, and flooding of the territory. Environmental hazards are caused by large amounts of gas emissions into the atmosphere and releases of contaminated water into canals and collectors, as well as by erosion washout and

Fig. 12.3 The depression of the earth's surface (July 28, 2007) in the minefield of a potash mine in Berezniki

filtration through salt and tailings dumps of surface and atmospheric waters, with the formation of technogenic hydrochemical streams of high concentration and aggressiveness that pollute surface water bodies and rivers.

Underground development of a deposit of potassium salts in the cities of Solikamsk and Berezniki was started in 1934–1936 In particular, the operation of the largest first Berezniki mine (BPIOM-1) was carried out since 1954. Its mine field extends under the industrial and residential areas of Berezniki. In 2006, a large technogenic accident occurred on this mine, which led to the flooding of all the developed underground space, with a volume of more than 80 million m^3. Flooding of the mine lasted 2 years and led to the formation in July 2007 of a huge depression of the ground with a finite size of 300 × 400 m and a depth of about 90 m. The amount of damage from this accident amounted to more than 3 billion rubles. After flooding the mine and restoring surface water levels in rivers, ponds, and wells, suffusion-karst processes in the rock strata continue, which led to the formation of other depressions in 2010, 2011, and 2015.

Deformation and destruction of undermined rock masses are caused by the two groups of factors: technical (man-made) and natural (geological). The main ones involved in the formation of geological dangers and risks in the city of Berezniki are in the Table 12.1.

Complex geological research and monitoring on the territory of Berezniki established complex engineering-geological conditions of the site, undermined

Fig. 12.4 The general plan of the industrial zone of Berezniki with three earth surface depressions

underground mines, as well as revealed a great range of natural and man-made factors in the formation of techno-natural (geological engineering) processes and the dangers caused by them. These factors need to be taken into account in the current assessment of the natural and man-caused dangers of the territory of Berezniki and the analysis of the causes of the formation of earth surface gaps.

The most developed, widely used, visual method of generalizing and analyzing data for assessment and forecasting of processes forming geological dangers and risks is targeted thematic mapping, expressed in the compilation of a series of engineering-geological maps (engineering-geological conditions, engineering-geological zoning, and schemes zoning of the territory according to the degree of danger of development of dangerous techno-natural processes). Based on the analysis of these maps and the combined surface plan of the city's territory, methods are developed for drawing up and mocking up the vulnerability map of industrial and civil objects in Berezniki, which will show important industrial, transport, administrative, and residential objects for the city, in relation to which the vulnerability is assessed. These objects differ in hazard classes, scale of production, and capitalization of funds.

In general, vulnerability (potential damage) is the degree of possible loss (damage) for a given object or a combination of objects from different elements of risk that may occur under the influence of a natural or techno-natural process of a certain value.

Vulnerability or the degree of potential damage attributed to a particular territory (zone) and expressed in value form is the main content of risk maps that are planned to be developed for the researchable object.

Table 12.1 Natural and technogenic factors of formation of geological dangers and risks in the territory of Berezniki in the Perm region

Natural	Technogenic
Variegated petrographic composition of layered horizontally lying rocks of terrigenous, carbonate, and halogen formations with actively developing karstic processes of all genetic and geomorphological types	The territory of the underground working part of a large volume, arranged stacked one above the other, complicated by a significant delay, and low-volume filling operations
Complex structural-tectonic plan of the territory with a fault-block structure of rock massifs that were updated in modern times, disturbed by multiscale and multidirectional systems of tectonic faults and cracks	The impact of technogenic-induced seismicity from rock hits, cave roofs of underground excavations, and earth surface failures with the strength of seismic events up to 3–4.5 points on the scale MSK-64
Active hypergenesis of strata of rocks of Permian age, occurring from the surface, caused by unloading of stresses and fracturing, weathering, and leaching of rocks; the thickness of the core rock hypergenesis zone established by drilling and geophysical studies is 80–90 m, and over highly fractured zones of tectonic disturbances, it can extend deep into the first hundreds of meters	The creation in 1954 of the Kama Reservoir, the hydrological regime of which contributes to the annual cyclic flooding of the earth strata of Quaternary sediments and weathering zones and high fracturing of bedrock waters. This accelerates the process of formation of voids in massifs of karst rocks by dissolving, leaching, and mechanically removing small ground particles with increasing volumes and rates of groundwater filtration
Relatively high natural seismicity of the Perm region, which is a zone of high seismic activity for the East European Platform and the Urals with the force of seismic events to 5–6 (rarely 7 points) for MSK-64 scale	The same negative role is played by artificial reservoirs and reservoirs of the dam type
Tectonic activity of the region caused by compression of rock massifs in the west Ural folded zone and the Ural pre-Ural marginal trough, renewal of folded structures, and movement of crustal blocks along zones of major tectonic disturbances	The impact of daily pumping of large volumes of fresh groundwater (up to 200 thousand m^3/day) from the supersaline aquifer on the change in the hydrodynamic regime of groundwater and activation of suffusion-karst processes in the upper near-surface strongly weathered the bedrocks
Complex hydrogeological conditions of the territory, characterized by different factors, including very high (up to 300 g/dm^3) mineralization and variegated chemical composition of groundwater, interaction of complexes and separate horizons of groundwater, close hydraulic communication with the underground water level of Kama Reservoir, and unloading groundwater in overdeepened Kama River valley and its tributaries	Exploration and exploitation of oil deposits in the researchable territory, which cause additional gas saturation of rock massifs lying above, including salt and over-salt strata. The saturation of rock strata with gas emanations of methane, carbon dioxide, hydrogen sulfide, hydrocarbon components, and other gases present in dispersed, coherent, and dissolved species causes increased aggressiveness of groundwater and accelerates the processes of dissolution, leaching, and mass transfer of individual elements tens of times
Activation of suffusion-karst processes during a 2-year period of accidental flooding of underground excavations of the largest potash mine (BPIOM-1), which formed in the strata of rocks	Geochemical pollution of the environment. The industrial zone of Berezniki with a total area of about 140 km^2 includes more than 900 economic entities.116 enterprises have a

(continued)

Table 12.1 (continued)

Natural	Technogenic
the filtration flows of large volumes and velocities with a vertical washing regime of filtration	total of about 1400 stationary emission sources that pollute the atmosphere. They throw into the atmosphere about 160 pollutants, chemical elements and their compounds, which causes an increased aggressiveness of groundwater and speeds up the processes of dissolution and leaching of rocks tens of times
A complex of dangerous natural (geological) processes: flooding of territories; linear erosion along valleys of large erosion infiltrations (ravines) and lateral erosion of river valleys and karst; suffusion, dissolution, mass transfer, and redeposition of soluble rocks; prolonged filtration of aggressive waters in solids, leading to an increase in their porosity and permeability; and the formation of a complexly constructed network of karst channels and large cavities	The same concerns discharges and releases of insufficiently treated sewage from industrial and communal enterprises, as well as mine waters of potash mines
A complex of dangerous natural (geological) processes: flooding of territories; linear erosion along valleys of large erosion infiltrations (ravines) and lateral erosion of river valleys and karst; suffusion, dissolution, mass transfer, and redeposition of soluble rocks; prolonged filtration of aggressive waters in solids, leading to an increase in their porosity and permeability; and the formation of a complexly constructed network of karst channels and large cavities	
A change of the stress-strain state of the rocks in the construction and long life of mines located there and having large areas and volumes	

This work was supported by a grant from the Russian Science Foundation (project No. 16–17-00125 "Risk Assessment of dangerous natural processes in urban areas").

References

1. Kudryashov, A. I. (2013). The salts deposits in Verkhnekamsk. 2nd edition. M.: Epsilon plus, Uralkali Funds. 368 p.
2. Mamaev, Y. A., & Yastrebov A. A. (2015). Integrated protection of the mined potassium mine in the Perm region of the Russian Federation. *Engineering Protection 3*(8) May–June, 26–35.

Chapter 13
Geo-environmental Sustainability of the Arctic Area of the Russian Federation

Vyacheslav Zaikanov, Irina Zaikanova, Lidia Matveeva, and Tatiana Minakova

Abstract The natural systems in the Arctic zone of the Russian Federation have a high vulnerability. The most significant environmental problems in the Russian Arctic are mainly associated with the availability of impact territories of intensive economic activity. One of the main conditions of geo-environmental stability in this area is to maintain the existing permafrost conditions, which excludes the development of adverse cryogenic engineering-geological processes. The solution to the problem of security of the population in the special conditions of the Russian Arctic requires the use of environmentally safe and economically feasible innovative technologies and rehabilitation areas.

13.1 Introduction

Strategy of formation of the ecologically sustainable development of the Arctic zone of the Russian Federation (AZRF) includes the provision for current and future generations of a healthy environment through the efficient usage of natural resources, preventing harm to the natural environment and humans, reducing risks of emergency situations, preservation of valuable objects of the nature, and development of a system of protected natural territories.

The rate of development of the north, which increased in the last years (oil and gas development in combination with the deposition here of polluting the natural environment of substances, accumulated from past economic activities, as well as coming from neighboring regions), is posing a serious threat of environmental degradation, given the high vulnerability of natural Arctic systems. The consequences of negative anthropogenic impact in the Russian Arctic include pollution by oil and oil products, geochemical pollution by heavy metals, persistent organic pollutants, solid waste, and radiation pollution.

V. Zaikanov (✉) · I. Zaikanova · L. Matveeva · T. Minakova
IEG RAS, Moscow, Russia

V. Svalova (ed.), *Natural Hazards and Risk Research in Russia*,
Innovation and Discovery in Russian Science and Engineering,
https://doi.org/10.1007/978-3-319-91833-4_13

The territory of the Russian Arctic is characterized by extreme climatic conditions. Nowadays the problem of stability of natural systems, associated with the rapid increase of man-made stress in conditions of the warming climate, has risen. The most important distinctive feature of the Russian Arctic, determining the conditions for the sustainability of natural-man-made systems, is the presence of permafrost.

The main man-made loads are confined within the impact zone. A considerable area is occupied by carriers and dumps of mining and processing enterprises. The total area of lands that are transformed as a result of economic activities is up to 3% of the total area of the mainland of the Russian Arctic. The results of the previous economic activity are providing the negative impact on humans and nature: unreclaimed pits, tailings, geochemical pollution, and area of radioactive contamination. The annual growth of disturbed land in the oil industry is 5000–6000 ha, in the natural gas industry is 2,5–3 thousand ha, and in the construction of pipelines is 0,4–0,5 thousand ha.

13.2 Geo-environmental Sustainability of Geosystems

Due to the high vulnerability, the natural and natural-man-made systems of tundra and Arctic have a low environmental potential and thus are easily destroyed under the influence of a person and recovered a long time afterward. In the Russian Arctic, there are dozens of points in which the degradation of natural and man-made systems has reached critical values. Pollution of atmospheric air, water objects, and soils, emissions and discharges of enterprises of the mining and metallurgical industry and enterprises of production and transportation of oil and gas, accumulation of industrial waste, chemical reagents for drilling, etc. are all dangerous to the health of the local population and geosystems. Environmental management in the Arctic zone is unstable because of the low technical level and high depreciation of production assets, the crisis in the reproduction of mineral-raw material base for many types of natural resources produced.

At the same time, the Arctic remains one of the least polluted regions of the Earth and plays an important role in maintaining the biological balance on the planet. There are unique natural complexes that do not have analogues in the world, like the objects of natural heritage UNESCO (coastal areas of the Barents Sea, Wrangel Island, etc.), and their contribution in maintaining the global ecosystem balance is estimated at 12% of the world. However, the Arctic zone of Russia is a poorly protected area of natural biodiversity. The total area of all the federal parks and protected areas, which are distributed very unevenly, is only 1.5–5% of the territory and waters of the Russian Arctic (in other countries there are territories-analogues of 20–50%). Due to large-scale commercial development and application of new technologies in the last decades, remote regions of the Arctic become more accessible to humans. As a result, the impact is increasing on biosystems, which reduced the number of rare species of plants and animals and transformed their habitats.

A crucial role in maintaining the stability of the natural-man-made systems is their ability of self-purification, which depends on the ecological potential of the system and the scale of the technogenic impact on it.

Geo-environmental sustainability of geosystems and natural-man-made systems can't be saved and secured, if there is a violation of the law of internal dynamic balance. The essence of this law is that the natural system has an internal energy, substance, information, and dynamic quality. They are so interconnected that any change to one of these results causes other related functional quantitative changes. On the whole are retained amount of material and energy, information and dynamic characteristics of the entire natural or natural-anthropogenic system. This provides the system of maintaining balance, closing the cycle in the system and its "self-repair" and "self-cleaning."

This law acts as a regulator of impacts on the environment, provided that "component balance" and "balance of large territories" are not broken. These "balances" are the norms of environmental management, and they should underpin the development of measures for the protection of the environment.

Geo-environmental sustainability – the ability of natural and natural-man-made system to maintain their properties and mode parameters in terms of existing internal and external disturbances, including the following three types:

1. The ability of an geosystem to retain under external influence original state for some time – the inertia of the system (resistant stability, buffering)
2. The ability of geosystem to move from one state of balance to another, while maintaining internal communication – flexibility of the system
3. The ability of the landscape to return to its original state after a temporary external influence – the recoverability of the system (elastic stability, elasticity).

The first two concepts are interpreted as adaptive repair the third as regenerating. If one or the other function of geosystems under the influence of perturbations deviates from the "norm," the degree of this deviation indicates the relative stability of the system and the time required to restore the "norm" – its relative elasticity.

These all types of geoecological stability are widespread in the territories of the Russian Arctic, but unlike other zones, their provision is due to specific properties of the cryolithozone. The main condition of geo-environmental stability in this area is to maintain the existing permafrost conditions of the cryolithozone, which excludes active development of adverse cryogenic engineering-geological processes, adversely affecting natural and natural-man-made geosystems.

Environmental sustainability in natural-man-made systems is contributed to the following environmental events:

– Improving the efficiency of use of resources, through the introduction of environmentally friendly and advanced technologies, alterations in the structure of economy, natural resources, science, secondary use, and consumption of waste
– Reduction of man-made impact on the environment by reducing emissions, cleaning of territories from pollutants, waste management, prevention of environmental emergencies, and improved management of environmental protection

on the basis of implementation of effective economic mechanism ("green investments," in particular) and the principle of ecosystem trans-regional implementation of sustainable development programs
- Restoration and preservation of the natural environment (landscapes, ecosystems, and biodiversity)

13.3 Zoning of Geo-environmental Sustainability

Zoning of the Arctic zone in geo-environmental sustainability is impossible without accounting and assessment of natural diversity and the specificities of the territory. As the basis physical-geographical zoning was selected, with the rank of big taxa as a system of territorial division of the earth's surface. It's the most useful system, which has an inner unity and distinctive individual features of nature. Along with the physical-geographical zoning, the socioeconomic characteristics of the territory and its administrative division were taken into account. In accordance with Decree No. 296 on the land border of the Russian Arctic, which entered into force on May 2, 2014, the territories of the eight regions are included.

The Arctic zone is characterized by extreme natural conditions, which determine the weak geo-environmental stability of geosystems. These include low temperatures throughout the year, the long polar night and polar day, strong winds and blizzards, dense fogs, and the monotony of Arctic deserts and tundra; the dynamics of climate change is in recent decades. The Arctic is characterized by two zonal "easily vulnerable" vegetation types: tundra and polar desert.

The most important distinctive feature of the Russian Arctic from other natural areas in the country, defining the conditions of sustainability of natural-man-made systems, is the presence of permafrost. Virtually any types of activities in the Russian Arctic, including the construction, development of transport system, agriculture, mining, etc., and also rational use of nature, are impossible without taking into account the permafrost conditions and operating in specific geological processes and phenomena. In this regard, there is a need to take into account characteristics such as power and temperature of permafrost for zoning of the Russian Arctic, their distribution, etc. A more objective assessment was made of geo-environmental stability of natural-man-made systems. The main condition of geo-environmental stability in this area is the preservation of existing geocryological conditions of the cryolithozone which eliminates the development of adverse cryogenic engineering-geological processes, adversely affecting natural and natural-man-made systems. The almost ubiquitous distribution of permafrost ensuring geo-environmental sustainability in the development of the site requires strict control over changing environmental conditions, mode of operation of buildings, and land.

The most significant environmental problems in the Russian Arctic are mainly associated with the availability of impact territories of intensive economic activity. At the same time, the territory is the focal character of industrial economic development and low population densities (one to two persons per 10 km2) and high vulnerability of natural systems from anthropogenic impact of industrial activity, among which the most common is mining (Fig. 13.1).

Zoning was carried out allowing to identify the eight types of integrated areas for the characteristic combinations of natural and economic characteristics and to develop criteria for determining their geo-environmental sustainability in the specific conditions of the Russian Arctic. The result of this regionalization was the network of habitats, reflecting river wave feature space and the inherent territorial organization hierarchy of the natural and natural-man-made systems subject to assessment of sustainability.

The geo-ecological criteria of sustainability of natural and natural-man-made systems and their assessment are reflected in the legend to this map. The geo-environmental stability changes from stable to dangerously unstable, depending on the environmental capacity of the systems, anthropogenic impact, ability to cleanse themselves from pollutants of organic and mineral substances, and taking into account the climatic characteristics and presence of permafrost. At the same time, less sustainable is the natural-man-made system.

Forecast maps of development of the Russian Arctic are called to warn and to give advice under which you can avoid loss of geo-environmental stability of geosystems and natural-man-made systems.

In the forecast period, in accordance with the "Strategy of development of the Arctic zone of the Russian Federation and national security for the period up to 2020," the foundations of the new unified national system of monitoring of condition and pollution of the natural and industrial environments of the Russian Arctic are laid.

Geo-environmental sustainability is one of the most important components of national security, including natural, social, technical, and other conditions, ensuring the comfort of life constantly living (or working in shifts) in extreme conditions of the Russian Arctic population.

Modern geo-environmental situation in the old industrial Murmansk region is determined mainly by the action of the enterprises of mining and metallurgical and mining industry, housing, and transport. The largest sources of waste are mining and smelters "Pechenganikel," JSC "Apatite," JSC "Kovdor GOK," JSC "Olencgorsky GOK," and JSC "Kola GMC." Of particular concern is the accumulation of the tailings and waste dumps of toxic waste, including carcinogenic substances. One of the main geo-environmental problems in the Murmansk region is the organization of safe handling of radioactive waste and spent nuclear fuel. Temporary storage areas for solid radioactive waste are filled to 90–100%. Established on the Kola Peninsula, geo-environmental situation is presenting a real threat to public health, biodiversity, and ecosystems.

Sources of negative impact in Severodvinsk impact area are pulp and paper mill and other timber production, shipbuilding, and engineering. For the region it is typically the pollution of coastal waters from ships and port facilities and critical level of water pollution in estuaries of the rivers Northern Dvina and Vychegda.

The economy of the Nenets Autonomous Okrug is defining the mineral resources sector (hydrocarbons). Having developed 16 fields, it is planning to develop additional 10 hydrocarbon deposit areas prepared for industrial development and 8 oil fields.

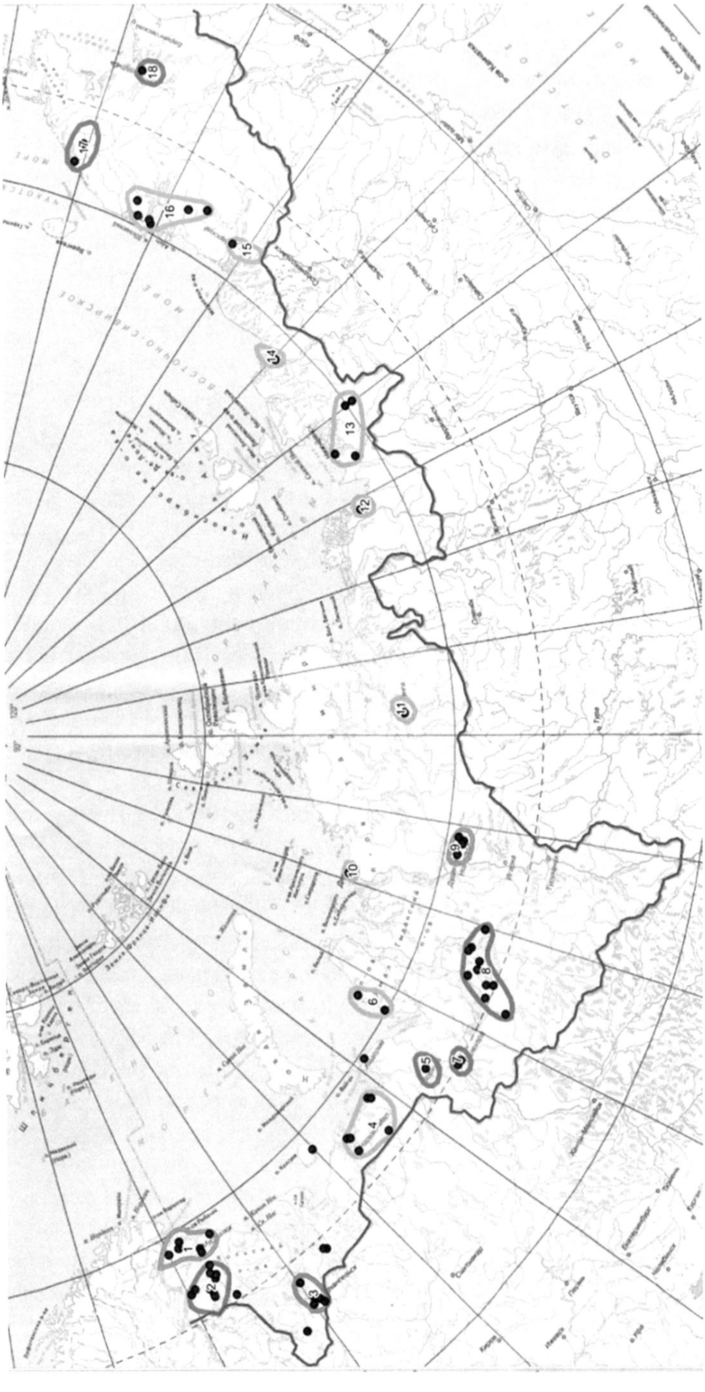

Fig. 13.1 Main mined deposits (marked by dots) and the impact areas of the Arctic zone of the Russian Federation

The Vorkuta District impact area of the Russian Arctic is defined by coal mining. Coal mining is conducted in 1937 by underground method. Despite the closure of some mines and environmental measures, this district is in the first place in terms of gross emissions of polluting substances into the atmosphere among others.

The leading place of industrial production is the Yamalo-Nenets Autonomous Okrug, which consists of gas, timber, woodworking, and fishing industry. The developed nine gas fields on the territory of the Lower-Ob District impact area are major traffic flows of gas to the Urals, Central Russia, and Eastern and Western Europe. On the territory of the Taymyr (Dolgano-Nenets) AO lies the largest plant in Russia belonging to Norilsk Mining and Metallurgical Combine, which mines and processes copper-nickel ore. In the area of the plant within a radius of tens of kilometers of disturbed geological conditions, soil cover, resulting in aktiviziruyutsya the process of transformation of natural landscapes. Norilsk is number one for years in the list of Russian cities with the highest levels of air pollution.

The territory of the Republic of Sakha (Yakutia) within the boundaries of the Russian Arctic is focusing on the extraction and enrichment of mineral raw materials. There are developing ports in the Tiksi and Green Cape (Chersky). In the Anabar Ulus, there are three mining operations and seasonal concentrating factory. The basis of the economy of Ust-yansky District is tin mining and gold. The largest enterprise is JSC's mining and processing complex "Deputy," responsible for the mining and beneficiation of tin ore.

In Chukotka, the main industry is mining (gold, tin, tungsten, mercury, coal, and brown coal). There is developed production of construction materials, power engineering, and fishing industry. Environmentally hazardous facilities are waste storage Bilibino nuclear power station and concentrator in the Chaun District in the tailings storage facility where radioactive waste is accumulated.

According to the zoning, a large part of the Russian Arctic is of natural-type and relatively weakly sustainable systems. Natural-man-made systems with different lower types of stability are mainly confined in the impact area.

Currently we are implementing many projects on the elimination of past damage on the territories of settlements, which are dominated by point maloplastichnye objects, or projects to clean up the uninhabited islands. The solution to the problem of security of the population and of nature in the special conditions of the Russian Arctic requires the use of environmentally safe and economically feasible innovative technologies for rehabilitation areas.

13.4 Conclusion

A new stage of development of the Russian Arctic may lead to new environmental damage. At the same time, there are no further development of Russia without economic development of the Arctic. There are formulated priorities for the development of this zone and national security in the "Strategy of development of the

Arctic zone…" [1]. The creation of a reserve fund of deposits is being planned, guaranteeing the energy security of Russia and sustainable development of fuel and energy complex in the long run. Developing deposits of chromium, manganese, tin, alumina, uranium, titanium, and zinc is also being planned on the islands of the Arctic Ocean, the Kola Peninsula, and in the mountainous massifs of the Polar Urals. In the eastern regions of the Arctic zone will be developed indigenous gold deposits. The Timan-Pechora oil and gas province and hydrocarbon deposits on the continental shelf will be developed. It's to provide for the development of high-tech marine service complex, including offshore exploration.

Judging by the perspective and scale of the projects, with the aim of nature conservation and natural resource potential of the Russian Arctic, its development should be based on rational use of natural resources.

Reference

1. Arctic development strategy of the Russian Federation and national security for the period up to the year 2020, approved by Russian President Vladimir Putin February 20, 2013. The text of a document posted at the Russian Government pursuant to the relevant instructions of the President of Russia. http://www.government.ru/docs/22846/.

Chapter 14
Geo-environmental Zoning for Urban Planning and Design

Ekaterina Buldakova, Vyacheslav Zaikanov, and Tatiana Minakova

Abstract Geo-environmental problems are caused by natural and anthropogenic factors, especially true for urbanized areas. The modern growing cities, with limited space, are accompanied by a steady trend of new construction "up and down" and increase the density of housing and reconstruction of the already developed city areas, which gives rise to conflict in geo-environment. This predetermines the new requirements to the planning of urban space. Geo-environmental zoning of the urban areas is based on geosystemic principle, is reflecting the territorial differentiation of urban environment condition in dynamics for different scenarios of urban design, and should be conducted at an early stage of design. This technology will help to choose the best option and justify the investment. Maps of geo-environmental zoning of the urban territory should be the basis for justifying design decisions and the placement of geo-environmental monitoring. Its use in urban design will contribute to the formation of environment favorable for human life and sustainable development of the city.

14.1 Introduction

Rapid urbanization and high level of construction with the wide implementation of investment projects are caused by the necessity to adapt to existing spatial planning system to modern realities. Urban growth drives to the forced involvement the uncomfortable land under the construction, accompanied by the construction "up and down", with the higher building densities. This increases the pressure on natural components and activates dangerous natural and man-made processes. As a result, the geo-environmental situation becomes worse in the city. Currently the geo-environmental problems stand in one row by severity and urgency of their decisions with social and economic development of the urban space.

E. Buldakova (✉) · V. Zaikanov · T. Minakova
IEG RAS, Moscow, Russia

V. Svalova (ed.), *Natural Hazards and Risk Research in Russia*,
Innovation and Discovery in Russian Science and Engineering,
https://doi.org/10.1007/978-3-319-91833-4_14

To ensure an enabling human, plant, and animal environment, as a basis for sustainable development, the geo-environmental situation must require safety in the cities, where about 75% of the population lives. It depends not only on environmental indicators and reliability of technical systems, but also on the conditions of influence the last one on the geological environment, its compensatory possibilities.

The important tool in making managerial decisions of planning and designing urban space should become a single geo-informational platform with analytical information unit of geo-environmental assessment of the territory. The main principles of this assessment are geo-systemic approach and a single criterion for zoning the city territory according to the degree of danger of geo-environmental constraints. The geo-environmental constraints are understood as conditions that complicate the development of the urban area in accordance with current regulations, regulatory documents, design, urban planning, and environmental management.

The geo-environmental map of the urbanized area should become the basis for the justification of design decisions, and its practical use will contribute to the formation of an environment favorable for human life and sustainable development of the city.

14.2 Geo-environmental Constraints and Their Systematization

An important role is played by the existence of natural and man-made processes and phenomena that limited the development of the urban area, and their ignoring by geo-environment assessment leads to a violation of its stability. In this case, it is absolutely not necessary that the unsuitable urban area presented the "inconvenience" (wetland, erosion dissected, etc.) [1]. This can be the territory, for example, with rare and specially protected species of flora or fauna.

Among the many constraints for the development and exploitation of urban area were identified separately geo-environmental constraints.

Since the 1960s of the last century in the different countries and Russia too [2], geologists have intensively studied not only the geological structure of urban areas but also their natural and man-made hazards. In recent years, the growth of cities often comes at the expense of developing the areas with weak and man-made soils. Large areas in 135 Russian cities are exposed to exogenous geological processes, which the economic damage from them for the last 15 years amounted to 23 billion rubles [2].

The consequences of development the cities are lead for increasing such a processes as flood, karst, suffusion, landslides, ravines, etc. Hydrogeological conditions are especially important in the sphere of interaction of buildings with the geological environment. In many cities are the zones of tension in tectonic dislocations, areas of development of geodynamic processes. About 40% of Russian territory, with a population of over 20 million people, is seismically active.

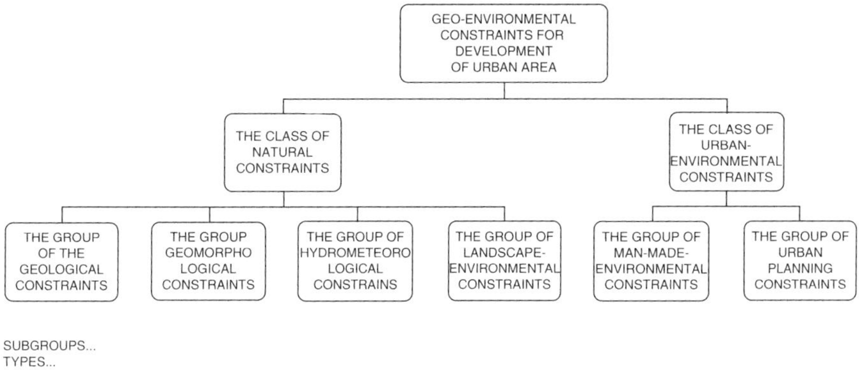

Fig. 14.1 The scheme of systematization of geo-environmental constraints

The zones of geological tension are allocated in many cities by overlaying the zones of geological constraints that are unfavorable for construction, associated with the presence of specific soils, exogenous and endogenous processes, and hydrogeological conditions.

Geo-environmental situation of the city also depends on the geomorphological, hydrometeorological landscape and man-made environmental characteristics of the urban area, as well as the kind of recipient under the impact of these constraints.

The effect of negative consequences increases dramatically by combining natural and man-made environmental constraints (pollution depositing environments and geophysical effects of noise, vibration, electric and magnetic fields, ionizing radiation from technogenic sources, etc.)

Here are some of geo-environmental constraints for the development of the urban area in urban planning: size (border) of land, environmental requirements (pollution, etc.), and the protection zones of individual natural objects.

The study of natural and anthropogenic processes and phenomena essentially reflects the interaction of nature and society, allowed to carry out systematization, with regard to their parameters and distribution [3]. Currently, in the systematization of the geo-environmental constraints, two classes are allocated: natural and urban environmental (Fig. 14.1). The first one refers to four groups, while the second are two groups of constraints. In each group, we can identify several subgroups; among the subgroups are from one to nine species.

14.3 The Algorithm of Zoning

Zoning of urban area based on geo-environmental constraints is carried out in several stages (Fig. 14.2) [4, 5].

The first step is gathering of basing information. In accordance with the selected scale, the source maps are chosen for analysis and visualization of spatial data

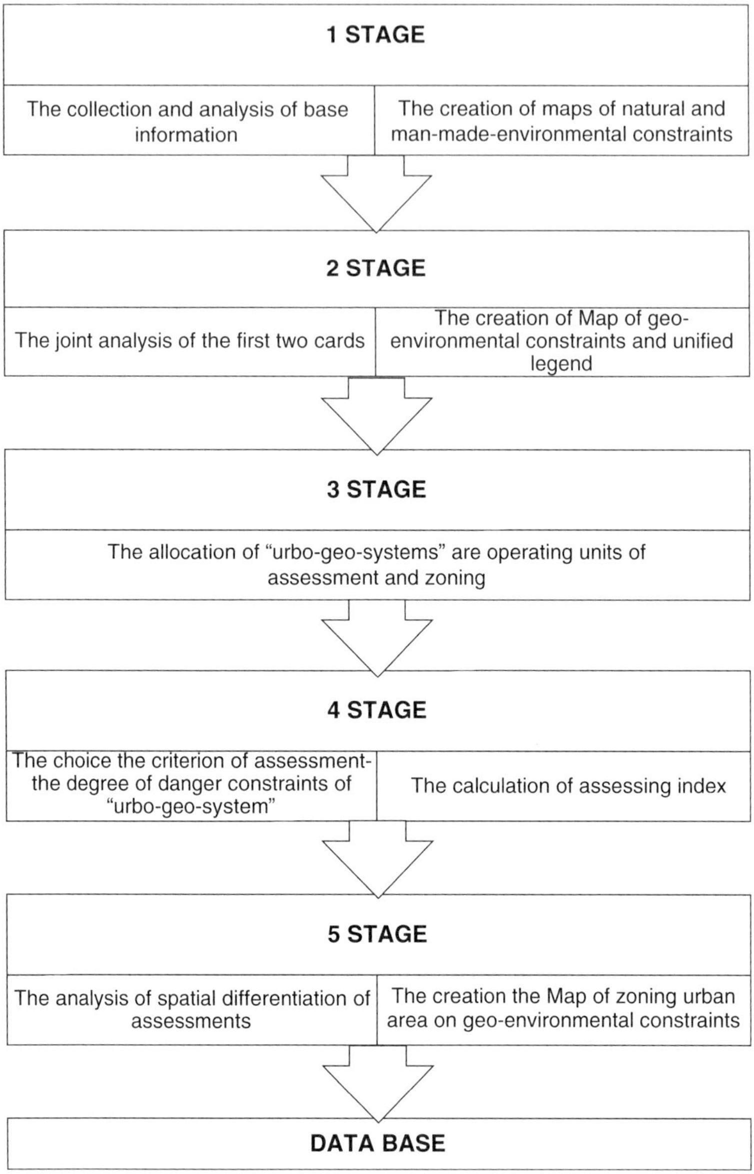

Fig. 14.2 The algorithm for the zoning of the urban area on geo-environmental constraints

(vector and/or raster); designed and maintained the contents of the primary databases, which are presented in a geographic information system (GIS). Each spatial layer is associated with an attribute table containing qualitative and/or quantitative data. The scale of the maps depends on the intended target of the rendering and style of the city. As a base map sources can be used map 1:25,000 and 1:10,000. For big cities, technologically, more convenient maps are specific scale (1:20,000 and 1:5000).

The task of this stage is to identify in the study area the geo-environmental constraints and to develop original maps: landscape; geological-geomorphological, hydrometeorological and man-made environmental constraints are characterized by parameters reflecting the intensity of the event (in physical units). Based on their analysis a map of "geo-environmental constraints" is created. This is a result, at the same time, it is the basis for selection of operating units for the next assessment of spatial differentiation of the city.

Functional zoning is due to restrictions on the development of construction, using the site for different types of activities, accommodation and recreation, and environmental protection. The presence/absence of natural and man-made environmental constraints is one of the indicators of the quality of the site, which largely determines the form of land use.

Various recipients (the subject of the economy, people, green spaces, functional formation) have unique vulnerability to the effects of the same constraints. The impact of the recipient on constraints is reflected when combining maps of geo-environmental constraints and the functional zoning of the urban area. The result is a map of new smaller sites of "urbo-geo-systems" (UGS) which is a natural functional unit and the basis of allocation which is the results of interaction of economic entities with the components of nature that are determined in the process of consistent differentiation of the territory. In this approach, the boundaries of the UGS, on the one hand, due to the natural boundaries, on the other, consider the impact of the city on natural components and are distinguished by set of constraints [6]. UGS is selected as an operating unit to assess danger and zoning in the urbanized areas.

The next stage is to assess the degree of danger of certain types of geo-environmental constraints and urbo-geo-system, in general. The calculation module zoning database includes a definition of the areas of certain types of constraints in each UGS, the development of a system of indicators, and the algorithm for calculating the degree of danger of geo-environmental constraints for a particular territory.

Today is the absence of standards for calculating (per unit area) cost values at neutralizing certain types of constraints; therefore, for calculating we used specific cost values for the base scale (estimation of parameters of identified constraints). The value of the weighting factor of the recipients is determined as a value proportional to values of the official cadastral valuation of lands for the permitted uses for specific areas. The feedback coefficients (the expected responses of the recipients to the constraints) reflect the expected damage, if you are ignoring the boundary conditions. Justification of the commensurability of those indicators is the adoption as a single basis of assessment value of all the parameters given to this point in time and normalized to unit area. The attraction of monetary units enables the mapping in the single dimension of all considered factors.

The calculation can be carried out by software implementing the mathematical assessment model, for example, in Excel:

- Columns of table contain the types of geo-environmental constraints which are adopted in relation to a particular territory, base points.
- The rows contain for each urbo-geo-ecosystem: area distribution of individual species limits in the UGS and the total area of UGS and the coefficients of the reverse reaction for each constraint and a single factor of the importance of the recipient

This approach enables to present spatial differentiation of geo-environmental constraints and their relative assessment to compare alternative options of the development of urban areas.

Maps of danger assessment are developed for different scenarios. Changing the type of the recipient, according to the scenario, changes the coefficients of its significance and back reaction on the impact of the constraints and sometimes the boundaries of the UGS. These maps are the basis for the zoning of urban areas, determining current problematic situations, developing recommendations for urban design, and assessing of efficiency of investment project for redevelopment of territory.

According to the basic criterion of division into districts – the same degree of danger of geo-environmental constraints in urbo-geo-systems – we conducted zoning (current and future state scenarios). Map legend is presented as a table, in which the final assessment of geo-environmental danger reflects the patterns of their spatial differentiation. It contains recommendations on necessary actions to neutralize (warning) dangerous geo-ecological limitations for the development or the reorganization of the territory.

Analysis of landscape-environmental conditions is needed to identify natural areas' special status and develop recommendations for the conservation of rare flora and fauna, developing environmental frameworks and others, and for the solution of planning and design tasks.

Zoning database is constantly updated and corrected in the process of all stages of zoning. It contains the data with their spatial reference to a single coordinate system on the main parameters of and limitations to functional and planning formations which provide estimates and maps. Each urbo-geo- and ecosystem have their own set of database attributes.

Geo-environmental zoning aimed at ensuring safe operation of the economic facilities and favorable for the life of geo-environmental situation due to prevent and address the adverse environmental and geological factors (geo-environmental constraints). The zoning map will help in the solution of practical problems of urban planning and design.

Spatial differentiation of danger geo-environmental constraints reflects the relationship of urban and geo-environmental zoning, so the proposed algorithm should be made mandatory in the planning and design of urban space.

So, zoning the urban area on geo-environmental constraints is based on the use of the following principles:

- Cartographic visualization of spatial data

- Sequential mapping in GIS of urban areas for certain types of constraints and complex geo-environmental assessment
- Functional zoning on different scenarios of land use
- Allocation of urbo-geo-systems and assessment of the degree of their danger to the use by different scenarios
- Integration of all layers of maps for final zoning
- Creating and maintaining the functional state of the database

Thus, the spatial analysis of the city on geoinformational platform allows you to identify and determine the location and structure of geo-environmental constraints, relationships of natural conditions, and economic facilities.

14.4 The Example of Geo-environmental Zoning

Currently, one of the modern trends of urban space is redevelopment – reconstruction of old industrial and commercial properties. More than 17% of the territory of "old Moscow" was occupied by industrial zones, of which 70% is proposed to reorganize [7]. The idea of this renewal is the establishment of integrated urban development with housing, social facilities, and new jobs.

By 2035, it is planned to make the Moscow River the linear center city. The concept of renovation of coastal areas covers almost 11,000 ha, considering the trends of recent years a more active participation of these territories in the structure of the city.

For the practical implementation of the process of zoning on geo-environmental constraints, the area of the metropolis was chosen as model area, planning to reorganization. By natural the chosen site is a part of the valley complex on the right Bank of the river of Moscow, including areas of high floodplain the three terraces. The main natural limiting factors are the presence of anthropogenic soils and exogenic geological processes (flooding, potential of suffusion-karst hazard).

Map of geological-geomorphological constraints was compiled using a map of "Engineering-geological conditions of Moscow", scale 1:10,000 [8], and data from field surveys of the territory. With a rather complex geological structure, there are 17 contours with different combinations of natural constraints on the map.

This site has a real distribution of geochemical pollution of soil and vegetation, noise and vibration impacts along the open metro lines and railways, local contamination of sediments, etc. Here were allocated four districts with three types of man-made environmental constraints. Statutory (environmental standards) urban planning limits water protection and sanitary protection zones, protected natural territories.

Functionally planning structure of the territory is not homogeneous. For each functional purpose (recipient) was installed a set of priority constraints of the existing danger: for buildings and facilities, geological and geophysical (vibration), and for a person, geochemical and geophysical (noise pollution) types of limitations.

As a result of joint analysis of the maps (natural and man-made environmental constraints), the spatial differentiation of the main types of geo-environmental constraints is represented by the 29 contours on the new map of geo-environmental constraints. Based on the analysis of implicative relations between functional-planning formations and geo-environmental constraints, 52 urbo-geo-systems have been allocated. Each UGS has its own index, reflecting the sequence of its formation.

UGSs are merged into 16 subareas of the four types of districts that are similar in value danger degree but due to different reasons. The recommendations are presented in the legend, reflecting the measure prevention and the negative reaction of the recipients to the constraints. The ratio of natural and man-made environmental constraints varies from 0/100 to 71/29 in subareas.

According to the Moscow General Plan, the 20% of area of the territory is planned to reorganize: industrial enterprises will be driven out, communal-storage areas will be exempted, and coastal area will be landscaped. The project provides for the development of these areas for multi-flat residential and mixed-use public area.

The result of a joint analysis of the maps of existing geo-environmental constraints and planned functional zoning of the reorganized areas was the allocation of 12 new UGS. In the case of placing objects without accounting the constraints, the danger for their presence are increased five times. This indicates the need for significant investment in preparation for the development of the land. The developed recommendations for the preparation and rehabilitation of the territory, which are subject to reorganization (reduction in the level of groundwater, the replacement of man-made soils, reclamation, the use of specific technologies in the areas potentially dangerous for the existence of karst, etc.).

After the implementation of the recommendations, the degree of danger of geo-environmental constraints had decreased substantially. Further, on the hazard assessments, geo-environmental constraints were carried out zoning of the territory after its reorganization. While maintaining the typology of regionalization (four types), the number of allocated sub-districts was increased to 22, compared with the current state of area with the minimum danger.

The area ratio of groups of constraints has changed in the direction of increasing the role of natural constraints. However, a part of the reorganized areas is still dominated by man-made constraints due to the negative effects of "neighbors." There are saved the necessity of conducting monitoring observations at sites of potential manifestations of suffusion-karst processes and the elimination of the effects of the nearby industrial enterprises and transport.

At the reorganization of the urban area, an important role plays the stepping of submission of geo-environmental information in the maps and efficiency of the calculations of assessing index, which are made in the GIS. The integrated assessment of geo-environmental constraints in the development phase of the project will increase the geo-environmental safety of the territory through a more objective evaluation of alternatives.

14.5 Conclusion

Testing of methodology of zoning in practice on this and other cases has proved its effectiveness. In the future, the types of restrictions can be filled up by a new one after implementation of the method in practice, but using the application of unified legend of zoning is already possible for various cities now. To the lack of a common regulatory framework, methods of danger assessment of environmental constraints will differ necessity of an individual approach to each territory, not only because of the differences of the original data but when determining the valuation ratios.

Undoubtedly, the method including identifying, assessing, zoning, and visualizing the geo-environmental constraints will help to optimize the planning and management of urban space. Virtually any urban project must be geoecologically safe when theoretically provided by balance natural resource potential and damage (accumulated or annual) in the zone of interaction of technical facilities with nature. The modern concept of safe development of urban areas should be based on the introduction of "forward" strategy of safe spatial development. The basis of such a strategy should be urban planning and design with the obligatory accounting of geo-environmental constraints.

For justification of town planning regulations, optimization of planning, and prioritizing the territory development, it is necessary to study the geo-environmental features of the urban space on the principle of geo-system and GIS technologies. The proposing approach for zoning is based on geological constraints and, as a result, can reflect both modern danger geo-environmental constraints, promising formulation of master plans of cities, projects of reorganization, and development of new territories.

The information model of the assessment and mapping of geo-environmental constraints is made on geographic information platform; can be integrated easily as a separate unit into the municipal GIS; presents flexible, quickly updating system; has the function of support of acceptance of administrative decisions on the choice of an optimum variant of development of urban space; and justifies the investment. This will contribute to the formation of environment favorable for human life and sustainable development of the city.

References

1. Abaturova, I. V., Emelianova, I. A., & Koviazin, I. G. (2015). Litho-technical system "the city – geological environment" and its malfunction as a result of development of natural and natural-technogenic. In *Sergeevskie chtenia* [*The seventeenth conference in commemoration of academician E.M. Sergeev "Engineering geological and geoenvironmental problems in urban area*] (pp. 463–468). Moscow: RUDN. (in Russian).
2. Osipov, V. I., Eremina, O. N., & Kozlyakova, I. V. (2017). Ozenka ekzogennikh opasnostei I geologicheskogo riska na urbanizirovannikh territoriaiakh [Assessment of exogenous geohazards and geological risk in urban areas (review of foreign publications)]. *Geoecologiya*, *3*, 3–15.

3. Zaikanov, V. G., & Minakova, T. B. (2015). Geo-environmental constraints and zoning for urban design. In *Sergeevskie chtenia [The seventeenth conference in commemoration of academician E.M. Sergeev "Engineering geological and geoenvironmental problems in urban area"]* (pp. 549–554) Moscow: RUDN. (in Russian).
4. Zaikanov, V. G., Minakova, T. B., & Buldakova, E. V. (2016). Teoriticheskie osnovi I metodicheskie podkhodi k raionirovaniu urbanizirovannoi territorii po geoecologicheskim ogranicheniam [Theoretical fundamentals and methodical approaches to urban area zoning by geoecological restrictions]. *Geoecologiya*, *3*, 272–281.
5. Zaikanov, V. G., Minakova, T. B., Buldakova, E. V., & Savisko, I. S. (2017). Geoecologicheskie ogranichenia pri proectirovanii reorganizatcii gorodskogo prostranstva [Geoenvironmental constrains for planning rearrangement of urban space]. *Geoecologiya*, *4*, 82–96.
6. Buldakova, E. V., & Minakova, T. B. (2013). Prinzipy i kriterii vydeleniai urbogeosystem pri geoecologicheskoi ozenki regionov [The principles and criteria for selection urbogeosistems for geoenvironmental assessment of regional]. *Geoecologiya*, *4*, 376–384.
7. https://stroi.mos.ru/renovaciya-promzon?from=cl
8. Map Engineering-geological conditions of Moscow (1:10,000), IEG RAS, Moscow, 2010.

Chapter 15
Objective Methods for Compiling Integral Maps

O. K. Mironov

Abstract Synthetic maps of integral natural and technogenic-natural hazard are used in various management procedures. The main possible goals are strategy planning for the territory development, decision-making for a placement or protection of infrastructure objects, and risk management.

A traditional approach provides map compiling as follows. At first, experts in every type of hazard compile maps for each hazard. As a rule, these maps show zones of various hazard intensities. When the total number of considered hazards is small, then it is possible to demonstrate all the hazards in one map. In other cases, some algorithm is applied to compute the integral index. The final map shows the space distribution of this index.

A popular method of numerical scores calculates the integral index as a sum of scores that are attached to each intensity level of each hazard. The main defect of this method is an arbitrariness and voluntarism of attached scores. Methods of agreement of experts' private opinions do not solve the problem of objective content of the result. Another problem is the possible correlation of various hazardous processes. For example, if a group of source hazards has a similar genesis and, consequently, the positive correlation, the procedures described above will give a preference to this group.

The paper considers the following methods developed in IEG RAS:

1. The method of the unified legend for hazard maps provides a possibility of comparing maps of various hazards from the united point of view, especially for risk analysis. The economic criteria are used in this method instead of inner properties of the processes.
2. The method of Pareto optimum provides integral hazard maps with easy understandable legends. Multicriterion optimization finds the best or the worst areas according to source maps.

O. K. Mironov (✉)
Sergeev Institute of Environmental Geoscience, Russian Academy of Sciences (IEG RAS), Moscow, Russia
e-mail: geoinf@geoenv.ru

V. Svalova (ed.), *Natural Hazards and Risk Research in Russia*,
Innovation and Discovery in Russian Science and Engineering,
https://doi.org/10.1007/978-3-319-91833-4_15

3. Method of optimal marks attaches to each point of map a set of digital marks that explain in the optimal way differences in the set of source zonal maps. The task is dual to the cluster analysis: to find the best digital estimation for many divisions of points into clusters instead of finding the best division of points with many digital characteristics into clusters.

The analysis of results is like the analysis of principal components. From the computation point of view, the procedure calculates optimal marks as a sum of scores that are attached to each intensity level of each hazard, but the scores reflect objective dependencies between data and not an arbitrariness of an expert.

The described approaches were applied by IEG RAS in various programs on safety of Russia, its regions, and Moscow megacity. The algorithms do not depend on the map's scale. The resulting maps had scales from 1:10,000,000 to 1:2000

15.1 Introduction

Integral maps of natural and/or technogenic-natural hazard are used in various management procedures. The typical goals are strategy planning for territory development, selecting a placement or protection for infrastructure objects, and risk management.

Many other applications use geographical analysis to make integral synthetic maps to summarize influence of various factors for decision-making in various areas of investigation. Many publications and conference thesis may be referenced as examples.

A traditional approach to synthetic map compiling provides as follows. At first, an expert in the individual factor (e.g., a type of a hazard) compiles the corresponding map. The main thematic layer in this map is territory zoning according to the factors' value. Other layers may give additional thematic information and may be used for extra analysis.

After all factor maps are compiled, some procedure of geographical analysis is used. If the total number of considered factors is small, then it is possible to demonstrate all the data in one map. In other cases, some algorithm is applied to compute the integral index. The final map shows the space distribution of this index. The integral index is a primary object in this method, and the map is a secondary object.

The thematic layer of zoning may use a nominal, range, or numerical scale. Hazard maps use range scale as a rule.

We consider the following problem. Let several zoning thematic layers be given. What reasonable procedure should be used to compile the synthetic map?

The scientific approach requires the following conditions for making the integral map:

1. *The reproducibility of results*. If various groups of scientists use the same source data, then the results should be analogous, if not the same.

2. *The reasonable interpretation of the result.* The legend of the synthetic thematic layer should have an informal semantics. (The explanation in the map legend "a green zone corresponds to index 1, whereas red zone corresponds to index 5" is not sufficient. Authors must explain the substantial interpretation for values 1 and 5.)

The computing procedure for compiling synthetic maps may be applied to both vector and raster format of source data. For the vector format, we take the full overlay of all zoning layers. Each taxon at the resulting layer inherits attribute values from all source layers. The set of attributes of the resulting layer is the union of attributes of source layers.

In raster format we assume that all source layers are defined at the same grid. We attach the values of source layers to each point of the grid as a set of attributes. For uniformity, below we shall call each point of a grid a taxon too and distinguish formats only when it is necessary.

A popular method of numerical scores attaches some digital score to each gradation of each factor (e.g., an intensity level of each hazard). The integral index of a taxon is a sum of scores that are attached to its attributes. The main defect of this method is an arbitrariness and voluntarism of attached scores and thresholds for the resulting map.

Methods of concordance of experts' private opinions [5] do not solve the problem of the objective content of the resulting map. A change of the expert group may cause the change of the result. Moreover, the Saati method contains inside arbitrary scores for pairwise comparisons. From the declared above point of view, the resulting map should present objective data and not a public opinion.

Another problem for the numerical scores method is a possible correlation of source data. For example, if a group of hazards has a similar genesis and, hence, a positive space correlation, the procedure described above will give a preference to this group. If some factors have a negative space correlation, then the result will be blurred or biased depending on the score values.

Compiling the risk maps often is reduced to the numerical score method with all its inner specifics and defects.

This paper considers two methods of compiling synthetic maps:

1. The method of Pareto optimum based on the unified legend for hazard maps provides a possibility of comparing maps of various hazards from the united point of view, especially for risk analysis. The economic criteria are used in this method instead of inner properties of the processes.
2. Method of optimal marks attaches to each point of map a set of digital marks that explain in the optimal way differences in the set of source zoning thematic layers.

The computing procedure in both cases coincides with a numerical score method. But in the first case, attached scores are just a technical tool, and in the second case, scores are the solution of the mathematical problem.

The described approaches were applied by the Institute of Environmental Geoscience Russian Academy of Sciences (IEG RAS) in various programs on safety of Russia, its regions, and Moscow megacity. The algorithms do not depend on the map's scale. The resulting maps had scales from 1:10,000,000 to 1:2000.

15.2 The Method of Pareto Optimum for Synthetic Maps

The Project "Safety of Russia"

Natural hazards are serious obstacle for economical development and threat to human civilization. Yokohama strategy declared the mitigation of losses from natural hazards to be an important part of government regulations and sustainable development. One of the main strategic ideas is the transfer of works to an alert of a hazard from accident eliminations.

In Russia the problem of natural hazards is very important due to large territory and a great variety of natural conditions. The Russian Academy of Sciences carries out very wide research in this area in collaboration with the government, regional authorities, and industrial and exploring companies.

At the end of the twentieth century, the project "Safety of Russia" was fulfilled. The project included many scientific and practical works on mitigation of natural hazards. The results of the project are published in the monograph "Natural Hazards in Russia" [4]. The publication was initiated by the Russian Academy of Sciences and the Ministry of the Russian Federation for Civil Defence, Emergencies and Elimination of Consequences of Natural Disasters (Russian Emergency Ministry). The monograph consists of six volumes:

1. Natural hazards and society
2. Seismic hazards
3. Exogenous geological hazards
4. Geocryology hazards
5. Hydro- and meteorology hazards
6. Natural risks: estimation and management

More than 120 leading scientists from various scientific institutes and centers took part in the monograph. Chief editors are academician V.I. Osipov (Director of IEG RAS) and S.K. Shoygu (Minister of Russian Emergency Ministry).

The monograph considers various aspects of natural hazards. Below we demonstrate the principal decisions in hazard mapping.

The Typification of Natural Hazard and Unified Legend

The project "Safety of Russia" included the series of natural hazard maps. There are many variants of a typification of hazards from various points of view. In the project the unified scale was adopted for all hazard maps according to the possible size of hazard consequence [6, Chap. 2].

Six levels of hazards were defined (see Table 15.1).

For each type of hazard, the zoning thematic layer was developed according to this classification. This procedure provides the same interpretation for similar zones

Table 15.1 Hazard levels

Level	Name of level	Description	Possible human victims	Possible economic losses, $mln	Score
6	Extra high	Mass destructions at large areas	More than 25	More than 500	10^{10}
5	Very high	Mass destructions at restricted areas	Less than 25	150–500	10^8
4	High	Large or medium destructions at local areas	Accidental unique	15–150	10^6
3	Medium	Medium, rarely large destructions	No	1.5–15	10^4
2	Small	Medium, rarely large destructions	No	0.5–1.5	10^2
1	Very small	Small destructions	No	<0.5	1

at different maps. Usually specialists in concrete fields apply their inner criteria for zoning. The problem of concordance notions from several maps is rather hard and nonobvious (e.g., how to compare “intensive” karst and “medium” seismicity).

All the maps use the same cartographical projection, coordinate system, and basic cartographic layers at scale 1: 5,000,000. It provides the possibilities of geographical analysis with GIS tools.

According to this methodic, the maximal hazard level at the corresponding map depends on a hazard type. For the list of hazard maps and their maximal hazard levels, see Table 15.2.

We shall use the scores from the last column of Table 15.1 to compute the resulting index of a taxon at the resulting map. The single cause of such score selection is that given a resulting index, we can define how many of hazard processes of each level are attached to that taxon. For example, the index 40510 means 4 hazards of the medium level, 5 hazards of the small level, and 10 hazards of very small level. Concrete values of scores are selected only for simplicity. For example, we could receive the same result attaching the score 20^{k-1} to the level k.

Pareto Optimum for Synthetic Maps and Its Modifications

Intuitively, the task of compiling a synthetic map consists in ordering of all taxons according several criteria: $\mathbf{F_1}, \mathbf{F_2}, \ldots, \mathbf{F_p}$, where p is a number of thematic layers. One of a partial task is to find the “best” or “worst” regions.

Let us recall that a set M of arguments of multicriteria optimization task is called the Pareto optimum set if for any $y \notin M$and any $x \in M$, there exists a criterion $\mathbf{F_i}$ such that $\mathbf{F_i(x)} > \mathbf{F_i(y)}$. It means that if we go off optimum set, then at least one of criteria will be worse.

Table 15.2 The list of hazard maps and maximal hazard levels

#	Hazard	Maximal hazard level
1	Flooding	6
2	Hurricanes and strong winds	6
3	Tsunami	6
4	Earthquakes	6
5	Landslides and rockfalls	6
6	Avalanches	6
7	Mudflows	6
8	Lateral erosion	5
9	Karst	5
10	Surface and ravine erosion	4
11	Groundwater uprising	4
12	River erosion	4
13	Suffosion	4
14	Swelling	4
15	Loess subsidence	4
16	Thermokarst	3
17	Ice mound	3
18	Thermoerosion	3
19	Solifluction	2

Application of Pareto optimum to integral map compiling proceeds as follows. Let the set of first rank M_1 be the maximal Pareto optimum set of taxons. That means that $M_I = \cup_{k=1}^{p} M_{1k}$, where M_{1k} is the union of taxons for which the criterion $\mathbf{F_k}$ takes on its maximal value. The union of taxons from M_1 is the region of the first rank.

After deleting the set M_1 from all taxons, we define the set M_2 to be the maximal Pareto optimum set of resting taxons. Repeating this procedure we get Pareto ranking for the resulting synthetic map.

We may use Pareto minimum set instead of Pareto maximum. In this case we start from the best regions instead of worse regions. Generally speaking, the results should not coincide.

It is possible to define subregions inside regions of a fixed rank according the number of maximums of taxon's attributes.

When all source maps are compiled according to the unified scale, we use the following modification of optimum Pareto method. Let M_6 be the set of taxons such that at least one attribute of each taxon is equal to 6. It means that in each taxon of M_6, at least one hazard of maximal level may take place.

Continuing, let M_5 be the set of taxons such that at least one attribute of each taxon is equal to 5 and each attribute is less than 6.

Similarly, we define sets of taxons M_4, M_3, M_2, M_1. The union of taxons from M_k is the region of the rank k, $k = 1, 2, 3, 4, 5, 6$. Maximal possible hazard level in the region of rank k is equal to k.

We may use scores from Table 15.2 to get this division into regions. The thresholds T_k should be set to $10^{2(k-1)}$, where the set M_k is defined by the inequality $T_k \leq$ score (taxon) < T_k. Here scores and thresholds are used as purely technical tools.

It is easy to see that the described methodic satisfies two principles declared in introduction.

This methodic was used for compiling the resulting synthetic maps in the project "Safety of Russia" and other regional projects.

15.3 Optimal Marks for Maps Analysis

This section is based on the author's papers [2, 3].

Motivation and Areas of Application

The objectives of compiling a synthetic map is to find some characteristic parameter for each point of mapping area and show the space distribution of this parameter. Below we formulate the optimization task for computing such a parameter.

We shall work with raster format of maps. We assume that the same square grid is placed over all the source maps and consider grid points as our objects. Each source zoning thematic layer defines a partition of objects into disjoint classes. We propose to attach an optimal numerical mark $\mathbf{z}$ to each object (grid point or pixel) of the source area.

We begin from some denotations. Suppose $\mathbf{n}$ is the total number of objects; $\mathbf{z_i}$ is the sought optimal mark for $\mathbf{i}$th object, $\mathbf{i = 1,...,n}$; $\mathbf{p}$ is the total number of thematic layers; $\mathbf{k_l}$ is the number of classes at $\mathbf{l}$th layer, $\mathbf{l = 1,...,p}$; $\mathbf{G_{lj}}$ is the set of objects in $\mathbf{j}$th class at $\mathbf{l}$th layer; and $\mathbf{n_{lj}}$ is the number of objects in $\mathbf{G_{lj}}$.By definition, we put the *correspondence coefficient* of a numerical mark $\mathbf{z}$ with $\mathbf{l}$th layer

$$\boldsymbol{\rho}_1 = \mathbf{n}\left(\sum_{j=1}^{k_l} \mathbf{n_{lj}}\left(\overline{\mathbf{z_{lj}}} - \overline{\mathbf{z}}\right)^2\right) / \sum_{i=1}^{n} \left(\mathbf{z_i} \quad \overline{\mathbf{z}}\right)^2,$$

where $\overline{\mathbf{z}} = \left(\sum_{i=1}^{n} \mathbf{z_i}\right) / \mathbf{n}$ is the total mean value of $\mathbf{z}$ and $\overline{\mathbf{z_{lj}}} = \left(\sum_{m \in G_{lj}} \mathbf{z_m}\right) / \mathbf{n_{lj}}$ is the mean value of $\mathbf{z}$ within $\mathbf{G_{lj}}$.

We assume that $\mathbf{z}$ is centered $\sum_{i=1}^{n} \mathbf{z}_i = 0$ and weighted $\sum_{i=1}^{n} \mathbf{z}_i^2 = \mathbf{n}$.

The *correspondence coefficient* is the ratio of the interclass square deviation of a numerical mark to the common square deviation of mark values, where partition into classes is defined by the thematic layer.

The *optimization problem* is to maximize the sum of correspondence coefficients of mark $\mathbf{z}$ and all source layers: $\mathbf{K} = \sum_{l=1}^{p} \rho_l \Rightarrow \max$.

This problem is dual to the cluster analysis. In both cases we want to find the best correspondence between quantitative and qualitative data. The criteria of correspondence of one quantitative and one qualitative layer in both cases are defined by the same formula. In the cluster analysis, many numerical marks are given and one optimal partition into classes is required. In our case we have many partitions into classes and one quantitative mark is required.

As usual in data analysis, it is reasonable to consider a multidimensional solution of the problem. We propose to attach a set of optimal numerical mark $\mathbf{z}^{(\mathbf{r})}$, $\mathbf{r} = \mathbf{1,...,R}$, to each object. Marks are supposed to be centered, weighted (see above), and orthogonal: $\sum_{i=1}^{n} \mathbf{z}_i^{(\mathbf{r})} \mathbf{z}_i^{(\mathbf{t})} = \mathbf{0}$, when $\mathbf{r} \neq \mathbf{t}$.

Let $\mathbf{K_r}$ be the criterion defined as above for $\mathbf{r}$th mark. The global optimization criterion for multidimensional problem is the sum of all partial criterions: $K = \sum_{r=1}^{R} K_r \Rightarrow \max$.

The solution of this problem is formulated in the following

Theorem There exists a family $u_{lj}^{(r)}$, $\mathbf{r} = \mathbf{1,...,R}$, $\mathbf{l} = \mathbf{1,...,p}$, $\mathbf{j} = \mathbf{1,...,k_l}$, of optimal factors defined for each class $\mathbf{j}$ at each layer $\mathbf{l}$ such that:

(i). Optimal mark $\mathbf{z}_i^{(\mathbf{r})}$ at a point of a map equals to the sum of the factors, defined for classes realized at this point at all layers: $\mathbf{z}_i^{(\mathbf{r})} = \sum_{z_i \in G_{lj}} \mathbf{u}_{lj}^{(\mathbf{r})}$.

(ii). Optimal factor $\mathbf{u}_{lj}^{(\mathbf{r})}$ for jth class of lth map equals to average value of rth optimal marks at points of region $\mathbf{G_{lj}}$: $\mathbf{u}_{lj}^{(\mathbf{r})} = \frac{1}{\mathbf{n_{lj}}} \sum_{z_i \in G_{lj}} \mathbf{z}_i^{(\mathbf{r})}$.

The theorem states that computing procedures for the optimal marks at points of a map are the same as in the method of numerical scores. But in the optimal procedure, marks come from the solution of the optimization problem based on a collocation of layer data and not from a voluntarism of experts.

The statement (ii) says that optimal factors can be used for classification of data on thematic layers. If optimal marks for some classes have small difference, then the nearness of corresponding regions is very likely. We can find the correlated or anomalous classes too. Application examples confirm this hypothesis [2].

The analysis of optimal marks should be considered in the context of exploratory data analysis. The main objectives for application to maps data are:

1. To get a compact representation of data (data compression)
2. To check a hypothesis of links between data
3. To frame a new hypothesis of links between data
4. To define some generalized characteristics of data
5. To formulate recommendations how to compile a synthetic map

There are two main fields for an application of this method.

First, the analysis of optimal marks for classes at thematic layers is similar to the well-known analysis of principal components. We can consider scatterplots of several marks and try to interpret the closeness of points on them as the closeness of corresponding objects (taxons at various thematic levels). The estimation of number of optimal marks that is sufficient for analysis is also similar to principal component method.

Second, if we find a reasonable interpretation for some optimal mark, then its distribution can be mapped as a sketch for a synthetic map.

Mathematical Background

In this section we prove the theorem stated above. Readers that are not interested in mathematical details may pass this section.

It is easy to see that the one-dimensional problem $\mathbf{K} = \sum_{l=1}^{p} \rho_l \Rightarrow \max$ is equivalent to $\mathbf{K}' = \sum_{l=1}^{p} \sum_{j=1}^{k_j} \overline{\mathbf{z}_{lj}}^2 \Rightarrow \max$.

Let $\mathbf{m} = \sum_{l=1}^{p} \mathbf{k}_j$ be the total number of classes at all layers. We define $(m \times n)$ matrix $\mathbf{A}$ in which rows correspond to classes and columns correspond to objects as follows. We set $\mathbf{a_{ji}} = \mathbf{1}$, if $\mathbf{i}$th object belongs to $\mathbf{j}$th class and $\mathbf{a_{ji}} = \mathbf{0}$ otherwise. Let $\mathbf{D}$ be a diagonal $(\mathbf{m} \times \mathbf{m})$ matrix with $\mathbf{d_{jj}}$ equals to the number of objects in $\mathbf{j}$th class. We consider values of mark $\mathbf{z}$ as a column vector. Then $\mathbf{K}' = \mathbf{z}^*\mathbf{A}^*\mathbf{D}^{-1}\mathbf{Az}$, where * denotes a transposed matrix.

Let $\mathbf{z}$ be the eigenvector of symmetric positive defined matrix $\mathbf{A}^*\mathbf{D}^{-1}\mathbf{A}$ that corresponds to maximal eigenvalue. It is well known that $\mathbf{z}$ is the solution of our one-dimensional optimization problem. The eigenvectors corresponding the next eigenvalues in decreasing order give us the solution of multidimensional optimization problem.

Matrix $\mathbf{A}^*\mathbf{D}^{-1}\mathbf{A}$ has the trivial maximal eigenvalue 1 because vector $\mathbf{z}$ with $\mathbf{z_i}=\mathbf{1}$ is its eigenvector. We consider the eigenvalues that are less than 1. In spite of matrix $\mathbf{A}^*\mathbf{D}^{-1}\mathbf{A}$ having colossal size ($\mathbf{n} \times \mathbf{n}$), we can solve the eigenvalue problem in the following way.

Consider a matrix $\mathbf{B} = \mathbf{D}^{-1/2}\mathbf{A}$, and then $\mathbf{A}^*\mathbf{D}^{-1}\mathbf{A} = \mathbf{B}^*\mathbf{B}$. The matrices $\mathbf{B}^*\mathbf{B}$ and $\mathbf{B}\mathbf{B}^*$ have the same set of equal positive eigenvalues. If $\mathbf{z}$ is an eigenvector of $\mathbf{B}^*\mathbf{B}$ corresponding to eigenvalue $\boldsymbol{\lambda}$, then $\mathbf{v} = \mathbf{Bz}$ is an eigenvector of $\mathbf{BB}^*$ corresponding to the same eigenvalue $\boldsymbol{\lambda}$:

$$\mathbf{BB}^*(\mathbf{Bz})=\mathbf{B}(\mathbf{B}^*\mathbf{Bz})=\mathbf{B}(\boldsymbol{\lambda}\mathbf{z})=\boldsymbol{\lambda}\mathbf{B}(\mathbf{z}).$$

Similarly if $\mathbf{v}$ is an eigenvector of $\mathbf{BB}^*$ corresponding to an eigenvalue $\boldsymbol{\lambda}$, then $\mathbf{z} = \mathbf{B}^*\mathbf{v}$ is the eigenvector of $\mathbf{BB}^*$ corresponding to the same eigenvalue $\boldsymbol{\lambda}$.

Let $1 > \boldsymbol{\lambda}_1 > \boldsymbol{\lambda}_2 > \ldots > \boldsymbol{\lambda}_\mathbf{R}$ be $\mathbf{R}$ maximal nonzero eigenvalues of matrix $\mathbf{BB}^*$ and $\mathbf{v}^{(1)},\ldots,\mathbf{v}^{(\mathbf{R})}$ be corresponding eigenvectors. Then $\mathbf{z}^{(\mathbf{r})} = \mathbf{D}^{-1/2}\mathbf{Au}^{(\mathbf{r})}$ are optimal marks for objects.

The matrix $\mathbf{BB}^*$ has size ($\mathbf{m} \times \mathbf{m}$); therefore its eigenvalues and eigenvectors may be calculated without specific difficulties.

The matrix element $\mathbf{b_{uv}}$ of matrix $\mathbf{BB}^*$ is equal to $\frac{N_{uv}}{\sqrt{N_u N_v}}$, where N_{uv} is the number of points in the intersection of regions u and v and N_u, N_v are the numbers of points in regions u and v, respectively. As size of grid cell tends to zero, the matrix element $\mathbf{b_{uv}}$ of matrix $\mathbf{BB}^*$ tends to $\frac{S_{uv}}{\sqrt{S_u S_v}}$, where S_{uv} is the area of the intersection of regions u and v and N_u, N_v are the areas of regions u and v, respectively. This formula shows the way of calculation if source maps are given in a vector format.

The optimal marks method is an application of multivariate correspondence analysis [1] to map data.

The Algorithm for Calculating Optimal Marks

1. The source data for the optimal marks calculation may be given in raster or vector formats. Each thematic layer contains a division of total regions into subregions according to values of some attribute data. Let $\mathbf{m}$ be the total number of all values of classifying attributes and $\mathbf{R_k}$ be the region, corresponding the $\mathbf{k}$th value, $\mathbf{k} = \mathbf{1},\ldots,\mathbf{m}$.

2a. If the source data are given in raster format, calculate:

$\mathbf{N_u}$, be the number of points in the region $\mathbf{R_u}$, $\mathbf{u} = \mathbf{1},\ldots,\mathbf{m}$.
$\mathbf{N_{uv}}$ be the number of points in the intersection of regions $\mathbf{R_u}$ and $\mathbf{R_v}$, $\mathbf{u} = \mathbf{1},\ldots,\mathbf{m}$, $\mathbf{v} = \mathbf{1},\ldots,\mathbf{m}$.

$b_{uv} = \frac{N_{uv}}{\sqrt{N_u N_v}}$, **u** = **1,. . .,m, v** = **1,. . .,m.**

2b. If the source data are given in vector format, calculate:

$\mathbf{S_u}$, be the area of the region $\mathbf{R_u}$, **u** = **1,. . .,m.**
$\mathbf{S_{uv}}$ be the area of the intersection of regions $\mathbf{R_u}$ and $\mathbf{R_v}$, **u** = **1,. . .,m, v** = **1,. . .,m.**
$b_{uv} = \frac{S_{uv}}{\sqrt{S_u S_v}}$ **u** = **1,. . .,m, v** = **1,. . .,m.**

3. Calculate eigenvalues and eigenvectors of the symmetric $(m*m)$ matrix (b_{uv}). Order eigenvalues that are less than 1 descending and select several first eigenvalues in this row.

Components of corresponding eigenvectors are optimal marks for regions $\mathbf{R_u}$, hence for classes at source maps.

4. Calculate optimal marks for points at a map by formulae

$$\mathbf{z}_\mathbf{i}^{(\mathbf{r})} = \sum_{\mathbf{z_i} \in \mathbf{G_{lj}}} \mathbf{u}_{\mathbf{lj}}^{(\mathbf{r})}$$

It is easy to see that algorithms can be realized by GIS technologies both in vector and raster formats of data.

Applications

We present an example of optimal mark computation [2].

The source information consists of eight thematic geology layers for Ladoga lake. List of classes and their optimal marks is presented in Table 15.3.

We see that values of the first optimal mark may be associated (nonlinear) with the depth of the lake. In particular, the isobath 150 m at first approximation does not separate meaningful data in this consideration.

For the detailed analysis and other applications, see [2]. Here we discuss only the principal and visual results.

The scatterplot for first two optimal factors is shown in Fig. 15.1. Each point at the scatterplot corresponds to one class.

We see that points MORF_1, MORF_2, and MORF_3 are vertexes of almost regular triangle. It means that the partition into morphostructural regions has a good correspondence with optimal marks. By the scatterplot we can suppose that the first optimal mark can separate South region from two others and the second optimal mark can separate Central and North regions. It is easy to see this effect in Fig. 15.2.

Table 15.3 The results of computing optimal marks for Ladoga lake data

Thematic layers and classes	Symbol	% of area	Optimal marks	
			*F*1	*F*2
Depth, m				
<25	Bati_1	38.8	−1.039	0.288
25–50	Bati_2	18.7	−0.178	−0.730
50–100	Bati_3	32.1	0.936	−0.442
100–150	Bati_4	8.8	1.303	1.480
150–200	Bati_5	1.5	1.303	2.178
>200	Bati_6	0.1	1.371	3.205
Regimes of modern sedimentation				
Exceptional erosion	Regi_1	4.6	−0.592	1.813
Erosion	Regi_2	29.4	−0.470	0.654
Exceptional accumulative	Regi_3	2.4	−1.233	0.093
Accumulative	Regi_4	6.0	−0.961	−0.419
Undercompensation of flexure with sedimentation	Regi_5	8.6	1.298	0.527
Transitional denudation – accumulative	Regi_6	7.8	−0.500	−0.606
Transitional erosion – denudation	Regi_7	25.1	0.939	−0.439
Mixed	Regi_8	16.1	−0.348	−0.872
Morphology-structural regions				
Northern	MORF_1	27.4	0.670	1.184
Central	MORF_2	25.3	0.894	−0.845
Southern	MORF_3	47.4	−0.864	−0.233
Depth of the roof of pre-quaternary deposits, m				
<50	P_Qu_1	33.2	−1.091	0.420
50–100	P_Qu_2	39.0	0.184	−0.781
100–200	P_Qu_3	25.1	1.022	0.423
200–300	P_Qu_4	2.6	1.232	2.132
> 300	P_Qu_5	0.1	1.418	2.970
Thickness of Holocene deposits, m				
<5	Holo_1	83.1	−0.240	−0.126
5–10	Holo_2	9.2	1.094	0.306
10–20	Holo_3	7.1	1.283	0.965
>20	Holo_4	0.6	1.470	1.467
Thickness of quaternary deposits, m				
<10	Quat_1	16.3	−1.015	0.835
10–25	Quat_2	33.9	0.103	−0.146
25–50	Quat_3	36.3	0.408	−0.182
50–100	Quat_4	12.8	−0.184	−0.242
100–200	Quat_5	0.7	0.908	1.400
>200	Quat_6	0.1	0.288	2.091

(continued)

Table 15.3 (continued)

Thematic layers and classes	Symbol	% of area	Optimal marks *F*1	*F*2
Granulometric composition of modern deposits				
Boulder and blocks	Gran_1	1.7	−1.310	0.668
Pebbles and gravel	Gran_2	14.3	−0.981	0.540
Sand	Gran_3	31.1	−0.795	−0.087
Mictite	Gran_4	13.7	0.162	−0.748
Siltstone	Gran_5	15.7	0.768	−0.612
Siltstone pelite	Gran_6	22.6	1.122	0.533
Pelite	Gran_7	1.1	1.209	1.550
Geological zoning of bottom deposits				
Lower Cambrian deposits	Geol_1	3.4	−1.297	0.559
Kotlin horizon of Vendian deposits	Geol_2	35.0	−0.920	−0.184
Redkin horizon of Vendian deposits	Geol_3	10.3	0.187	−1.084
Middle Riphean gabbro – diabase	Geol_4	3.3	−0.034	0.900
Middle Riphean sandstone, siltstone, and mudstone	Geol_5	41.3	0.933	0.099
Middle Riphean basalt	Geol_6	1.0	−0.593	−0.022
Lower and upper Proterozoic granite	Geol_7	0.4	−0.713	−0.865
Lower Proterozoic granite, gabbronorite, granodiorite	Geol_8	1.5	−0.488	1.759
Lower Proterozoic amphibolites, gneiss, and shale	Geol_9	3.7	−0.568	1.730

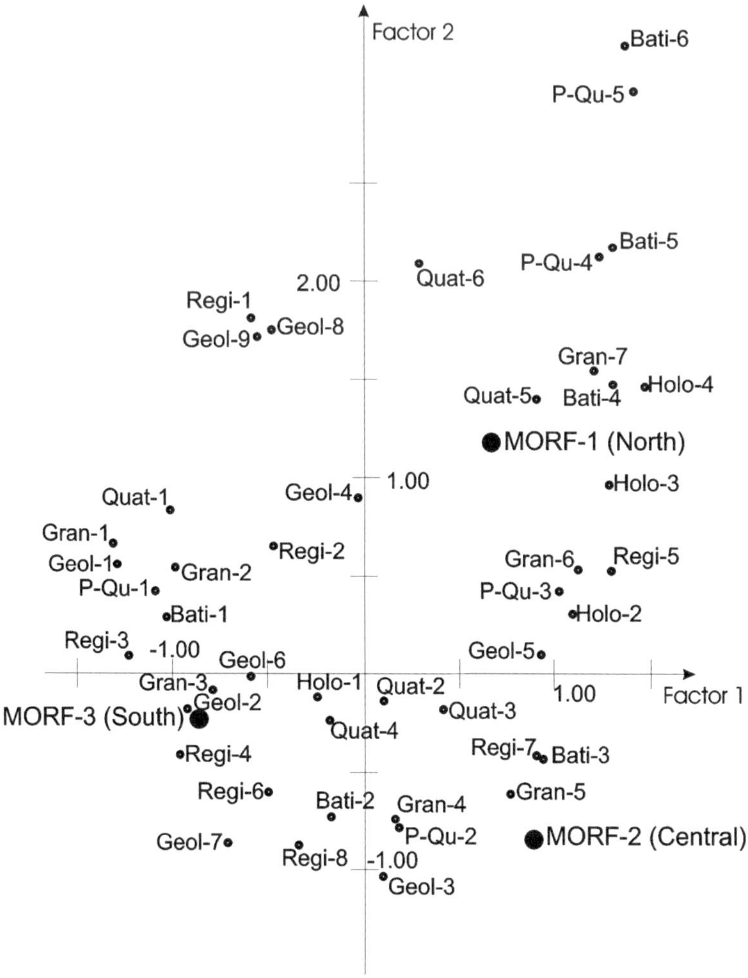

Fig. 15.1 Scatterplot of optimal factors for Ladoga lake geology layers

Fig. 15.2 Morphostructural regions and isolines of optimal marks for Ladoga lake

References

1. Lebart, L., Morineane, A., & Warwick, K. M. (1985). *Multivariate descriptive statistical analysis. Correspondence analysis and related techniques for large matrices*. Chichester: Wiley.
2. Mironov, O. K. (1999). Optimal marks for compiling synthetics maps (in Russian). *Geoekologiya*, (3), 277–285 (in Russian).
3. Mironov, O. K. (2002). *Optimal estimations for compiling synthetic maps*. 8th Annual Conference of the International Association for Mathematical Geology (IAMG), Berlin, 5 pp.
4. Osipov, V. I., & Shoygu, S. K. (Eds.). (2002). *Natural hazard in Russia. In 6 volumes*. Moscow: KRUK (in Russian).
5. Saaty, T. (2008). Decision making with the analytic hierarchy process. *International Journal of Services Sciences, 1*, 83–98.
6. Vladimirov, V. A., Vorobjov, Y. L., & Osipov, V. I. (Eds.). (2002). *Natural hazards and society*. Moscow: KRUK (in Russian) 248 pp.

Chapter 16
Suture Zones as Upper Crustal Seismogenerating Structures

Sergei Nesmeyanov and Olga Voeykova

Abstract For the upper crust seismicity, the tectonic nature of earthquakes plays an important role. The allocation of zones of origin of earthquake focus (OEF zones) should be based on the tracing and typification of seismogenerating structures (SGS), the main of which are the suture zones. Dangerous for the construction, SGS zones belong to the upper seismically active layer.

16.1 Geological Criteria of Seismicity

Seismologists consider the focus of the earthquake to be a fault or a fracture with a geologically instantaneous displacement. The main element of the focus is combined with the fracture or deep fault plane. The SGS zones include disjunctive structures that meet the main geological criteria for seismicity. Such criteria are disjunctive nature, scale, age, the degree of new growth, activation character, and the depth of penetration [1].

The disjunctive nature is determined by the generally accepted mechanism of earthquake focus.

The scale is the connection of the strongest earthquakes with the largest disjunctive structures.

The age estimates the duration of the formation of the SGS. The considerably ancient (several geological periods) SGS succeed in increasing its length and at the same time complicating its structure. The main and feathering planes of displacement and tectonic wedges are formed in the fractures, and marginal sutures with their disjunctive plumage as well as the marginal uplifts are formed in the suture zones (faults) in addition.

The degree of new growth is determined by the fact that not all SGS elements are ancient. Usually there is an alternation of ancient and modern elements. The

S. Nesmeyanov · O. Voeykova (✉)
IEG RAS, Moscow, Russia

V. Svalova (ed.), *Natural Hazards and Risk Research in Russia*,
Innovation and Discovery in Russian Science and Engineering,
https://doi.org/10.1007/978-3-319-91833-4_16

occurrence of earthquakes is often associated with the latest conversion of ancient structures.

The activation character is related to the degree of modern mobility of seismogenerating structures. This mobility is determined by geodetic and geological-geomorphological methods.

Depth of Penetration of the SGS The SGS should reach the main seismically active layers of the region.

By the sum of the criteria, the largest seismicity is associated with large disjunctive structures capable of intense pulsed discharge of accumulated tectonic stresses in them.

The uneven distribution of earthquakes in the SGS is treated in two ways. The seismotectonic method of seismic zoning suggests the appearance of new strong earthquakes in the "windows" between previous earthquakes. The method of seismoactive nodes takes into account the timing of earthquakes to the "nodes" of the intersection of the SGS.

"Nodes" can have a different structure. Disjunctive nodes correspond to articulation and intersection of disjunctives and deep faults (suture zones), and seismoactive disjunctive-plicative nodes are considered as intersections of suture zones with regional transverse uplifts.

Large SGS usually include kinematically different disjunctives. Therefore, SGS *producing focus with a different set of mechanisms* prevail. A simultaneous seismogenic activation of a number of different types of disjunctives in the focal zone is possible. A good example of such revival is the strongest Olyutorsky earthquake of 2006 in Koryakia ($M_s = 7.8$, I = 9–10 points). Its focus came to the surface in the form of a segregated seismic system with a total length of about 140 km. Vertical and horizontal discontinuous displacements were revealed. At different sites, seismic disjunctives are represented by primary dislocations of the disjunctive uplift, uplift-shear, and shear type.

16.2 The Main Types of Suture (Fault) Zones

The main zones of the OEF are represented in the upper crust of the earth by suture zones. The structure of the suture zones is determined by the nature of the displacements along the deep disjunctives. Usually depth shifts, faults, uplifts, thrusts, and spreadings are distinguished. Varieties of the largest suture zones are rift zones and Benioff or subduction zones.

Typical examples of seismically active *deep shifts* are the disjunctives of San Andreas in California and the Thalasso-Fergana fault in Tien-Shan.

The general crustal spreadings include rifts (deep grabens) and cracks made by deep magmatic material. E.E. Milanovsky [2] distinguishes two main types *of rift structures* on platforms: ancient – avlakogeny – and young (which began to form not earlier than the late Paleozoic). He proposed a complex classification of near-surface rift structures (i.e., rift suture zones). It reflects the cycles and stages of development of riftogenesis in the polycyclic and monocyclic development of these structures with the presence or absence of inversion.

The Great Dike of Rhodesia – an intrusion of the main and ultrabasic rocks with a width of 5–10 km and a length of 500 km – can be used as an example of *deep spreading cracks*.

The Benioff zones are usually typical examples of *deep thrusts*. They are traced according to the distribution of focus of earthquakes and are interpreted by plate tectonics as the subduction zone. However, many strong earthquakes are caused by uplifting along steep disjunctives in these zones. Such gaps do not match to subduction models. Recently there has been a trend in favor of recognizing the thrust and not the underthrust nature of these zones on the eastern periphery of Asia. Suture zones of deep thrusts penetrating the mantle usually include deep-sea trenches, island arcs, and adjoining parts of marginal seas.

The classification of discontinuous violations used equally satisfies both the fixed and mobilistic constructions.

Less large suture zones are studied mainly for intracontinental orogenic (not rift) areas.

16.3 Types of SGS of Orogenic Areas

The structural basis of the OEF zones is the SGS (suture zones). Among the suture zones of some of the newest orogens (Caucasus, Central Asia), three main types exist: suture-block, flexure-rupture, and suture-depression (Fig. 16.1). For all of them, uniform stages of development have been traced. The main elements of all types of suture zones can be considered sutural depression (suture grabens or graben-synclines), edge rupture sutures, and marginal uplifts. In the flexure-rupture zones, such sutures and uplifts are located on one side of the depression (graben-synclines), while in the rift or ramp suture-depression zones on both sides of the depression (graben). In the suture-block zones, there are also axial uplifts framed by depressions and further marginal uplifts [1, 3, 4].

A significant part of the focus of large crustal earthquakes is associated with marginal regional rupture sutures. Some focus tends to rupture of the plumage of these sutures. Pleisto-seismic areas of catastrophic earthquakes often completely overlap individual segments of the suture zones [1].

Detailed studies make it possible to identify a much greater variety of suture zones in individual orogenic zones. This is clearly seen in the example of the Caucasus.

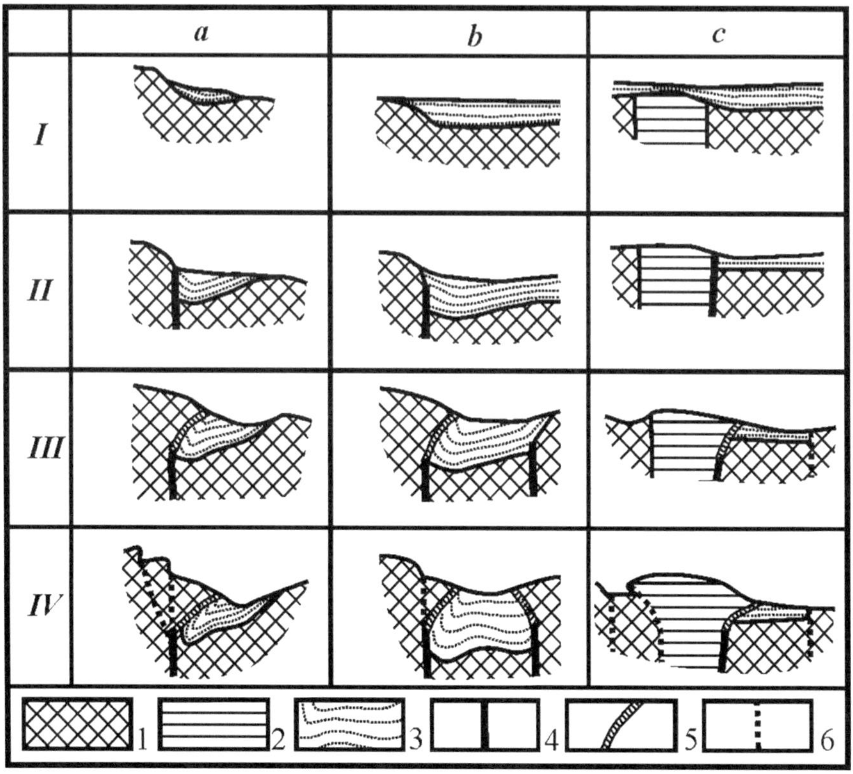

Fig. 16.1 Stages of formation of the main types of the Tien Shan suture zones [3]. (***a***) Flexure-rupture, (***b***) suture-depression, (***c***) suture-block stages. Stages: ***I***, primary flexure; ***II***, structural step; ***III***, frontal uplift thrust and a rupture depression; ***IV***, rear fault. 1, Paleozoic foundation undivided; 2, the most ancient Paleozoic rocks of the axial horst of the suture-block zone; 3, Meso-Cenozoic cover; 4, axial rupture, which caused the formation of the structural step; 5, frontal uplift thrust; 6, faults and rear uplifts

Suture Zones of the Greater Caucasus

Along the periphery and inside the new megaarch of the Greater Caucasus, there are the suture zones longitudinal and transverse to the axis of the megaarch (Fig. 16.2, Table 16.1). Among those and others, there are flexure, suture-fault, and suture-shift zones. Only suture-depression, suture-shift, and suture-block are longitudinal structures, and various zones of transverse crushing with activation of faults, upsets, or shifts are transverse ones.

The suture zones of the flexural type (Western Caucasian, Buinakskaya, Eastern Caucasian) are large inflections in the outsole of molasses, up to 20 km wide and with a vertical amplitude of up to 6–7 km.

Flexure-rupture suture zones (Northern and Southern Crimean-Caucasian, Vladikavkazskaya, Sulakskaya) are similar to flexural ones but complicated by

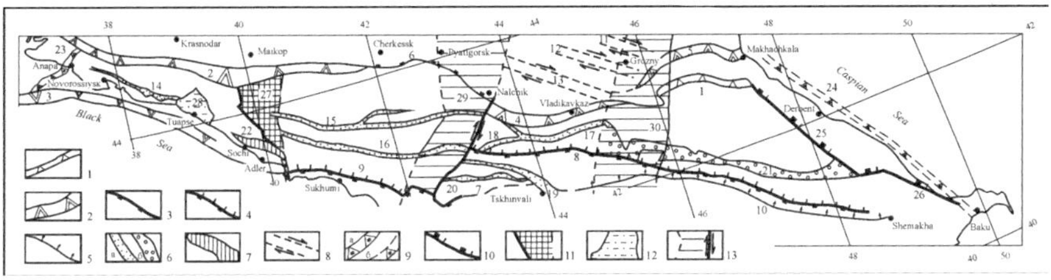

Fig. 16.2 Suture zones of the Greater Caucasus [1]. 1–8, longitudinal (hereinafter, (**a**) reliable, (**b**) assumed): 1, flexural; 2, flexure-rupture; 3, suture-fault; 4, up-and-thrust; 5, suture-thrust; 6, suture-depression; 7, suture-block; 8, suture-shift in the basement of the marginal trough; 9–13, transverse; 9, flexural; 10, suture-fault; 11–13, zone of transverse crushing; 11, with active fault uplift; 12, with active graben deflections; 13, with a shift component (edge shifts) (for zone numbers, see Table 16.1)

large rupture systems in a predominantly thrust type. In some cases thrusts deform the upper bend of the regional flexure (Akhtyrsky subtype) and, in others, its lower bend (Sulaksky subtype).

Suture-fault zones (Essentukskaya, Eastern Dagestanskaya, supposedly Sachkerskaya, and Sumgaitskaya) are a system of steep step faults. Sometimes they form a steep ledge in the relief with a graben at the foot (e.g, the Eastern Dagestanskaya zone).

The fault-thrust or *fault-upthrow suture zones* (the Main Ridge, Sukhumskaya) are marked by large ledges in the relief, which are formed imbricate faults, upthrows, and thrusts. In the case of attenuation of activity, such zones turn into flexures [5].

The suture-thrust zone (Alazano-Shemakhinskaya) is a system of flaky thrusts. Its formation is explained by the underfoot of the Kura microplate under the southern slope of the Eastern Caucasus [6].

Linear *suture-shift* zones are revealed mainly by geophysical methods in the foundation of the marginal trough (Gudermesskaya, Benaisko-Eldarovskaya, Dattykh-Akhlovskaya). They are marked by shift deformations in the overlying molasses.

The suture-depression zones (Mikhailovskaya, Pshekish-Tyrnyauzskaya, Mzimtinskaya, Sadonskaya, Verkhneurukhskaya, Mele-Javskaya, Racha-Lechkhumskaya, and alleged Bezhitinskaya) are represented by chains of grabens, marginal horsts, and separating faults or up-and-thrust [1, 5]. In the feathering of large discontinuities, circus faults are formed.

In the suture-block zones, the axial chain of the horsts from the side of the adjacent depression is limited by a system of uneven-aged up-and-thrust and, on the side of the adjacent uplift, by a chain of suture depressions (Vorontsovskaya zone).

The zones of transverse crushing are characterized by various trends of recent development. For Pshehsko-Adlerskaya depression, uplifting is characteristic, for Tuapsinskaya zone, of active graben formation, and for Nalchikskaya and Groznenskaya, the shift component is obvious. The Pshehsko-Adlerskaya zone

Table 16.1 The main new suture zones of the Greater Caucasus

Type	№ in Fig. 16.2	Name	Length (km)
Longitudinal zone			
Flexural	1	Buinakskaya	150
Flexure-rupture	2	Northern Crimean-Caucasian	>400
	3	Southern Crimean-Caucasian	>500
	4	Vladikavkazskaya	200
	5	Sulakskaya	150
Suture-fault	6	Essentukskaya	170
	7	Sachkerskaya (?)	160
Fault-thrust	8	Main Ridge	>500
	9	Sukhumskaya	150
Suture-thrust	10	Alazano-Shemakhinskaya	400
Suture-shift	11	Gudermesskaya	70
	12	Benaisko-Eldarovskaya	100
	13	Dattykh-Akhlovskaya	150
Suture-depression	14	Mikhailovskaya	120
	15	Pshekish-Tyrnyauzskaya	270
	16	Mzimtinskaya	250
	17	Zadonskaya	150
	18	Verkhneurukhskaya	60
	19	Mele-Javskaya	120
	20	Racha-Lechkhumskaya	110
	21	Bezhitinskaya (?)	400
Suture-block	22	Vorontsovskaya	80
Transverse zones			
Flexural	23	Western Caucasian	80
	24	Eastern Caucasian	450
Suture-fault	25	Eastern Dagestanskaya	200
	26	Sumgaitskaya (?)	100
Transverse crushing	27	Pshehsko-Adlerskaya (with fault uplifts)	>100
	28	Tuapsinskaya (with graben depression)	50
	29	Nalchikskaya (with a shift component)	>200
	30	Groznenskaya (with a shift component)	200

defines the western limitation of the central segment of the megaarch. It distinguishes the western frontal fault scarp with a vertical amplitude of up to 1 km, a belt of advanced horsts, and a chain of rear grabens. The Nalchikskaya and Groznenskaya zones limit the different width segments in megaarch. Transverse grabens of flexure, and flexure-rupture zones (Groznenskaya), shifts (Nalchikskaya) and other structures are widely developed [1].

The neotectonic suture zones of fault-thrust type are large, in places subvertical structural bend, which are complicated by faults, upthrows, or thrusts. The Sukhumskaya zone with a bend of up to 5 km is the neotectonic neogenesis. The

Table 16.2 Characteristics of the main OEF zones of the Northern Caucasus and pre-Caucasus

Name			Depth, km	
OEF-SGS zones	Segments	M_{max}	*h*	Seismically active layer
Southern Crimean-Caucasian	Gelendzhik	6.5	20	10–40
	Sochi	6.0	10	3–15
	Sukhumi	7.0	15	3–35
Western Caucasian		6.5	20	10–45
Mikhailovskaya		6.0	15	10–45
Mzimtinskaya		7.0	15	3–40
		6.5	5	
Pshekish-Tyrnyauzskaya		6.5	10	3–40
Essentukskaya		6.0	20	15–35
Gudermesskaya		6.0	15	9–20
Benoisko-Eldarovskaya	Northwest	6.0	25	15–50
	Southeast	5.5	10	8–20
Dattykh-Akhlovskaya		6.0	10	3–25
Groznenskaya		6.5	25	20–50
Northern Crimean-Caucasian	North Krymsky	6.5	20	15–50
	Akhtyrsky	6.0	20	15–40
	Chercessky	5.5	10	5–25
	Pyatigorsky	6.0	20	5–35
	Naursky	5.5	10	8–20

zone of the long-lived ancient Main Caucasian overthrust for the most part was renewed weakly. It partially turned into the neotectonic flexural bend [5], but in places it retained some ability to the neotectonic thrust formation.

Most of the examined suture zones are the SGS, on the basis of which the OEF zones are allocated. The latter are usually wider than the corresponding SGS. The width of the OEF zones is measured by the width of the focus, which is often taken to be half its length. In the western sector of the Caucasus (the Maikopskaya, Akhtyrskaya, Anapskaya, Gelendzhikskaya, Mikhailovskaya, and Sochinskaya zones of the OEF), the width is $M = 5.5$–5 km, $M = 6.0$–11 km, and $M = 6.5$–18 km [1]. Seismic characteristics of the OEF zones require taking into account the deep seismically active layers (Table 16.2).

Suture Zones of Different Orogenic Areas

The majority of suture zones represented in the Caucasus have analogs in other active orogenic areas, for example, in Central Asia and Sakhalin (Table 16.3) [1, 3, 7]. Consequently, the same suture zones are characteristic for both the intercontinental Central Asian inland and the East Asian mobile belts transitional to the ocean. Therefore, the classification used is not only of regional importance.

Table 16.3 Examples of the main types of neotectonic suture zones in orogenic areas (the numbers in parentheses are the maximum magnitudes of possible earthquakes)

Main types of suture zones	Intracontinental orogens		The near continental island-arc medium active Sakhalin orogen
	Epiplatform high active orogen of Tien Shan	Epigeosynclinal medium active orogen of the Greater Caucasus	
Flexural		Buinakskaya (local) (6)	
Flexure-rupture	Central Ferganskaya (7?)	Southern Crimean-Caucasian (6,5)	Nyshsko-Nabilskaya (transverse) (7)
Suture-fault		Essentukskaya	Western Sakhalinskaua (7,5)
Suture-thrust	Zerafshanskaya (7?)	Alazano-Shemakhinskaya (7)	Central Sakhalinskaua (6,5)
Suture-shift	Talaso-Ferganskaya (8)	Gudermesskaya	Easttrn Sakhalinskaua (7,5)
Suture-depression	Predgissaro-Kokshaalskaya (8)	Mzimtinskaya (7)	Pogranichnaya (local)
Suture-block	Kansaiskaya (local)	Vorontsovskaya	

16.4 Types of SGS of the Ancient East European Platform

Platform SGS are poorly investigated. The materials on the ancient East European platform are most interesting.

As studies in the northwestern part of the Russian plate [8] have shown, it is important to take into account the glacioisostatic factor in the analysis of seismicity. Comparison of seismological materials with the direction of the front of the last continental glaciation has unambiguously shown that all the tectonic earthquakes detected are located exclusively in the territory of the fairly powerful cover of this glaciation. According to theoretical calculations, at a distance from its front about 50–100 km, the thickness of this glacier could reach 1 km and the amplitude of glacioisostatic deflections and subsequent uplifts up to 0.3 km. These movements over the total amplitude are commensurable with the amplitudes of vertical neotectonic movements, which were carried out for 35–37 million years [9]. Consequently, the glacioisostatic movements of the last glaciation, realized for no more than 20,000 years, were on the average three orders of magnitude more intense than neotectonic ones. With such amplitudes and velocities of vertical glaciostatic movements, it is quite probable that the recent (including Holocene) “revival” of discontinuous zones with block motions of amplitude tens of meters takes place. This circumstance explains the localization of modern seismicity, dangerous for construction, within the area of distribution of the last ice sheet. Consequently, data on glacioisostasy should be taken into account in the seismic zoning of the marginal northwestern part of the Russian plate. Therefore, the boundary of the last ice sheet is drawn on the scheme (Fig. 16.3). The entire Baltic (Fennoscandian) shield and the adjacent parts of the Russian plate fell into its limits.

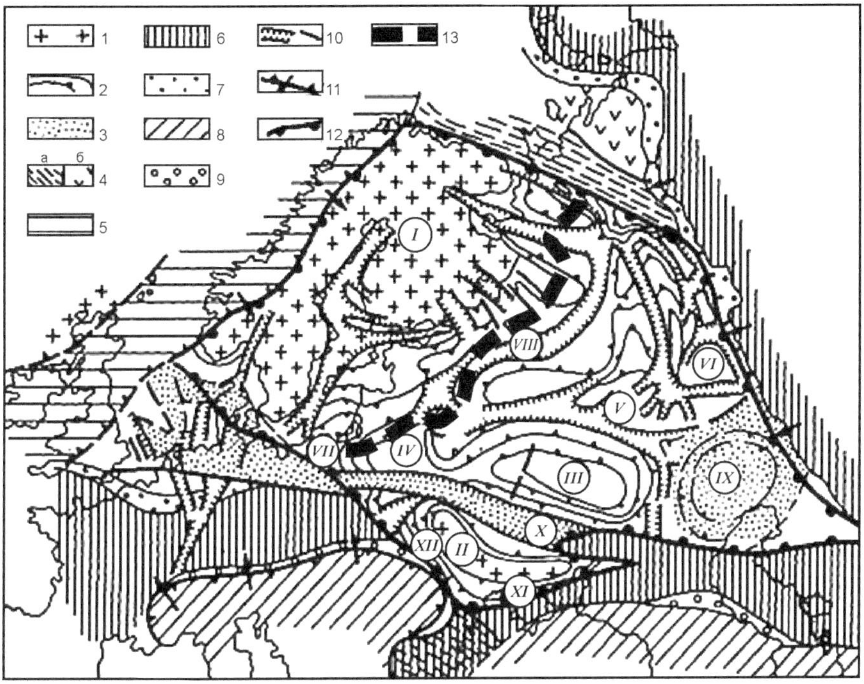

Fig. 16.3 The boundary of the western and eastern seismotectonic segments of the East European Platform in its tectonic scheme. 1, protrusions on the surface of the pre-Riphean basement (I, Baltic; II, Ukrainian shields); 2, isohypse of the foundation surface (km), depicting the main structural elements of the Russian plate (III, Voronezh; IV, Byelorussian anteclises; V, Tatar; VI, Tokmovsky arches of the Volga-Ural anteclise; VII, Baltic; VIII, Moscow; IX, pre-Caspian syneclises; X, Dnieper-Donetsk trough; XI, pre-Black Sea depression; XII, Dniester trough); 3, areas of development of salt tectonics; 4, epi-Baikal Timan-Pechora slab, external (**a**) and internal (**b**) zones; 5, Caledonides; 6, Hercinides; 7, Hercynian marginal troughs; 8, alpine; 9, alpine marginal troughs; 10, aulacogens; 11, thrusts, covers, and direction of the advancing masses of rocks; 12, modern boundaries of the platform; 13, boundary of the western and eastern seismotectonic segments (the boundary of the last glacial cover)

Baltic (Fennoscandian) Shield

Seismicity of Fennoscandia is characterized by a fairly clear linearity (Fig. 16.4a [10]). In the northwest of this territory, the earthquake zone belongs to the Caledonian zone of the North Atlantic belt. On the shield, such bands are associated mainly with grabens, for example, with the graben of the Gulf of Bothnia (Fig. 16.4b [9]), which E.E. Milanovsky [11, p. 53] refers to aulacogenes. In the area of the Kola Peninsula, linear seismogenerating structures are also distinguished (Fig. 16.4b [10].)

In general, there are two types of SGS: (1) suture-depression and (2) suture-fault. Undoubtedly, the Botnichesky, Kandalaksha-Arkhangelsk (№ II in Fig. 16.4b), and the Oslo Graben belong to the suture-depression zones. The Kola-Mezenskaya,

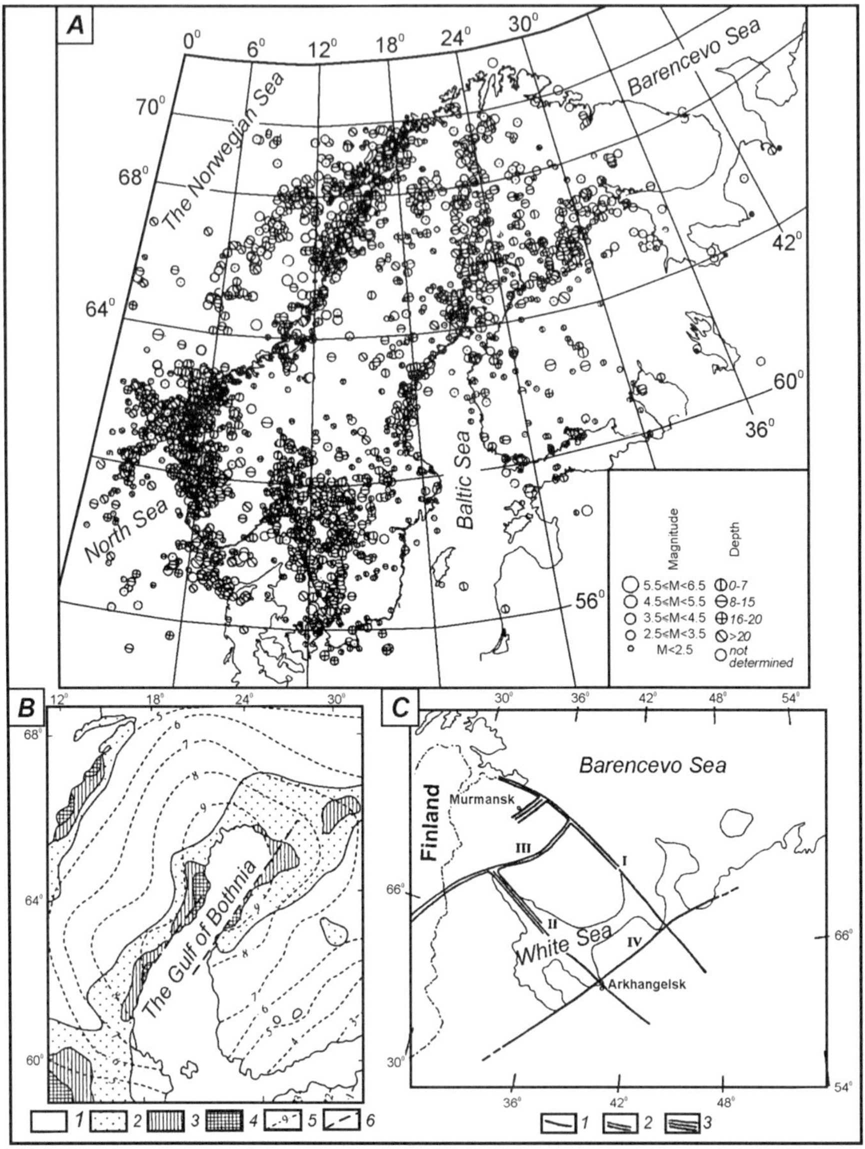

Fig. 16.4 Seismicity of the Baltic (Fennoscandian) shield. (**A**) Map of earthquake epicenters of Fennoscandia for 1900–1977 (according to Heikkinen et al. of [10]. (**B**) Relationship of seismicity (according to K.E. Saaltsrem), modern movements (according to A.A. Nikonov), and zones of discontinuous violations (according to A.G. Stremberg) in the center of the late Pleistocene Fennoscandian ice cover [9]: 1–4, the number of earthquakes in the period 1600–1925 (1, less than 5; 2, 6–11; 3, 12–17; 4, more than 18); 5, isobase, mm/year; 6, discontinuous violations. (**C**) Axes of seismogenic zones in the northeast of the Fennoscandian shield [10]. Maximum magnitudes of earthquakes: 1, up to 4; 2, up to 5; 3, 6 or more. Seismogenic zones: I, Kola-Mezenskaya; II, Kandalaksha-Arkhangelsk; III, Khibinskaya; IV, Onego-Czech

Khibinskaya, and, apparently, the Onego-Czech (№ I, III, IV in Fig. 16.4b) can serve as suture-fault zones.

Some researchers consider the main suture-depression zones (the Gotland-Bothnia and the Finnish graben zones), the structures of the initial (late Quaternary) stage of rifting [9]. But a complete set of signs of rifting did not appear in them. For example, there is no synchronous graben formation of basalt volcanism. Here, as in the suture-depression zones of orogens, a chain of seismic focus marks the marginal rupture graben and adjacent parts of their upper sides.

The Russian Plate

On the Russian plate, strong earthquakes are associated with aulacogens, grabens, and suture-fault zones, which have been particularly affected by repeated activation. Thus, south of the eastern end of Timan (Fig. 16.5), the seismic focus is related to the sides of the Kirov-Kazhim avlakogen [12, p 231]. Activation of this aulacogen occurred twice – on middle (?) Riphean and at the end of the middle – beginning of the late Devonian. And with each activation, up to 0.5–1.0 km of sedimentation accumulated in it [11, p. 67]. There are examples of the connection of earthquakes with other aulacogens and active ruptures.

Studies in the northwestern part of the Russian plate, taking into account the above-considered geological criteria of seismicity and the influence of the glacioisostatic factor, revealed a much greater variety of seismogenerating structures (Fig. 16.6) [13]. Here there are a number of suture-aulacogenic, suture-anticlinal small-block, rupture, flexure-rupture, and flexure zones [8].

General Characteristics of Platform SGS

1. The difference in the nature of the SGS and the level of seismicity for two commensurable segments of the ancient platform separated by the boundary of the spread of the last glacial cover was revealed.
2. The northwestern segment of the platform, which includes the Baltic (Fennoscandian) shield and adjacent marginal parts of the Russian plate, has increased seismicity. This segment underwent the last continental glaciation. Here, the main earthquakes are associated with suture-graben, suture-anticline, suture-fault, flexure-rupture, flexural, and, less often, suture-aulacogenic seismogenerating structures.
3. The southeastern segment of the platform, which corresponds to the main part of the Russian plate and the Ukrainian shield, is characterized by a weaker seismicity. Here, most of the local earthquakes are associated with renewed buried avlakogens, as well as with suture-graben and suture-fault seismogenerating structures.

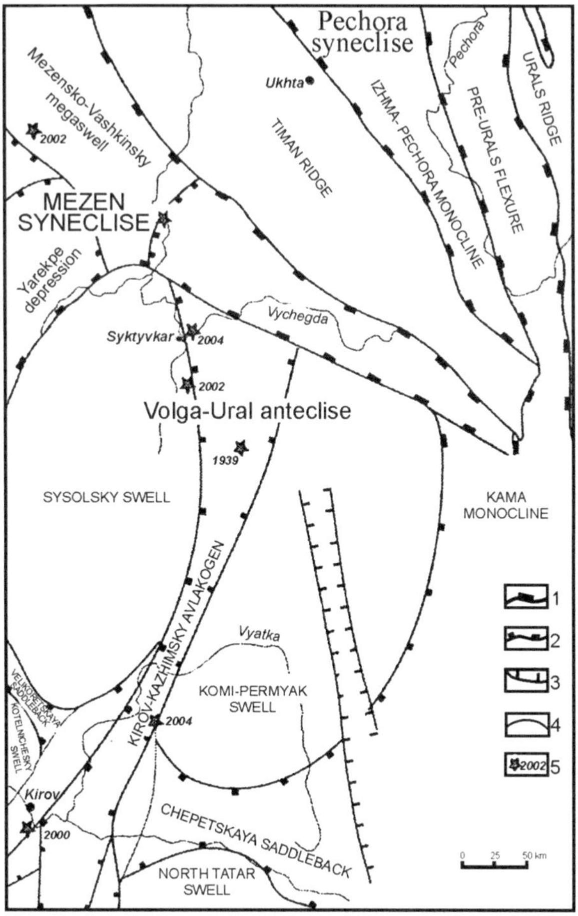

Fig. 16.5 Scheme of tectonic zoning along the foundation surface on the east of the junction of the Russian plate and Timan [12]. 1–4, boundaries of structures (1, superorder; 2, first order; 3, second order; 4, trough); 5, location of the earthquake focus, year

The Generality of the Set and the Regional Specificity of the SGS

Let us consider the following three aspects of this problem.

1. *Similarity of sets of basic types of SGS in different orogenic and ancient platform regions.* The main types of SGS can be considered: suture-flexure, suture-rupture (with variations in the kinematics of the prevailing displacements), suture-depression, and suture-block. There are particular regional differences. At the same time, the main geological criteria of seismicity remain relevant.
2. *The most varied is the suture-depression type of SGS.* It combines suture-aulacogenic and suture-graben zones of platforms, suture-depression zones of orogens, and suture-rift zones of riftogenes. Each of these species can be divided into a number of variants, differing in size (scale), complexity of structure,

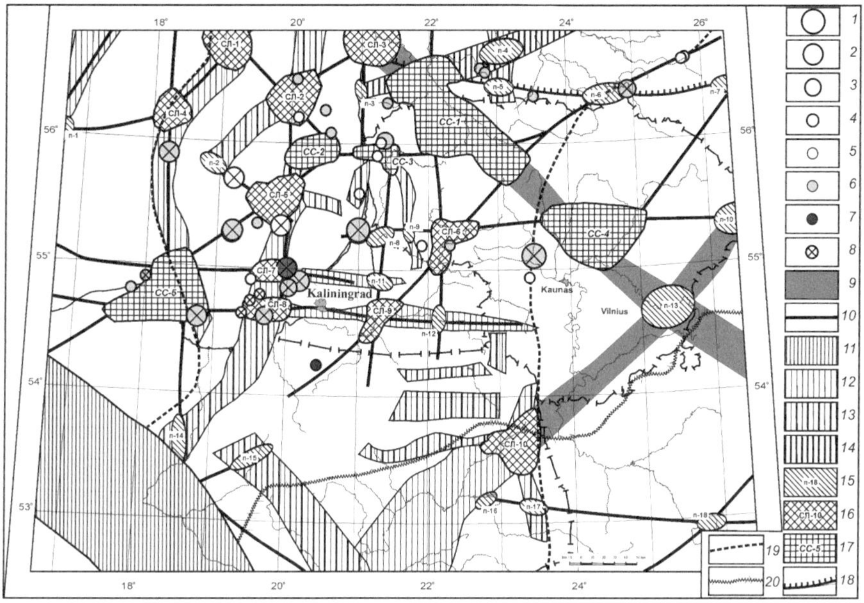

Fig. 16.6 Map of potentially seismogenerating structures of the northwest of the Russian plate [8]. 1–4, magnitude of earthquakes (*M*) (1, 4.8 $\leq M \leq$ 5.2; 2, 4.3 $\leq M \leq$4.7; 3, 3.8 $\leq M \leq$ 4.2; 4, 2.8 $\leq M \leq$ 3.7); 5–7, depth of hypocenter (*h*) in km (5, not defined; 6, 0–10; 7, 10–20); 8, earthquakes shown in the catalog; 9–10, rupture zones within the ancient platform (9, II order; 10, III order); 11, suture-aulacogenic zone of Teisseyre-Tornquist; 12–14, linear zones (12, flexural; 13, flexure rupture; 14, suture anticlinal); 15–17, types of structural nodes (15, simple; 16, complex; 17, composite); 18, the boundaries of the neotectonic structures of the fifth order; 19, boundary of distribution of the last ice sheet; 20, deep faults in the Moho surface

kinematics of marginal sutures, development history, etc. Usually, focuses of earthquakes are timed to the marginal sutures and their plumage.

3. *There is a connection between seismogenerating ability and the level of general new and modern tectonic mobility of the region.* The greatest mobility and practically the ultimate (up to M about 8) seismogenerating ability characterize island-arc systems, riftogens, and the most active orogens, and the least mobility characterize platforms.

Orogens can differentiate according to the level of neotectonic mobility and the level of seismogenerating ability proportional to it. Thus, the high active orogens include the mountains of the Pamirs and the Tien Shan, the middle active orogen is the Greater Caucasus, and the less active ones are the Middle and Southern Urals [1].

Two criteria were taken to evaluate the seismic generation capacity of the platform areas: (1) the antiquity of the initial consolidation, that is, division of platforms into ancient and young, and (2) the main trend of the newest vertical tectonic movements. The second criterion is more significant. So, among young platforms, the large seismicity is noted in the relatively uplifted platforms of the

Turan and Scythian plates in comparison with the active and prolonged submerged West Siberian plate. According to the first criterion, the Turan plate has a greater seismogenerating ability relative to the southeastern segment of the ancient Russian plate.

The presence of the glacioisostatic factor determines the increase in the seismogenic ability of the territories, regardless of their tectonic nature. It is important to take into account the thickness of the last ice sheet and the age of its deglaciation. The more power and the later the deglaciation occurred, the more significant the seismic capacity of the territory increases. Therefore, the seismicity of the northwestern part of the Baltic shield is higher compared to its southeastern periphery and the edge of the Russian plate.

Thus, the proposed classification of the SGS and the consideration of the glacioisostatic factor can substantially advance the question of the genetic nature of the seismogenerating ability of various parts of the continent. There is an opportunity not only of quantitative but also qualitative improvement of general and detailed seismic zoning.

References

1. Nesmeyanov, S. A. (2004). *Engineering geotectonics*. Russia: Science 780 p.
2. Milanovsky, E. E. (1983). *Riftogenesis in the history of the Earth (rifting on ancient platforms)*. Russia: Nedra 280 p.
3. Nesmeyanov, S. A., & Barkhatov, I. I. (1978). *The newest and seismogenerating structures of the Western Gissar-Alai*. Russia: Science 120 p.
4. Chedia, O. K. (1986). *Morphostructure and the newest tectogenesis of the Tien-Shan*. Russia: Ilim 314 p.
5. Milanovsky, E. E. (1968). *The newest tectonics of the Caucasus*. Russia: Nedra 484 p.
6. Lilienberg, D. A., Budagov, B. A., & Aliev, A. S. (1996). Morphotectonics of Azerbaijan and Eastern Transcaucasia from the standpoint of neo-mobility. *Geomorphology, 4*, 31–50.
7. Voeykova, O. A., Nesmeyanov, S. A., & Serebryakova, L. I. (2007). *Neotectonics and active ruptures of Sakhalin*. Russia: Science 186 p.
8. Nesmeyanov, S. A., Lyutikov, A. I., Voeykova, O. A., & Dontsova, G. Y. (2011). Seismicity of the northern part of Russian plate and its its glaciostatic nature. *Water Resources, 38*(7), 916–929.
9. Garetsky, R. G., Aizberg, R. E., Karabanov, A. K., et al. (1999). The newest tectonics and geodynamics of Central Europe. *Geotectonics, 5*, 3–14.
10. Yudakhin, F. N., Shchukin, Y. K., & Makarov, V. I. (2003). *Deep structure and modern geodynamic processes in the lithosphere of the East European platform*. Russia: UrB RAS 300 p.
11. Milanovsky, E. E. (1987). *Geology of the USSR. Part 1*. Russia: MGU 416 p.
12. Earthquakes and microseismicity in the tasks of modern geodynamics of the East European platform. (2007). *Book. 1: Earthquakes*. Russia. Petrozavodsk. KSC RAS. 391p.
13. *Seismotectonics of plates of ancient platforms in the field of quaternary glaciations*. (2009). Russia: Book and Business. 288 p.

Chapter 17
Stochastic Modeling of Human-Induced Thermokarst and Natural Risk Assessment for Existing and Planned Engineering Structures

A. S. Victorov, T. V. Orlov, V. N. Kapralova, O. N. Trapeznikova, S. A. Sadkov, and A. V. Zverev

Abstract The aim of the present research is a stochastic modeling of initiated thermokarst developing along a linear structure and its impact risk assessment by thermokarst basing on the mathematical morphology of landscape approaches. Two versions of the models of the initiated thermokarst for linear structures are considered: the model of the initiated thermokarst 1.0 (linear variant) and the model of the initiated thermokarst 1.1 (zone variant). The research allows to make the following conclusions: Models of initiated thermokarst developing along linear structures (a linear and a zone variants) are created, and their empirical verification is done. Distribution laws for initiated thermokarst foci are obtained for the area of the linear structure influence, including the lognormal distribution of the foci areas and the exponential distribution of distances between the projections of the centers of foci into a linear structure. The mathematical landscape morphology approach allows us to estimate the thermokarst impact probability for linear structures.

17.1 Introduction

There are a lot of research works devoted to natural thermokarst [1, 3, 7, 12, 13, 14, 19, 21, 22, 28], including some modeling attempts of the process [2, 4, 6, 11, 25–27, 34], such as approaches of the mathematical morphology of landscape [8, 9, 30]. The results of that very modeling are given in the previous section of this book. At the same time, there is another urgent problem concerning modeling of the thermokarst process initiated by the impact of an engineering structure itself [24]. The majority of researches deal with the thermodynamic approach [6, 26, 27] though it faces essential difficulties resulting from complexity of modeling from one hand, such

---

A. S. Victorov · T. V. Orlov (✉) · V. N. Kapralova · O. N. Trapeznikova · S. A. Sadkov
A. V. Zverev
Sergeev Institute of Environmental Geoscience RAS (IEG RAS), Moscow, Russia

V. Svalova (ed.), *Natural Hazards and Risk Research in Russia*,
Innovation and Discovery in Russian Science and Engineering,
https://doi.org/10.1007/978-3-319-91833-4_17

as account of phase transitions and three-dimensionality, and, on the other hand, a lack of information on many parameters of the background and their variability like ice content, soil temperature, particle size distribution, etc.

The aim of the present research is a stochastic modeling of initiated thermokarst developing along a linear structure and its impact risk assessment by thermokarst basing on the mathematical morphology of landscape approaches.

17.2 Data and Methods

The mathematical morphology of landscape is a base of the modeling [30–32, 33]. This is a new scientific brunch studying mathematical laws of mosaics formed by natural units including their patterns and developing as well as the methods of their mathematical analysis. A special feature of this approach is that the process is analyzed taking into account the positions and spatial patterns of its foci development on the earth's surface. Thus it makes it possible to use parameters of the foci spatial pattern such as location density, distributions of areas, diameters, etc. which are easily measured on the remote sensing images, instead of hard-to-measure parameters like ice content, soil temperature, granulometric composition, etc. At that this approach allows us to verify theoretical constructions using remote sensing data. The following general scheme was used in the work:

1. Formulation of basic assumptions for each type of the model.
2. Record the presented assumptions in the form of mathematical equations, based on the probabilistic approach.
3. Analytical transformations of obtained mathematical equations.
4. Obtaining the results of analytical transformations and formulating the consequences of the model validity.
5. Verification of the validity of models for key areas.

To test the model results, a wide range of remote sensing data was used:

- 1960–1970 – Corona (2–5 m / pix)
- 1976–1999 – Landsat, SPOT
- 2000–2018 – highly detailed survey of QuickBird, GeoEye, WorldView, RESURS-P

Modern images were provided by:

- DigitalGlobe Foundation (www.digitalglobefoundation.org)
- Scanex Group (http://www.scanex.ru/en/)
- Research Center for Earth Operative Monitoring Russian Space Systems (http://eng.ntsomz.ru/)

17.3 The Model of Initiated Thermokarst

Modeling the man-initiated thermokarst, we are talking about a situation where the creation of a linear structure was a trigger for the thermokarst development. It seems very interesting, what will be the laws of development for this type of thermokarst depressions.

We considered two versions of the models of the initiated thermokarst for linear structures:

The model of the initiated thermokarst 1.0 (linear variant)
The model of the initiated thermokarst 1.1 (zone variant)

The Model of the Initiated Thermokarst 1.0 (Linear Variant)

The case when thermokarst depressions develop in the immediate vicinity of the linear structure and stretch as a dotted line along one or two sides of the structure is what we call the linear variant of initiated thermokarst depressions (Fig. 17.1).

In this case development of thermokarst stems from either an impact of the functioning structure or a consequence of soil-vegetation disturbance directly near

Fig. 17.1 Satellite image (**a**) and a scheme (**b**) of the initiated thermokarst in the immediate vicinity of a linear structure (linear variant); the explanation is in the text

the construction line of the structure. The model is suitable for a uniform nature environment including permafrost, as well as design of the linear structure. The nature uniformity implies, as in the model of the natural thermokarst process in another section of the book, the uniformity of the surface topography, manifested in a uniform pattern of the remote sensing data, the constant composition of the surface sediments, and the underlying deposits, including their thickness, absence of buried troughs and tectonic disturbances, location within the same tectonic structure, and relatively homogeneous permafrost conditions. Naturally, the model does not assume absolute homogeneity but only statistical homogeneity that allows random oscillations. The homogeneity of a linear structure implies the uniformity of a structure (e.g., the preservation of the diameter and material of pipes in gas pipelines, the width of the road, the material composition of the soil in the embankment of the road, etc.). Such homogeneity ensures the uniformity of the impact of the linear structure and of the process of its construction on the environment.

The model of initiated thermokarst 1.0 (linear variant) is based on the following assumptions:

1. New thermokarst holes appear directly in the strip adjacent to the linear structure, they appear independently from each other, and the occurrence probability within a given strip depends only on the length of the strip (ΔL):

$$p(\Delta L) = \gamma \Delta L + o(\Delta L),$$

   where γ is the parameter.
2. The thermokarst holes (water bodies) grow due to thermoabrasion independently from each other, and it is directly proportional to the density of heat waste through the side surface of the water body.

Thus, we deal with the variant of the synchronous start when appearance of the initial holes takes place during the short period of time after the structure building has started. The hole at this scheme looks like an oval object, elongated along a linear structure (Fig. 17.1).

The first assumption seems to be natural because it results from the homogeneity of the area under consideration and the linear structure and means that any limited strip contains a terminate number of thermokarst holes (particularly their centers).

The second assumption is also valid. It reduces to the proportionality between the growth rate of the thermokarst hole and the average density of heat losses through a side surface covered with water, while the growth is influenced by a lot of random factors such as average annual air temperature, ice content of adjacent deposits, etc. This assumption was checked for natural thermokarst and found its confirmation, as shown in another section of the book.

These assumptions simplify the situation and do not reflect the whole variety of nature interrelations; nevertheless the flexible nature of the assumptions, based on probabilistic representations, gives them a certain "margin of safety."

Using transformations similar to those used in the previously developed model of nature thermokarst [30], it can be obtained that the distribution of the number of centers of thermokarst depressions at a given length of the linear structure (L) obeys the Poisson distribution [10]:

$$P_0(k, L) = \frac{[\gamma L]^k}{k!} \mathrm{e}^{-\gamma L} \quad (17.1)$$

where γ is the average linear location density of thermokarst foci.The second assumption, as a result of transformations similar to those used in the model of nature thermokarst [30], leads to the distribution of the area of depressions which correspond to the lognormal distribution.

$$f_r^0(x, t) = \frac{1}{\sqrt{2\pi}\sigma x\sqrt{t}} \mathrm{e}^{-\frac{(\ln x - at)^2}{2\sigma^2 t}}$$

where a, σ are the distribution parameters and x is the area of the holes after time t.

The analysis of the first equation gives us the distribution of distance ξ between the centers of thermokarst lakes along the linear structure.

$$F_\xi(x) = 1 - P\{\xi > x\} = 1 - P_0(0, x)$$

And taking into account (17.1), we get

$$F_\xi(x) = 1 - \mathrm{e}^{-\gamma x}$$

Thus the distance distribution for centers of thermokarst holes along a linear structure should obey the exponential distribution in case of the model validity.

The linear variant of the model can be easily used for those thermokarst foci that appear directly under the linear structure.

The Model of the Initiated Thermokarst 1.1 (Zone Variant)

The case when thermokarst foci are found not only in the direct vicinity of a linear structure but in a certain zone surrounding the construction is what we call the zone variant of initiated thermokarst (Fig. 17.2). These thermokarst foci also stem from the linear structure impact or a consequence of soil-vegetation disturbance in a rather wide zone along the construction line of the structure.

The model of the initiated thermokarst 1.1 (zone variant) is based on the following assumptions:

1. New thermokarst holes appear directly within a limited zone with width (a) adjacent to the linear structure; they appear independently from each other, and the occurrence probability within a given strip depends only on the size of the zone (ΔS) and distance (r) from the linear structure:

Fig. 17.2 The satellite image of initiated thermokarst along a linear structure (zone version); the explanation is in the text

$$p(r, \Delta s) = \lambda(r)\Delta s + o(\Delta s),$$

where $\lambda(r)$ is the coefficient.

2. The thermokarst holes (water bodies) grow due to thermoabrasion independently from each other, and it is directly proportional to the density of heat waste through the side surface of the water body.

Thus, we deal with the variant of the synchronous start when appearance of the initial holes takes place during the short period of time after the structure building has started. The hole at this scheme looks like an oval object, elongated along a linear structure.

The first assumption seems to be natural because it results from the homogeneity of the area under consideration and means that any limited area contains a terminate number of thermokarst holes (particularly their centers). Besides, this assumption takes into account those disturbances within the zone adjacent to the linear structure, which initiate thermokarst. Peculiarities of the disturbance vary depending on the distance from the linear structure and generally remain the same along the structure, including disturbance of soil and vegetation cover, underfeature, and permafrost. In other words the main direction of thermokarst condition variability is the direction perpendicular to the structure. Function $\lambda(r)$ depending only on the distance from the structure takes this into account.

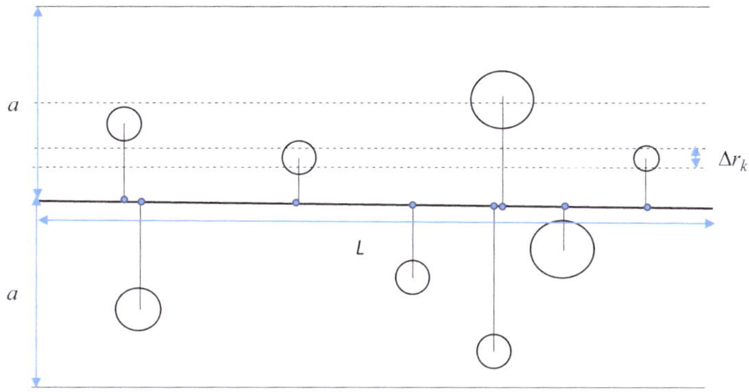

Fig. 17.3 A scheme of the initiated thermokarst in the zone of a linear structure (zone version); rounds, thermokarst foci; dots, projections of centers of foci on the line of a linear structure; dotted lines, boundaries of partitioning into relatively homogeneous strips; the explanation is in the text

The second assumption should be also fair. Similar to the first model, it reduces to the proportionality between the growth rate of the thermokarst hole and the average density of heat losses through a side surface covered with water, while the growth is influenced by a lot of random factors such as average annual air temperature, ice content of adjacent deposits, etc.

The model assumptions make it possible to obtain the distribution of the number of thermokarst centers projecting onto a given length of a linear structure (L) (Fig. 17.3).

Let us split the zone where thermokarst develops into several strips Δr_k. Every strip can be regarded as approximately homogeneous, and accordingly a number of centers of initiated thermokarst holes should [29] obey the Poisson law with average value equal to the product of the density of the center location according to the distance of the strip from the structure to the area of the strip.

$$\gamma_k \approx \lambda(r_k)\Delta r_k L.$$

An average number of thermokarst centers, projecting onto a given length of a linear structure (L), is approximately the sum of these average numbers of all the strips. Equality goes with the exact width of strips tending to zero and an infinite increase in the number of strips; in this case, we get the following equation for the average

$$\gamma = L\int_0^a \lambda(r)dr;$$

correspondingly, the average number of thermokarst centers per square unit is equal to

$$\mu = \int_0^a \lambda(r)dr.$$

The distribution law for centers of thermokarst foci projecting on a given length of a linear structure (L) remains the Poisson one as a sum of a certain number of independent random variables according to a number of centers of thermokarst foci within every strip distributing by the Poisson law.

$$P(k, L) = \frac{[\mu L]^k}{k!} e^{-\mu L}.$$

Applying the course of analysis similar to the linear case, it is not difficult to find out that under the condition of model validity, the distribution of the distances between the projections of the centers of thermokarst foci into a linear structure should also correspond to the exponential distribution.

After the transformations of the second assumption, similar to those used in the natural thermokarst model, we get to know that the area distribution of the foci must also correspond to the lognormal distribution.

Generally the linear variant of the model can be regarded as a particular case of the zone version if $\lambda(r) = \lambda_0$.

Empirical Testing of the Initial Thermokarst Model

Some conclusions of the considered models were empirically tested. The first stage of the testing dealt with key areas, where the testing was done (Fig. 17.4).

The empirical testing includes:

- Detecting thermokarst lakes using remote sensing data
- Detecting centers of thermokarst lakes and their projections on a linear structure
- Measuring distances between neighboring projections of lake centers
- Measuring lake areas
- Comparison of the empirical distribution of the measured distances with the theoretically inferred exponential distribution and the empirical distribution of the areas – with theoretically obtained lognormal distribution according to the Pearson criterion

Figure 17.5 demonstrates the comparison between empirical and theoretical (lognormal) distribution functions for areas of thermokarst depressions, and Fig. 17.6 demonstrates the comparison between empirical and theoretical (exponential) distribution functions for distances between neighboring projections of the centers of thermokarst depressions.

The results of the testing show that all samples correspond to verifiable hypotheses at a significance level of 0.99 (Tables 17.1 and 17.2).

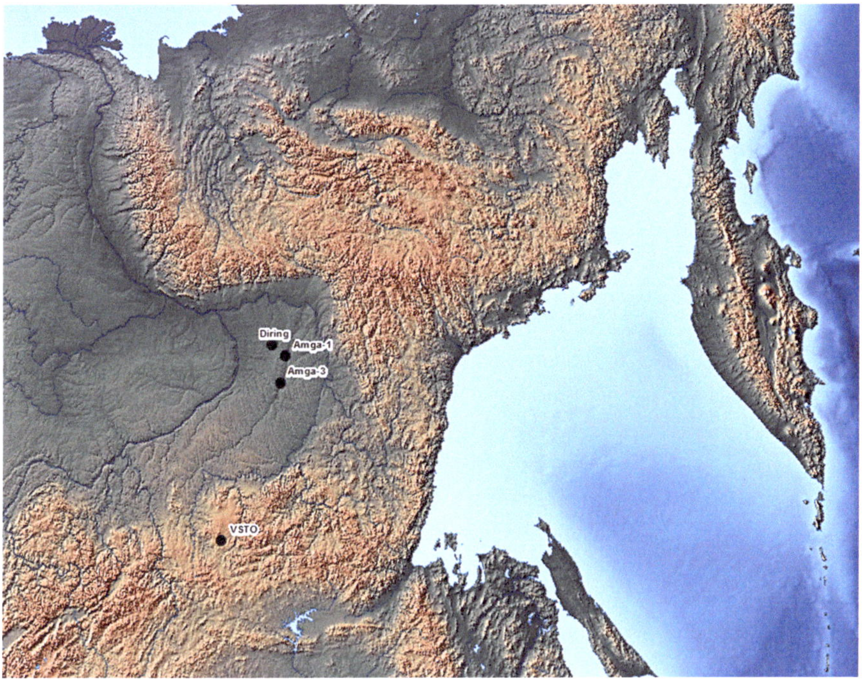

Fig. 17.4 The location scheme of sites 1, Diring; 2, Amga –1; 3, Amga –3; 4 VSTO. Background map – http://www.naturalearthdata.com

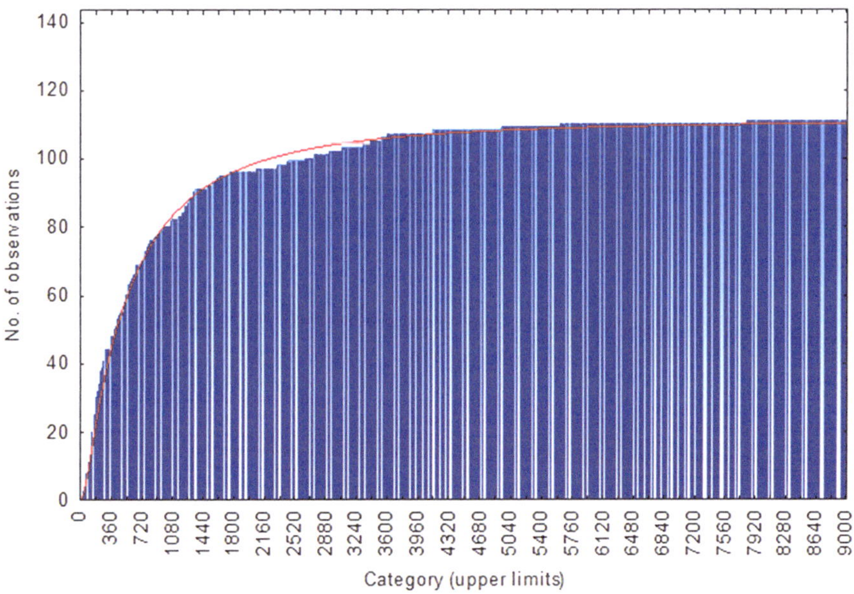

Fig. 17.5 Comparison between empirical and theoretical (lognormal) distribution functions for areas of thermokarst depressions (site Amga 1)

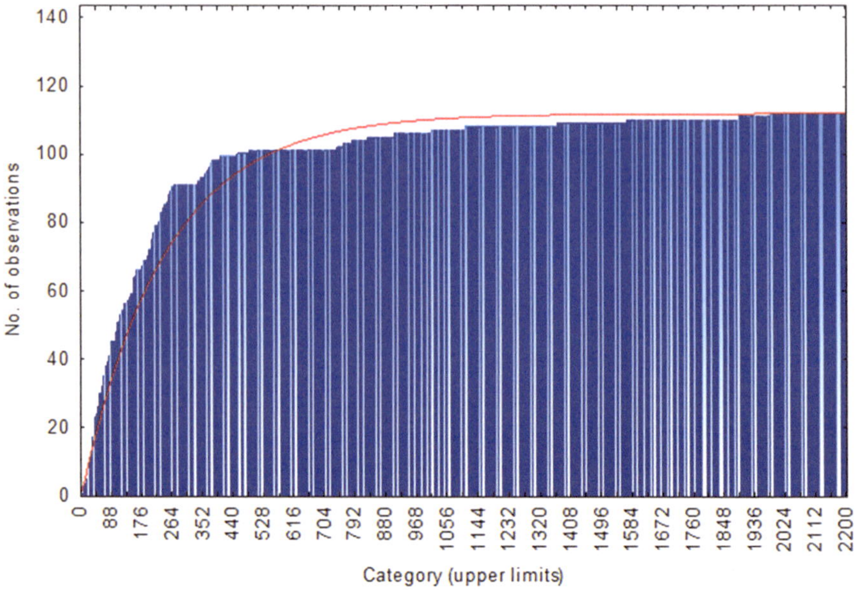

Fig. 17.6 Comparison between empirical and theoretical (exponential) distance distribution functions for neighboring projections of the centers of thermokarst depressions (site Amga 1)

Table 17.1 Correspondence of empirical and theoretical lognormal distributions of areas of thermokarst depressions (initiated thermokarst)

Site	Key site	Volume of sample	Length of the site (km)	Average area of the foci (m^2)	Standard deviation (m^2)	p^a
The valley of Amga river	Amga –1	35	8.19	5277	6198	0.518
	Amga –3	113	7.01	1269	2921	0.668
Vicinity of the settlement of Diring	Diring	78	35.73	1109	1444	0.459
Eastern Siberia–Pacific Ocean oil pipeline (VSTO)	VSTO – 1	74	3.96	77	56	0.064
	VSTO – 2	54	2.25	101	89	0.078

[a]*p* – the quantile corresponding to the computed value of the chi-square test; the hypothesis is not rejected at the level of significance 0.95 if $p > 0.05$

From the exponential distribution of the distances between the projections of the centers of thermokarst depressions, it is not difficult to obtain that the distribution of the number of projections of the foci centers onto a random segment of the linear structure obeys the Poisson law.

Table 17.2 Correspondence of empirical and theoretical (exponential) distribution functions for distances between neighboring projections of the centers of thermokarst depressions (initiated thermokarst)

Site	Key site	Volume of sample	Length of the site (km)	p^a
The valley of Amga river	Amga –1	34	8.19	0.156
	Amga –3	112	7.01	0.087
Vicinity of the settlement of Diring	Diring	77	35.73	0.095
Eastern Siberia–Pacific Ocean oil pipeline (VSTO)	VSTO – 1	73	3.96	0.658
	VSTO – 2	53	2.25	0.826

[a] p – the quantile corresponding to the computed value of the chi-square test; the hypothesis is not rejected at the level of significance 0.95 если $p > 0.05$

17.4 Natural Risk Assessment

Natural risk assessment is among the urgent directions of the perspective use of mathematical models of morphological pattern of cryolithozone landscapes. The most common risk measure is the impact probability of a natural hazard on an engineering structure.

Many researches such as [5, 16–18, 20, 23, 35], and deal with this topic; nevertheless it is still actual. Extreme practical relevance of the problem results from the wide economic development of cryolithozone including oil and gas production and pipeline construction.

Let us consider an example of impact probability of natural thermokarst on a linear structure. Let us examine a lacustrine thermokarst plain homogenous by its geomorphology and permafrost. We can estimate the impact probability for a linear structure of a given size using mathematical model of lacustrine thermokarst plain landscape pattern. The model is based on the following assumptions:

1. The process of thermokarst appearance within a tested site is probabilistic and dependent exclusively on the site area. For smaller sites, the probability of a single thermokarst starting point appearance is significantly higher than that of multiple starting points.
2. The lake size growth goes independently under thermoabrasive action. In addition, the growth is proportional to the heat accumulation in the lake and is inversely related to the lakebed surface area covered with water.

These assumptions stem from the homogeneity of the environment and the current concept of the process mechanism as it was previously examined in detail.

Let us estimate impact probability for a linear construction (section) of a given length (L). We should consider a strip of a finite width R (Fig. 17.7), with analyzed linear structure as its axis. Let us build the coordinate axis perpendicular to the linear structure.

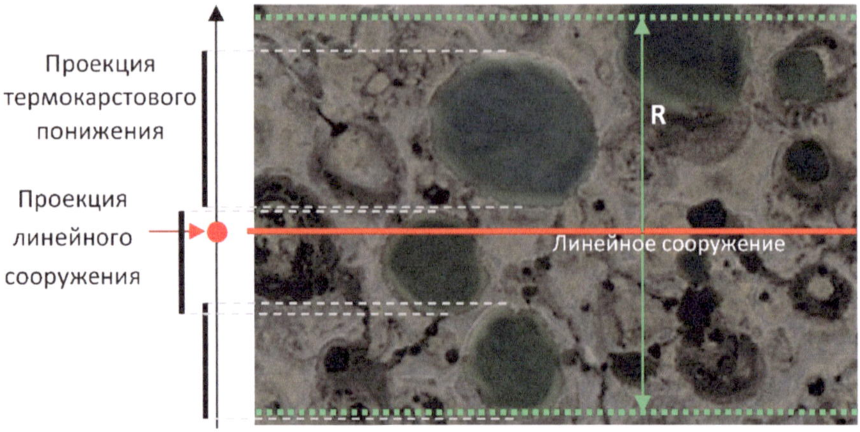

Fig. 17.7 The scheme illustrating the impact probability for a linear structure within the lacustrine thermokarst plain; the explanation is in the text

The structure becomes a dot in the projection on this axis; the thermokarst focus (a lake) becomes a section with the length squeal to the lake projection onto the axis. The lesion of the linear structure is a crossing of the section and a dot.

The developed model allows us to get laws of the morphological pattern of a thermokarst plain by means of the strict mathematical analysis. For instance, the distribution of the number of thermokarst lakes (centers) on a randomly chosen site obeys the Poisson law, that is,

$$P(k,t) = \frac{[\mu(t)s]^k}{k!} e^{-\mu(t)s},$$

where s is the test site area and $\mu(t)$ the average number of lakes per unit area at the time t.

Generally, location density of the lakes depends on time because lakes can merge. It is easy to show that Poisson distribution of thermokarst foci within a site provides for Poisson distribution of their projections on the coordinate axis and, hence, equal probability of the location of the foci center projections on certain sections of the axis and their independence from each other. In view of the said above, the probability (α) that any of the thermokarst foci affects the linear structure (i.e., the dot corresponding to the projection of the linear structure is inside the projection of the focus) stems from the ratio of the focus projection to the strip width. This, taking into account the probabilistic distribution of projection values, makes it possible to obtain an expression for the probability that one focus of the process among those named affects the linear structure.

$$\alpha = \int_0^{2R} \frac{x}{2R} f_{\mathrm{p}}(x,t)dx,$$

where $f_{\mathrm{p}}(x,t)$ is the distribution density of the focus projection at time t.

The probability that none of the thermokarst foci affect the linear structure on condition that there are k foci within the strip, and taking into account the independence of the foci and their Poisson distribution, is equal to

$$P^0(k,R,t) = (1-\alpha)^k \frac{[2\mu(t)RL]^k}{k!} \mathrm{e}^{-2\mu(t)\mathrm{RL}}.$$

The probability not to affect the linear structure with an arbitrary number of foci within the strip can be obtained by looking through all possible variants of a different number of foci within the strip. Therefore, summing over k, we obtain

$$P^0_{nl}(R,t) = \sum_{k=0}^{+\infty} (1-\alpha)^k \frac{[2\mu(t)RL]^k}{k!} \mathrm{e}^{-2\mu(t)RL} = \mathrm{e}^{-2\mu(t)\alpha RL}.$$

The exact value of the required probability is obtained by passing to the limit with unbounded expansion of the considered strip ($R \rightarrow +\infty$). Because of

$$\lim_{R\rightarrow+\infty} 2\alpha R = \int_0^{+\infty} x f_{\mathrm{p}}(x,t)dx = \overline{pr}(t),$$

where $\overline{pr}(t)$ is the expectation of the focus projection at time t, then after simplification the proofness (nonimpact probability) is given by the expression

$$P_{nl}(L,t) = \mathrm{e}^{-\mu(t)\overline{pr}(t)L}.$$

Hence, it is easy to find that the impact probability for a linear structure of length L with at least on focus is equal to

$$P_{dl}(L,t) = 1 - \mathrm{e}^{-\mu(t)\overline{pr}(t)L}.$$

In case of lakes of circular shape, the formula takes the form

$$P_{dl}(L,t) = 1 - \mathrm{e}^{-\mu(t)\bar{d}(t)L},$$

where $\bar{d}(t)$ is the average diameter of thermokarst lakes at time t.

Empirical testing of the obtained results was done. The research was concentrated on a set of sites which were conditionally called the reference ones (with size from 4 to 230 km^2 and average lake area within the sites from 0.01 to 0.6 km^2). We used the following criteria choosing the sites:

- Homogeneous geology and topography of the site including genetic and morphological homogeneity as well as the same surface sediment allover a site, proved by archive and literature data
- Uniform texture of the space image of a site including its inner structure and phototone, shape, and location of the lakes
- Availability of remote sensing data of the required resolution
- Different nature environment for different sites

Well-known interpretation criteria were used to regard the lakes as thermokarst ones at the space images [15].

The sites are situated in different nature environment including geology, permafrost, and climate. According to permafrost zoning, these sites are situated within the wide range of permafrost from rare discontinuous permafrost (sites 1 and 4) to massive discontinuous permafrost (site 5). The permafrost thickness also varies from a few tens to a few hundred meters. All sites are situated at river terraces within tundra zone and folded by quaternary deposits of different genetic types. Space images of the reference sites are shown in Fig. 17.8.

Space images from different satellites like Landsat (MSS, TM, ETM), SPOT, and IRS were used to get data about the morphological pattern of the area in question. Their spatial resolution varies from 60 and 14.5 m (Landsat) to 5.8 m (IRS) and 1.92 m (SPOT).

The empirical testing of impact probability assessment for linear structure results from the following logic. Let us suggest that we came to the testing area before thermokarst started. So we have no preferred location for future linear construction, and it can be placed with equal probability anywhere within the tested area. Then

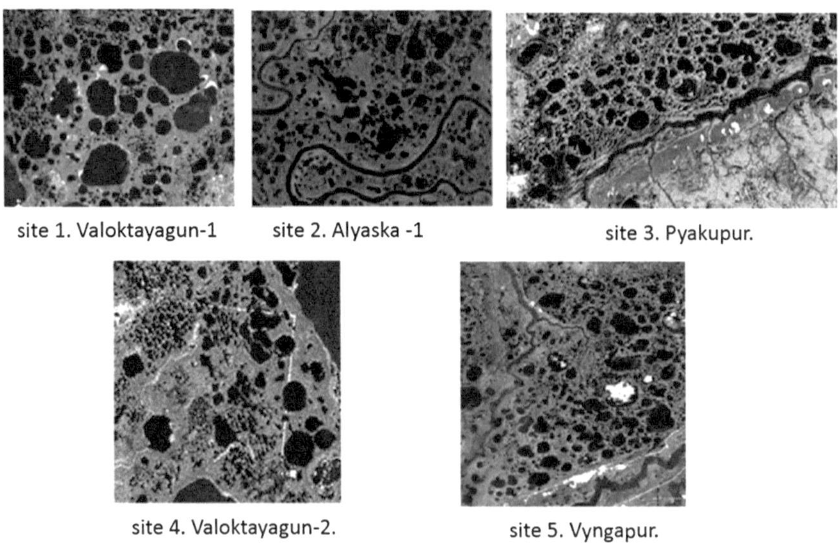

Fig. 17.8 Satellite images of the reference sites at lacustrine thermokarst plains

thermokarst lakes, which we can see now at the remote sensing imagery, could either affect or not the linear structure. Thus we can simulate the real situation: using software and a random number generator, we place a linear object of different length randomly within the testing area with already existing thermokarst lakes. Then we can estimate a number of lines (simulating linear structures at different locations) affected with the thermokarst and compare it with the theoretically calculated one using average diameters and average density of lakes. The following testing technique was used:

- Lake digitizing
- Measuring average diameter and average location density of lakes
- Detecting calculated impact probability for a linear structure (the length of the structure is given by the researcher)
- Random throws of linear objects of appropriate length and determination of empirical affect frequencies
- Comparison of calculated probability and empirical frequency for the linear structure affects

At first, the testing areas were digitized in the special software "Vectorizer" (author A.A. Victorov).

We have faced certain difficulties while testing because the lakes can merge in the course of their development and the equation was done without taking merging into account. At that the merging results in changing distribution parameters such as average location density of lakes, average lake area, and average diameter. We have made an attempt to take into account lake merging and recalculate their number after "separating" merged lakes according the visual peculiarities of their shapes (Figs. 17.9 and 17.10).

General results of the comparison are given in Fig. 17.11.

Thus the empirical testing confirms the proposed approach to the impact probability assessment for linear structures. The research in whole confirms the possibility of natural risk assessment for engineering structures basing on the mathematical model of the morphological landscape pattern of lacustrine thermokarst plains. Naturally, the actual cases are more complex than the analyzed ones. In a reality it is necessary to take into account the possibility of occurrence of thermokarst foci directly along the axis of the pipeline due to the thermal action. The problem can be solved by using the suggested model both for the whole area and the strip over the tube itself, assuming the same laws there but with different parameters. In this case the affect on the linear structures can stem under three factors:

- Due to expanding of the focus already existed before the construction started
- Due to the lakes appeared within the lacustrine thermokarst plain in course of linear structure functioning
- Due to foci appeared over the pipeline because of its thermal action

According to the results described at the beginning of this section, the conditional density of the radius distribution of thermokarst lakes is equal to

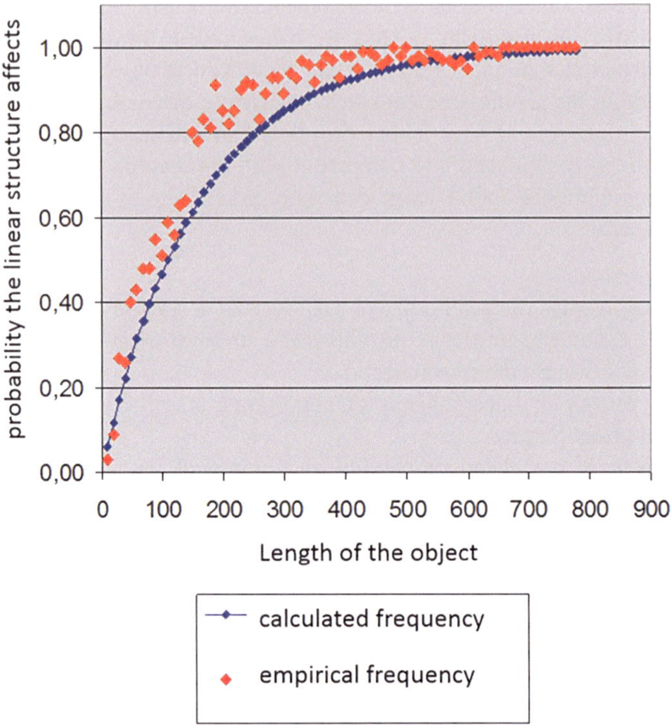

Fig. 17.9 Comparison of calculated probability and empirical frequency for the linear structure affects (site 1)

$$f(x,t|v) = \frac{1}{\sqrt{2\pi}\sigma x\sqrt{t}} \mathrm{e}^{-\frac{\left(\ln\frac{x}{v} - at\right)^2}{2\sigma^2 t}},$$

where v is the initial size of a lake, t is the time, x is the future size of a lake, and a, σ is the growth rate of the average logarithm of the radius and its standard in time.

From that we get the nonimpact probability for a linear structure in case of expanding i focus:

$$q_i(t) = \int_0^{r_i} \frac{1}{\sqrt{2\pi}\sigma u\sqrt{t}} \exp\left[-\frac{(\ln u - \ln v_i - at)^2}{2\sigma^2 t}\right] du,$$

where i=1,2,...,n is the focus number, n a number of foci, v_i initial radius, and r_i initial distance to the linear construction.

Finally, the nonimpact probabilities for a linear structure regarding the foci that arise directly at the axis of a linear structure, according to the model of the initiated thermokarst, are the following:

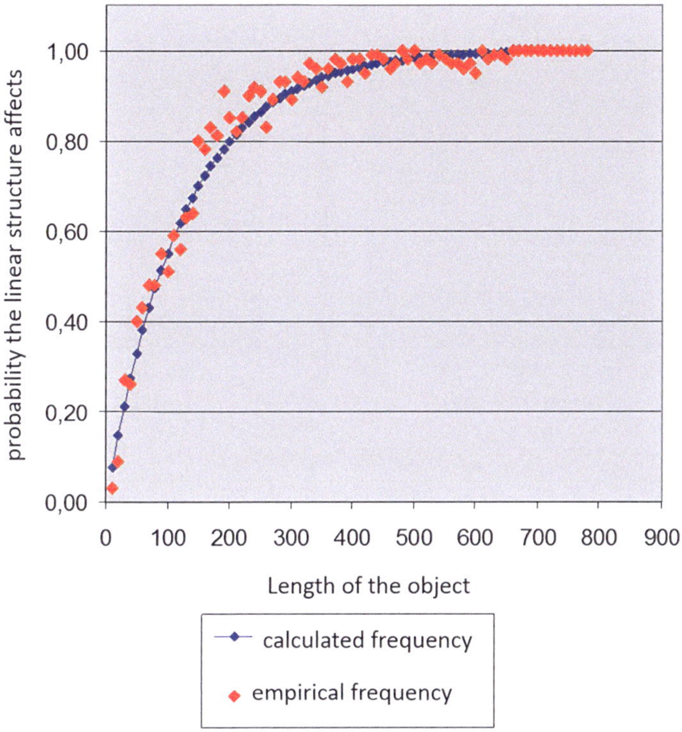

Fig. 17.10 Comparison of calculated probability and empirical frequency after "separating" merged lakes for the linear structure affects (site 1)

$$P_1(t) = e^{-\lambda t d L},$$

where d is the average lake diameter and λ average density of lake generation.

Finally, the nonimpact probabilities for a linear structure regarding the foci that arise directly at the axis of a linear structure, according to the model of the initiated thermokarst, are

$$P_2(t) = e^{-\mu t L},$$

where μ is the linear density of foci generation at the axis of a linear structure.

As a result the impact probability for a linear structure of the length L is

$$p(t) = 1 - e^{-(\mu+\lambda d)Lt} \prod_{i=1}^{n} \int_0^{r_i} \frac{1}{\sqrt{2\pi}\sigma u \sqrt{t}} \exp\left[-\frac{(\ln u - \ln v_i - at)^2}{2\sigma^2 t} \right] du,$$

where t is lifetime of a linear construction. The necessary parameters are determined from repeated remote sensing data:

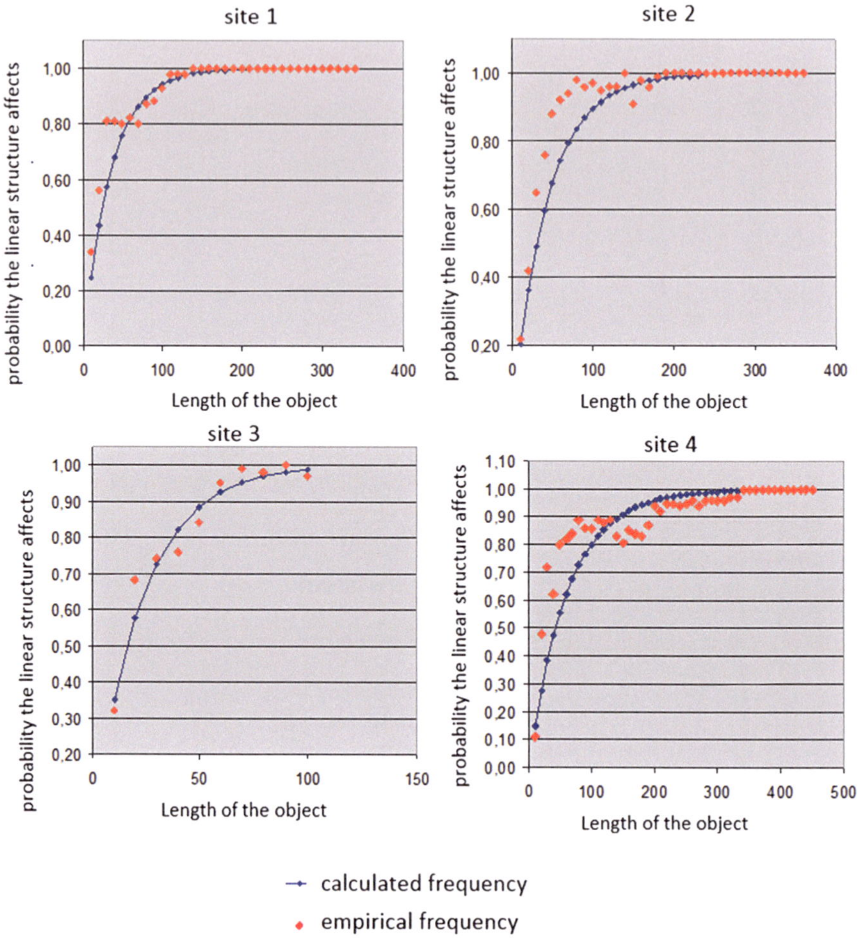

Fig. 17.11 General results of the comparison for sites 1–4

$$a = \frac{M(\ln \xi_2) - M(\ln \xi_1)}{t_2 - t_1}, \quad \sigma = \sqrt{\frac{D(\ln \xi_2) - D(\ln \xi_1)}{t_2 - t_1}}, \quad \lambda = \frac{n_2 - n_1}{(t_2 - t_1)S},$$
$$\mu = \frac{m_2 - m_1}{(t_2 - t_1)L_1}$$

where t_1, t_2 are the dates of the first and second space surveys, $M(\ln\xi_i)$ is the average logarithm of lake radii for the corresponding period, $D(\ln\xi_i)$ is the variance of the logarithm of lake radii for the corresponding period, n_i are a number of lakes within a given area S for the corresponding period, and m_i are a number of thermokarst foci at the axis of a linear structure of the given length L_1 for the corresponding period.

17.5 Conclusions

Thus, our research allows us to make the following conclusions:

Models of initiated thermokarst developing along linear structures (a linear and a zone variants) are created, and their empirical verification is done.

Distribution laws for initiated thermokarst foci are obtained for the area of the linear structure influence, including the lognormal distribution of the foci areas and the exponential distribution of distances between the projections of the centers of foci into a linear structure.

The mathematical landscape morphology approach allows us to estimate the thermokarst impact probability for linear structures.

Acknowledgments The research was supported by RGS-RFBR grant (project No. № 17-05-41141).

References

1. Arp, C. D., Jones, B., Bondurant, A., Grosse, G., Parsekian, A. D., Romanovsky, V., & Hinkel, K. (2016). Extreme sensitivity of shallow lakes and sublake permafrost to arctic climate change. In F. Günther, & A. Morgenstern (Eds.), *XI. international conference on permafrost – Book of abstracts*, 20–24 June 2016, Potsdam, Germany. Bibliothek Wissenschaftspark Albert Einstein. https://doi.org/10.2312/GFZ.LIS.2016.001.
2. Bogdanov, A. N., & Polishchuk, Y. U. M. (2014). Distantsionnoye issledovaniye prostranstvennogo raspredeleniya malykh ozer v zonakh aktivnogo termokarsta Zapadnoy Sibiri (Remote study of spatial distribution of small lakes in zones of active thermokarst of Western Siberia). *Vestnik Yugorskogo gosudarstvennogo universiteta, 3*(34), 8–12.
3. Bondurant, A. C., Arp, C., Jones, B., & Engram, M. (2016). Rates and mechanisms of expansion in thermokarst lakes with bedfast and floating ice regimes on the Arctic Coastal Plain of northern Alaska 2015. In *XI. International conference on permafrost – Book of Abstracts*, 20–24 June 2016, Potsdam, Germany. Bibliothek Wissenschaftspark Albert Einstein (pp 703–705). https://doi.org/10.2312/GFZ.LIS.2016.001.
4. Burn, C. R., & Smith, M. W. (1990). Development of thermokarst lakes during the Holocene at sites near Mayo, Yukon territory. *Permafrost and Periglacial Processes, 1*, 161–176.
5. Fell, R., Corominas, J., Bonnard, C., Cascini, L., Leroi, E., & Savage, W. Z. (2008). Guidelines for landslide susceptibility, hazard and risk zoning for land use planning. *Engineering Geology, 102*(3–4), 85–98. Fitzgerald, D., & Riordan, B. A. (2003). Permafrost and ponds. Remote sensing and GIS used to monitor Alaska wetlands at the landscape level. *Agroborealis, 35*(1), 30–35.
6. Grechishchev, S. Y., Chistotinov, L. V., & Shur, Y. U. L. (1980). *Kriogennyye fiziko-geologicheskiye protsessy i ikh prognoz. (Cryogenic physico-geological processes and their prognosis)*. Moscow: Nedra 384 p.
7. Grosse, G., Jones, B. M., Nitze, I., Lindgren, P. R., Walter Anthony, K. M., & Romanovsky, V. E. (2016). Massive thermokarst lake area loss in continuous ice-rich permafrost of the northern Seward Peninsula, Northwestern Alaska, 1949–2015. In *XI. International conference on permafrost – Book of abstracts*, 20–24 June 2016, Potsdam, Germany. Bibliothek Wissenschaftspark Albert Einstein (pp. 739–740). https://doi.org/10.2312/GFZ.LIS.2016.001.

8. Kapralova, V. N. (2014). *Zakonomernosti razvitiya termokarstovykh protsessov v predelakh ozerno-termokarstovykh ravnin (na osnove podkhodov matematicheskoy morfologii landshafta) (Regularities of the development of thermokarst processes within the lacustrine-thermokarst plains (based on the approaches of the mathematical morphology of the landscape))*. Author's Candidate's summery. Moscow: IGE RAN, 24 p.
9. Kapralova, V. N., & Victorov, A. S. (2009). Modelirovaniye morfologicheskoy struktury ozerno-termokarstovykh ravnin i yego geoekologicheskoye znacheniye. (Modeling of the morphological structure of the lake-thermokarst plains and its geo-ecological significance), Modelirovaniye pri reshenii geoekologicheskikh zadach. Sergeyevskiye chteniya. Is. 11. – Moscow: Geos (pp. 174–178).
10. Karlin, S. (1971). *Osnovy teorii sluchaynykh protsessov. (Fundamentals of the theory of random processes)*. Moscow: Mir 536 p.
11. Kirpotin, S. N., Polishchuk, Y. U. M., & Bryksina, N. A. (2008). Dinamika ploshchadey termokarstovykh ozer v sploshnoy i preryvistoy kriolitozonakh Zapadnoy Sibiri v usloviyakh global'nogo potepleniya (Dynamics of the areas of thermokarst lakes in the continuous and intermittent cryolithozones of Western Siberia under conditions of global warming). *Vestnik TGU, 311*, 185–189.
12. Kravtsova, V. I. & Tarasenko, T. V. (2009). Izucheniye i kartografirovaniye dinamiki termokarstovykh ozer na territorii Zapadnoy Sibiri po raznovremennym kosmicheskim snimkam // Vos'moye sibirskoye soveshchaniye po klimato-ekologicheskomu monitoringu. Materialy rossiyskoy konferentsii 8–10 oktyabrya 2009 (Study and mapping of the dynamics of thermokarst lakes in the territory of Western Siberia for various space images. The eighth Siberian conference on climate and environmental monitoring. Materials of the Russian conference October 8–10, 2009). – Tomsk: Agraf-Press (pp. 273–275).
13. Kravtsova, V. I., & Bystrova, A. G. (2009). Izucheniye izmeneniy rasprostraneniya termokarstovykh ozer Rossii po raznovremennym kosmicheskim snimkam. *Kriosfera Zemli, 15*(2), 16–26.
14. Makarycheva, E., Sergeev, D., Kapralova, V., & Jin, H. (2014). Water level regime of thermokarst lakes in the mountain areas. In *Book of Abstracts of EUCOP4 – 4th European conference on permafrost*. 2014, P. 206.
15. Metodicheskoye rukovodstvo po inzhenerno-geologicheskoy s"yemke masshtaba 1:200 000 (1:100000–1:500000). (1978). *(Methodical guidelines for engineering-geological survey of a scale of 1: 200 000 (1: 100000–1: 500000))*. Moscow: Nedra 391 p.
16. Osipov, V. I. (2009). Prirodnyye opasnosti i strategicheskiye riski v mire i v Rossii (Natural dangers and strategic risks in the world and in Russia). *Ekologiya i zhizn', 11–12*(96–97), 5–15.
17. Otsenka i upravleniye prirodnymi riskami. (2003a). *(Assessment and management of natural risks.) Mater. Vserop. konf. "Risk-2003"* (Vol. 1). Moscow: Izd. Rop. univer. druzh. narodov 412 p.
18. Otsenka i upravleniye prirodnymi riskami. (2003b). Tematicheskiy tom. (Assessment and management of natural risks. Thematic volume) Ragozin A.L. ed. – Moscow, Izd. firma. KRUK, 320 p.
19. Perl'shteyn, G. Z., Pavlov, A. V., Levashov, A. V., & Sergeyev, D. O. (2005). *Netemperaturnyye faktory teploobmena deyatel'nogo sloya s atmosferoy (Non-temperature factors of heat exchange of the active layer with the atmosphere), Materialy Tret'yey konferentsii geokriologov* (pp. 86–91). Rossii: MGU.
20. Ragozin, A. L. (1997). Osnovnyye polozheniya teorii opasnykh geologicheskikh protsessov i riskov (Basic theses of the theory of dangerous geological processes and risks), Novyye idei v naukakh o Zemle: Abstracts, Moscow (Vol. 4, p. 115).
21. Riordan, B., Verbyla, D., & McGuire, A. D. (2006). Shrinking ponds in subarctic Alaska based on 1950–2002 remotely sensed images. *Journal of Geophysical Research, 111*, G04002. https://doi.org/10.1029/2005JG000150.
22. Sejourne, A., Costard, F., Fedorov, A., Gargani, J., Skorve, J., Masse, M., & Mege, D. (2015). Evolution of the banks of thermokarst lakes in Central Yakutiya (Central Siberia) due to retrogressive thaw slump activity controlled by insolation. *Geomorphology, 241*, 31–40.

23. Sheko, A. I., & Krupoderov, V. S. (1994). Otsenka opasnosti i riska ekzogennykh geologicheskikh protsessov (Assessment of the danger and risk of exogenous geological processes). *Geoekologiya, 3*, 53–59.
24. Shiklomanov, N. I., & Nelson, F. E. (2013). Thermokarst and civil infrastructure. In *Treatise on geomorphology* (Vol. 8, pp. 354–373). Elsevier. https://doi.org/10.1016/B978-0-12-374739-6.00214-1
25. Tomirdiaro, S. V. (1972). *Vechnaya merzlota i osvoyeniye gornykh stran i nizmennostey. (Permafrost and development of mountainous countries and lowlands)*. Magadan publishing house: Magadan, 174 p.
26. Tumskoy, V. Ye. (2002). *Termokarst i yego rol' v razvitii regiona morya Laptevykh v pozdnem pleystotsene i golotsene. (Thermokarst and its role in the development of the Laptev Sea region in the Late Pleistocene and Holocene)*, Author's Candidate's summary. Moscow.
27. Tumskoy, V. Ye., Nikol'skiy, P. A., Basilyan, A. E., Kuznetsova, T. V., & Gavrilov, A. V. (2000). Evolyutsiya mnogoletnemerzlykh porod na poberezh'ye proliva Dmitriya Lapteva v pozdnem kaynozoye – Tezisy konferentsii "Ritmy prirodnykh protsessov v kriosfere Zemli" (Evolution of Permafrost on the Strait of Dmitrii Laptev in the Late Cenozoic – Theses of the conference "Rhythms of Natural Processes in the Earth's Cryosphere"), Pushchino, (pp. 123–125).
28. Veremeeva, A., & Gubin, S. (2009). Modern tundra landscapes of the Kolyma Lowland and their evolution in the Holocene. *Permafrost and Periglacial Processes, 20*(4), 399–406.
29. Victorov, A. S. (1995). Matematicheskaya model' termokarstovykh ozernykh ravnin kak odna iz osnov interpretatsii materialov kosmicheskikh s'yemok. *Issledovaniye Zemli iz kosmosa, 5*, 42–50.
30. Victorov, A. S. (2006). *Osnovnyye problemy matematicheskoy morfologii landshafta (The main problems of the mathematical morphology of the landscape.)*. Moscow: Nauka 252 p.
31. Victorov, A. S. (1998). *Matematicheskaya morfologiya landshafta (Mathematical morphology of the landscape)*. Moscow: Tratek 191 p.
32. Victorov, A. S. (2003). *An integrated mathematical model for diffuse exogenous geological processes*. Proceedings of the 9th annual conference of international association for mathematical geology. IAMG 2003 Portsmouth, UK, Sept 7–12.
33. Victorov, A. S. (2002). *Mathematical models of landscape patterns for geological interpretation of remote sensing data* (pp. 571–576). Proceedings of the 8th annual conference of international association for mathematical geology. IAMG 2002, Berlin, Germany.
34. Voskresenskiy, K. S. (1999). Sovremennyye rel'yefoobrazuyushchiye protsessy na ravninakh Severa Rossii, (Contemporary relief-forming processes on the plains of the North of Russia). Author's Doctor's summery.
35. Yelkin, V. A. (2004). *Regional'naya otsenka karstovoy opasnosti i riska (na primere Respubliki Tatarstan) (Regional assessment of karst danger and risk (on the example of the Republic of Tatarstan))*. Author's Candidate's summery. Moscow: IGE RAN. 24 p. 36.

Chapter 18
Stochastic Modeling of Natural Lacustrine Thermokarst Under Stable and Unstable Climate

A. S. Victorov, T. V. Orlov, V. N. Kapralova, O. N. Trapeznikova, S. A. Sadkov, and A. V. Zverev

Abstract Many researches are devoted to the problem of lacustrine-thermokarst plain development. The purpose of this study is a stochastic modeling of the thermokarst plains' pattern and development, primarily the morphological pattern of their landscapes. The bases of model creation are the approaches of mathematical morphology of landscape with wide use of random process theory. Studying territory type represents a slightly wavy subhorizontal surface with a predominance of different tundra or forest vegetation with the sprinkled lakes, without significant development of the erosion network. Several types of stochastic models were developed. These models were empirically verified at 16 sites all over cryolythozone. Theoretical statistical distributions of the lakes' number on the randomly chosen site within lacustrine-thermokarst plains correspond to the Poisson distribution on the plots relatively homogenous by the geomorphological, geocryological, and landscape conditions and under various assumptions about the nature of development of the lacustrine-thermokarst plains; this is largely confirmed by empirical data. Theoretical statistical distributions of the thermokarst lake sizes within the lacustrine thermokarst plains are different under various assumptions about the nature of development of the lacustrine-thermokarst plains on the plots relatively homogenous by the geomorphological, geocryological, and landscape conditions. The analysis of the empirical data shows that the observed distributions of the lakes' number at randomly chosen site correspond to the Poisson distribution and the distributions of the thermokarst lake sizes in the overwhelming majority correspond to the lognormal distribution and do not agree with other types of distributions being investigated. The analysis leads to a conclusion in favor of a model based on the proportionality of the growth rate of the thermokarst lake size to the average heat loss density through the side surface in case of the synchronous start of the processes of thermokarst lake appearance for every plot.

A. S. Victorov · T. V. Orlov (✉) · V. N. Kapralova · O. N. Trapeznikova
S. A. Sadkov · A. V. Zverev
Sergeev Institute of Environmental Geoscience RAS (IEG RAS), Moscow, Russia

V. Svalova (ed.), *Natural Hazards and Risk Research in Russia*,
Innovation and Discovery in Russian Science and Engineering,
https://doi.org/10.1007/978-3-319-91833-4_18

18.1 Introduction

Lacustrine thermokarst is widespread all over the Arctic and Subarctic. One can find vast plains (about 1.5–2 million km^2 of total area) [25], which topography is formed mainly under thermokarst processes, such as lacustrine thermokarst, thermokarst plains with fluvial erosion, etc. The thermokarst plains happened to become the territories with the most intense hydrocarbon extraction and pipeline infrastructure development.

Many researches are devoted to the problem of lacustrine-thermokarst plain development. The lacustrine-thermokarst plain dynamic is considered in the number of papers based on repeated remote sensing data [2, 3, 10, 18, 22–24, 27, 29, 30, 32]. Existing views of various researchers on the tendencies of lacustrine-thermokarst plain modern dynamic differ and are the subject of discussion. Some authors suppose that the sizeable shortening of the share of lakes occurs, and they associate it with climate change [5]. Another part of researchers consider that the changing of the share of lakes in different regions has different trends or does not have any one-way tendency [1, 14, 17, 18, 28, 37]. In some part of studies, the attempt of morphological pattern quantitative description of lacustrine-thermokarst plains on empirical base was made. Thus, in earlier publications of the authors of this paper, a variant of the model of morphological pattern development of lacustrine-thermokarst plain homogeneous plot in the case of constant climate and rounded lake is considered [12, 38]. Some researchers came to conclusion about lake areas' exponential distribution and uniform distribution of their centers [27]. The question of separate thermokarst lake development is associated with the problem under consideration. A significant number of research have been devoted to modeling of the thermokarst lake changes based on thermodynamic models [4, 9, 19, 34–36, 42].

The stationary observations of thermokarst lake dynamic were carried out in some research. Thus, the conclusion about uniform growth of the lake diameter was made during observation in one of the North America regions [4]. A number of studies detected the phenomenon of abrupt pulsation of the size of lake water surface, which was asynchronous for other lakes of studying plot [20]. Intra-soil channels' opening and locking conceivably explained this phenomenon. The question about thermokarst lake regime (generation) was discussed widely. A main part of researchers considered the appearance of thermokarst lakes for a relatively short time (in relation to the time of further development) (e.g., [35]). At the same time, another point of view on this problem was developed, based on the idea of a time-extended process of the lake appearance [33].

The purpose of this study is a stochastic modeling of the thermokarst plains' pattern and development, primarily the morphological pattern of their landscapes.

18.2 Data and Methods

The bases of model creation are the approaches of mathematical morphology of landscape [11, 39] with wide use of random process theory.

Mathematical morphology of landscape represents the new scientific direction devoted to the investigation of regularity of mosaics' pattern and development, which formed by natural units, and methods of their mathematical analysis. Thus, we are talking about the analysis of the landscape pattern – first of all, of the size and natural units' mutual location (lakes, marshes, moraine hills, etc.) on the Earth's surface. The theoretical core of this scientific direction is mathematical models of morphological patterns [40, 39]. Such an approach allows one to obtain process development regularities in landscape through analysis of its spatial pattern on the one hand and to verify theoretical constructions using the analysis of the Earth's remote sensing data on the other hand.

The following main scheme is used in the work at every model creating:

1. Formulation of the basic model assumptions.
2. Record the presented assumptions in the form of mathematical dependencies, based on the probabilistic approach.
3. Analytical transformations of obtained mathematical dependences and formulation of consequences.
4. Verification of the consequences of the model correctness in real territory.

   The wide range of remote sensing data was used for testing the model consequences:

   - 1960–1970 years – Corona (2–5 m/pix)
   - 1976–1999 years – Landsat and SPOT
   - 2000–2018 years – high-detailed survey QuickBird, GeoEye, WorldView, and ResourceP

18.3 Lacustrine-Thermokarst Model Set in Stationary State

Studying territory type represents a slightly wavy subhorizontal surface with a predominance of different tundra or forest vegetation (cotton grass tundra, sedge-cotton grass, etc.), with the sprinkled lakes, without significant development of the erosion network. Lakes are isometric, often round, sometimes angular or elongated, and randomly scattered across the plain. A typical image of such a plain is in Fig. 18.1. Such territory type is typical for a number of territories in Western and Eastern Siberia, Alaska, as well as for parts of the north of the East European Plain.

The different variants of the lake-thermokarst plain development hypotheses were considered in the study, based on the mathematical modeling of their morphological pattern.

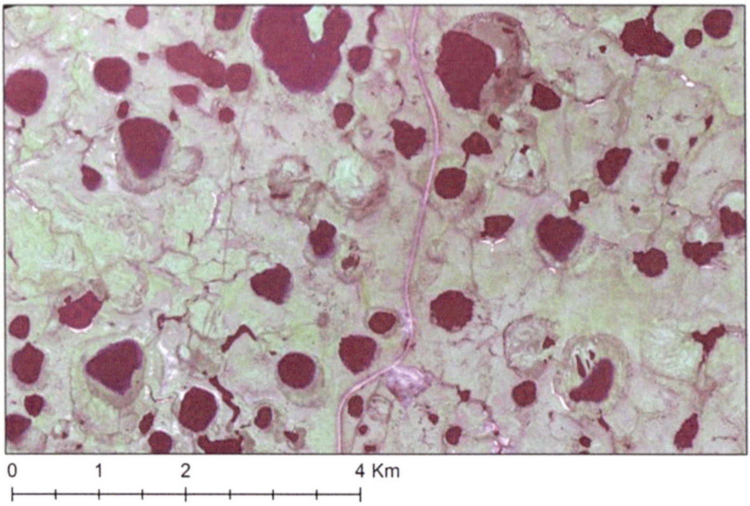

Fig. 18.1 Typical image of the lacustrine-thermokarst plain on the space image

Models of the morphological lake-thermokarst plain pattern belong to the case of the site with homogeneous physical-geographical and geocryological conditions developing under conditions of relative constancy of climatic circumstances. By homogeneity, we mean a combination of the following features: surface relief uniformity, which is manifested in a uniform image of the space survey materials, constant Earth surface and underlying sediments composing, uniform surface deposits thickness throughout all site, the absence of buried hollows, the absence of tectonic ruptures, location within the limits of one tectonic structure, and relatively homogeneous geocryological conditions. Naturally, the model does not assume absolute homogeneity, but only statistical homogeneity, allowing random oscillations. Thus, each model is not general and not covers a significant number of cases of thermokarst lake development, for example, the case of lakes located along a polygonal network dictated by re-veined ices. In such cases, the morphological pattern is different; in particular, the lakes are not characterized by a random arrangement but gravitate to the networks of lines (orthogonal or otherwise). In addition, the model does not cover the initiated thermokarst process.

In model creating, we adhere to the following concepts of thermokarst lake development, based on the number of geocryological research results (e.g., [26, 31]). Two unequal stages can be distinguished in the beginning of caused lake development.

At the first stage, it is depression caused by random factors triggering the thermokarst process – water accumulation in initial depression with thickness exceeding some critical value [26] of heat balance on account of snow accumulation, etc. Lake shape on this stage can be very diverse.

At the second stage, the lake already gets morphological features of the thermokarst lake, characteristic for the situation under investigation – it has circular or rounded outlines in the homogeneous environment considered by us [21]. At last, then the stages of lakes' merger and process attenuation come on.

The first stage is relatively short. The model belongs mainly to the second stage, which is the main one in duration, and the main characteristic features of morphology are manifested.

Let us consider the process of thermokarst form appearance and development on a homogeneous plot of the territory under conditions of relative constancy of climate. The developed complex of models includes the following elements:

- Mathematical model 1.0 of morphological pattern of lacustrine-thermokarst plains; "synchronous start" case
- Mathematical model 2.0 of morphological pattern of lacustrine-thermokarst plains; "synchronous start" case, quasi-uniform grows
- Mathematical model 1.1 of morphological pattern of lacustrine-thermokarst plains; "asynchronous start" case
- Mathematical model 2.1 of morphological pattern of lacustrine-thermokarst plains; "asynchronous start" case, quasi-uniform growth

Mathematical Model 1.0 of Morphological Pattern of Lacustrine-Thermokarst Plains; "Synchronous Start" Case

The model of base variant (model 1.0) is based on the following assumptions:

1. The process of thermokarst depression appearance took place for a short time ("synchronous start"); it was probabilistic and occurred independently on the two disjoint sites, while the probability of depressions occurrence on the trial plot depended only on its area; the probability of one depression occurrence was much greater than the probability of several depression occurrence for small plots.
2. Growth of the lake sizes took place independently one from another for the thermal abrasion,[1] and it was directly proportional to the heat reserves in the lake and inversely proportional to the area of the side surface of lake basin, flooded with water.

Thus, we consider the synchronous start variant, when the process of the appearance of initial lakes is relatively short as compared with their development period, and further the number of depressions remains constant. In this schematization, lake is considered as round-shaped object, as it is typical for many lakes.

The first assumption seems to be natural, as it follows from the homogeneity of the territory under consideration and reflects the fact that on any bounded plot, only a finite number of thermokarst depressions (more precisely, their centers) are observed.

[1]A low intensity of thermal abrasion is also possible while the process approaches "purely thermokarst."

The second assumption seems to be correct too. It reduces to the proportionality of the growth rate of the thermokarst center (lake) size to the average density of heat losses through the side surface of the lake, flooded with water. This is natural, because it is exactly these heat losses that are expended on the permafrost degradation in the sides of the lake, while the growth is influenced by a lot of random factors (average annual air temperature, iciness of rocks in the vicinity of the lake, etc.).

Such assumptions are simplified and do not reflect the whole variety of natural interrelationships; however, the flexible character of the assumptions, based on probabilistic representations, gives them a certain "margin of safety."

The created model bases allow us to obtain analytically the regularities of the thermokarst plain pattern. As it can be shown (e.g., [13]), it follows from the validity of the first assumption that the distribution of the number of thermokarst depressions (centers) on a randomly chosen plot obeys the Poisson law [39, 41], that is,

$$P(k,s) = \frac{(\gamma s)^k}{k!} \mathrm{e}^{-\gamma s},$$

where

γ – the average number of depressions per unit area
s – area of the test site

Let us consider the changing of the thermokarst depression sizes, which we schematically represent as cylinders. If V is the water value of the lake basin, which is the complex result of water concentration from catchment area, evaporation and flow, and other processes of water balance formation, then the area of the lateral surface flooded with water will be at the beginning of the j-th year

$$S = \frac{V}{\pi u_j^2} 2\pi u_j$$

or

$$S = \frac{2V}{u_j}, \tag{18.1}$$

where u_j is the lake radius at the beginning of the j-th year. Accordingly, the next expression implies from the second assumption of the model

$$\Delta u_j = \frac{\alpha c t^{\mathrm{o}} V}{S} \xi^0{}_j,$$

where Δu_j is the size increment for j-th year, c specific heat, t^{o} average temperature of water, α share of the heat amount in water mass, leaving through the side surface, and $\xi^0{}_j$ random variable, taking into account the impact of episodic factors. This expression after substituting the area from (18.1) and simplifying becomes

$$\Delta u_j = \xi_j u_j, \quad \text{where} \quad \xi_j = \frac{\alpha c t^0}{2} \xi^0{}_j. \tag{18.2}$$

The coefficient of proportionality is a random variable that takes into account the course of the degradation process in a particular year, due to summer and winter temperatures, the thickness of the snow cover, the volume of storm runoff, soil temperatures, the nature of precipitation and their quantity, etc. These coefficients are independent for different years, but they are equally distributed; denote their mathematical expectation and dispersion as follows:

$$M\xi_j = a, \quad D\xi_j = \sigma^2.$$

Transferring variables containing the depression size to the left part of expression (18.2) and summing and substituting the left sum by the integral

$$\int\limits_{x(q)}^{x(W)} \frac{du}{u} = \sum_{i=q}^{W} \xi_i,$$

we get

$$\ln x(w) - \ln x(q) = \sum_{i=q}^{w} \xi_i,$$

where $x(w)$ and $x(q)$ are the depression size in the time w and q accordingly.

As the sum of larger number of independent items (equally distributed in our case) is normally distributed according to the central theorem [15], then it follows that the growth process of thermokarst form radiuses, providing the significant $(w - q)$, can be considered as a Markov random process with continuous time with a transition function:

$$f(v, x, t) = \frac{1}{\sqrt{2\pi}\sigma x\sqrt{t}} e^{-\frac{\left(\ln\frac{x}{v} - at\right)^2}{2\sigma^2 t}},$$

where

a, σ – distribution parameters
v – initial size of thermokarst depression
x – depression size in the time t

If we assume, for the model simplicity, that initial thermokarst depressions have a unit radius at the beginning time (this corresponds to the radius normalization at the minimum value), then it follows that we should observe a lognormal distribution of the thermokarst form radius at any time, i.e.,

$$f_r^0(x,t) = \frac{1}{\sqrt{2\pi}\sigma x\sqrt{t}} e^{-\frac{(\ln x - at)^2}{2\sigma^2 t}}$$

where a, σ are distribution parameters and t is the time since the process beginning. Taking into account the quadratic relationship between area and radius, the area must also obey the lognormal distribution law. Expression (18.2), obtained from the model assumptions, is empirically confirmed in studies, devoted to the mechanism of development of Alaskan thermokarst lakes. A direct correlation was found between the value of changes in the linear lake sizes and the rate of their change in [3].

It is possible to show analytically that the circular lake shape in plan is not essential for the given constructions, if we assume that the geometric similarity of the outlines is persisted in the process of growth.

In this case, we consider the lake as cylinder, not with a circular directrix, but with a directrix of arbitrary shape. The question is also important, what is the influence of the ideas about activity degree of the thermoabrasion (or only thermokarst) process on the proposed model. Thus, the lakes are able to suddenly reduce the water surface area. This occurs independently even in nearby lakes, and one of the reasons seems to be the disappearance and subsequent recovery of ice blocks in the ground channels [20]. In these cases, the water surface area ceases to reach the side of the thermokarst depression, and there is no thermal abrasion of this side. Obviously, the growth rate decreases sharply at such times, and we can say that the lake practically ceases to grow. In another cases, the reason for a sharp growth slowdown is the fluctuation of the lake's water balance.

Analysis shows that an interruption in the lake growth does not affect the lognormal distribution. It can be considered as a feature of the distribution of the proportionality coefficients ξ_i in expression (18.2); this is because they can take a value of zero with some non-zero probability. Since we did not impose constrains on the distribution of these coefficients, and their independence was preserved, this means that the whole course of analysis remains valid.

Mathematical Model 2.0 of Morphological Pattern of Lacustrine-Thermokarst Plains; "Synchronous Start" Case, Quasi-uniform Growth

As a second variant (variant 2.0) of the model, let us consider a case based on the observations of a number of researchers on the uniform lake growth [4]. Since random fluctuations in the growth rate can always be embedded into this uniformity, the growth is more correctly called quasi-uniform, and the probabilistic analogue of this situation will differ by a second assumption having a somewhat different formulation:

2a. Lake size growth occurs independently of each other, due to the thermal abrasion, and has constant probability distribution.

In this case, the expression (18.2) is replaced by another

$$\Delta u_j = \xi^0{}_j,$$

where Δu_j is the size increment for j-th year and $\xi^0{}_j$ random variable, taking into account the impact of episodic factors (rainfall, precipitation regime, etc.), but in general the entire course of calculations remains valid. The corresponding analysis, which is exactly analogous to the previous one, leads to the following result: the process of growth of the thermokarst form radiuses can approximately be considered as a Markov random process for considerable t with uninterrupted time with a transition function:

$$f(v, x, t) = \frac{1}{\sqrt{2\pi}\sigma\sqrt{t}} \mathrm{e}^{-\frac{[(x-v)-at]^2}{2\sigma^2 t}},$$

where

a, σ – distribution parameters
v – initial size of thermokarst depression
x – depression size in time t

Following further and taking into account that the sum of normally distributed independent increments of radiuses is also normally distributed, we find that we should observe a normal distribution of the thermokarst form radius at any instant of time, that is,

$$F_r^0(x, t) = \Phi\left(\frac{x - \mathrm{at}}{\sigma\sqrt{t}}\right),$$

respectively; the diameter will also be distributed, according to Gauss's law.

Mathematical Model 1.1 of Morphological Pattern of Lacustrine-Thermokarst Plains; "Asynchronous Start" Case

Let us consider a slightly different situation – the situation of an asynchronous start. Let generation of new thermokarst lakes continue throughout all considered time $(0, t)$ and the generation intensity given by a function $\lambda(u)$ determine the average lake number appearing on a unit area at the instant of time u. In this case, model (model 1.0) is slightly modified, and its first assumption is replaced:

1a. Process of thermokarst depression appearance occurs constantly ("asynchronous start"); it is probabilistic and goes independently on disjoint sites and time intervals. In this case, the probability of depression appearances on the test plot for a time interval depends only on the area of the plot (Δs) and the interval value; the probability of one depression appearance for small plots and small time intervals is greater than the probability of several ones, that is,

$$p_1 = \lambda(t)\Delta s\Delta t + o(\Delta s\Delta t)$$
$$p_k = o(\Delta s\Delta t) \quad k = 2, 3\ldots,$$

where $\lambda(u)$ is a function that determines the average number of lakes appearing on a unit area at an instant of time u(lake generation density). Under this assumption, the distribution of the thermokarst lake areas (radiuses) at the instant of time t after process starting can be obtained by integrating taking into account the probability of thermokarst lake appearance at one time or another in the plot under consideration:

$$f(x,t) = \frac{\int_0^t f_0(x, t-u)\lambda(u)du}{\int_0^t \lambda(u)du}, \tag{18.3}$$

where $f_0(x, u)$ is the lake area (radius) distribution in time u after its appearance.

Depending on the function of the lake generation density of distribution, it is possible to obtain obviously different results of determining of the thermokarst lake areas' (radiuses) distribution. Let us consider the simplest case of a constant uniform generation of lakes, that is,

$$\lambda(u) = \lambda_0.$$

In this case, the base model 1.0 is transformed into model 1.1 with asynchronous start (assumption 1a), and the density distribution of the radiuses is given by the expression after simplification, as shown by the use of the expression (18.3):

$$f_r(x,t) = \frac{1}{t}\int_0^t \frac{1}{\sqrt{2\pi}\sigma x\sqrt{u}} e^{-\frac{(\ln x - au)^2}{2\sigma^2 u}} du.$$

For the big t ($t \to +\infty$), the last integral can be calculated, as it reduces to the Laplace transformation of the function $\frac{1}{\sqrt{u}} e^{-\frac{\ln x}{2\sigma^2 u}}$ with an argument equal to $\frac{a^2}{2\sigma^2}$. Using the above transformation, we obtain for the distribution density:

$$f_r(x,t) \approx \frac{1}{atx} \text{при } x \geq 1.$$

Thus, a good approximation of the distribution itself may be for normalized radiuses (and also areas) in model 1.1:

$$F_r(x) \approx \frac{\ln x - \ln a_0}{\ln b_0 - \ln a_0}, a_0 \leq x \leq b_0,$$

where a_0, b_0 are distribution parameters.

The distribution is called "logarithmic" for brevity in the framework of this paper.[2]

Mathematical Model 2.1 of Morphological Pattern of Lacustrine-Thermokarst Plains; "Asynchronous Start" Case, Quasi-uniform Growth

Similarly, the case of an asynchronous start in model 2.0 can be considered, while it is transformed into a model that can be called 2.1. The distribution of the thermokarst lake radiuses can be found in the way indicated above in this model, calculating the expression

$$f_r(x,t) = \frac{1}{t}\int_0^t \frac{1}{\sqrt{2\pi}\sigma\sqrt{u}} e^{-\frac{(x-au)^2}{2\sigma^2 u}} du.$$

Using the approximation described above, it is easy to see that the distribution of the thermokarst lake radiuses (and also the diameters) for a long time in model 2.1 should approximately correspond to a uniform distribution:

$$F_r(x) \approx \frac{x-a}{b-a}, a \leq x \leq b.$$

Thus, the results show:

- All four qualitatively different hypotheses for the development of the morphological pattern of the lacustrine-thermokarst plains (Table 18.1) lead to four very different conclusions about the shape of the distribution of thermokarst lakes radiuses (areas, diameters).
- In this case, the conclusion about the Poisson distribution of the number of thermokarst lake centers remains constant for a randomly chosen plot.

The main properties and consequences of the developed models are shown in Table 18.1.

[2]The name is purely conditional, since in general a logarithmic distribution is a distribution different from a given one.

Table 18.1 Model of the distribution of the thermokarst lake parameter under different lacustrine-thermokarst evolution scenarios

Thermokarst plain evolution scenario	Synchronous start	Asynchronous start
Lognormal growth of the lake radii	Model type 1.0	Model type 1.1
	Lognormal distribution of the lake areas (radii)	"Logarithmic" distribution of the lake areas (radii)
	$f(x) = \frac{1}{\sqrt{2\pi}\sigma x} e^{-\frac{(\ln x - a)^2}{2\sigma^2}}$	$F(x) \approx \frac{\ln x - \ln a}{\ln b - \ln a}, a \le x \le b$
Quasi-uniform growth of the lake radii	Model type 2.0	Model type 2.1
	Normal distribution of the lake radii	Uniform distribution of the lake radii
	$f_r(x) = \frac{1}{\sqrt{2\pi}\sigma} e^{-\frac{[x - a]^2}{2\sigma^2}}$	$F_r(x) \approx \frac{x - a}{b - a}, a \le x \le b$

Where α, σ are distribution parameters

18.4 Models of Morphological Pattern Development of Lacustrine-Thermokarst Plains Under the Climate Change

Mathematical Model 3.0 of Morphological Pattern Development of Lacustrine-Thermokarst Plains Under the Climate Change; "Synchronous Start" Case

The question of model for development of lacustrine-thermokarst plain morphological pattern in climate change conditions is worthy of special notice. The real situation assumes the presence of slow climatic trends, and it seems interesting to model the development of the morphological pattern of the lacustrine-thermokarst plains under these conditions.

Model, as a base model, belongs to the case of a plot with homogeneous physics-geographical and geocryological conditions.The next main assumptions lie into the basis of model (model 3.0):

1. The process of thermokarst depression appearance is probabilistic and goes independently on the disjoint sites, while the probability of depression appearance on the trial plot depends only on its area (Δs); the probability of one depression appearance for the small plots is greater than the probability of the appearance of some depressions, i.e.,

$$p_1 = \gamma \Delta s + o(\Delta s)$$
$$p_k = o(\Delta s) \quad k = 2, 3 \ldots,$$

where γ is the average number of depressions per unit area.

2. The growth of lake sizes occurs independently of each other due to thermal abrasion effect; it is directly proportional to the heat reserves in the lake and inversely proportional to the area of the lateral surface of the lake basin.

Thus, we consider the synchronous start variant, when the process of appearance of initial lakes is relatively short in comparison with its development period, and in the future, the number of depressions remains constant. The lake is considered as the cylindrical object in this schematization.

The first assumption follows from the homogeneity of the territory under consideration. The second assumption is reduced to proportionality of the growth of the thermokarst center (lake) size to the average heat flux density through the side surface.

It is important the model does not assume that the climatic characteristics are constant over time, but climate can change; obviously, this should influence the thermokarst lake growth.

Model allows us to determine analytically the regularities of thermokarst plain structure. As it was already shown in the previous part, it follows from the assumptions underlying the model that the distribution of the number (k) of thermokarst depressions (centers) on a randomly chosen plot obeys the Poisson law [38, 41], i.e.,

$$P(k,s) = \frac{(\gamma s)^k}{k!}\mathrm{e}^{-\gamma s}$$

where

γ – the average number of depressions per unit area
s – test's plot area

Let us consider the change of thermokarst depression sizes. We divide the entire time interval t into n intervals with length t_i, $i = 1, \ldots, n$, where the climatic situation can be regarded as approximately constant; consider i-th interval $\left(\sum_{k=1}^{i-1} t_k, \sum_{k=1}^{i} t_k\right)$. Let's analyze the size development of one thermokarst depression within i-th time interval.

The second thesis, underlying the model, causes the proportionality of the random change of the depression radius for the j-th year, belonging to the considered time interval, the already existing radius.

It follows, from the results of model 1.0, that the process of the thermokarst form growth can be approximately considered in the every time interval at sizable t as a Markov random process with continuous time with a transition function:

$$f(v,x,t) = \frac{1}{\sqrt{2\pi}\sigma_i x\sqrt{t}}\mathrm{e}^{-\frac{\left(\ln\frac{x}{v} - a_i t\right)^2}{2\sigma_i^2 t}}, t = r - s, \sum_{k=1}^{i-1} t_k \le s < r \le \sum_{k=1}^{i} t_k$$

where

a_i, σ_i – distribution parameters (in given i-th time interval)
v – initial size of thermokarst depression
x – depression size in time t (in given i-th time interval)

In other words, if the considering thermokarst depression radius is equal to $\zeta_{i\ -\ 1}$ at the beginning of the analyzed time interval, then the next expression is rightly

$$\zeta_i = \zeta_{i-1}\eta_i,$$

where η_i is the random value, which is equal to the ratio of the thermokarst form radius (ζ_i) at the end of time interval to the radius ($\zeta_{i\ -\ 1}$) at the beginning of the considered (j-th) time interval. This value, as it follows from the previous expression, has a lognormal distribution with parameters

$$M[\ln \eta_i] = a_i t_i, D[\ln \eta_i] = \sigma_i^2 t_i, i = 1, \ldots, n-1. \tag{18.4}$$

From the above, conditionally taking the initial size of the thermokarst form for unity, it is not difficult to obtain the ratio

$$\ln \zeta(t) = \ln \eta_1 + \ln \eta_2 + \ldots + \ln \eta_{n-1} + \ln \eta_n, \sum_{i=1}^{n-1} t_i \le t \le \sum_{i=1}^{n} t_i,$$

where$\zeta(t)$ is the thermokarst form size in the instant of time t, η_i ($i = 1, \ldots, n-1$) lognormally distributed independent random variables with the parameters indicated above (18.4), and η_n lognormally distributed independent random variables with the parameters

$$M[\ln \eta_n] = a_n\left(t - \sum_{i=1}^{n-1} t_i\right), D[\ln \eta_n] = \sigma_n^2\left(t - \sum_{i=1}^{n-1} t_i\right).$$

It means, taking into account the summation of independent normally distributed values, that the logarithm of the thermokarst form size at the end of the time interval (0, t) is a normally distributed random value with mathematical expectation:

$$M[\ln \zeta(t)] = \sum_{k=1}^{n-1} a_k t_k + a_n\left(t - \sum_{k=1}^{n-1} t_k\right)$$

and dispersion:

$$D[\ln \zeta(t)] = \sum_{k=1}^{n-1} \sigma_k^2 t_k + \sigma_n^2\left(t - \sum_{k=1}^{n-1} t_k\right).$$

Thus, even under climate change, the distribution of thermokarst form radiuses, and therefore areas, obeys the lognormal law, not with linear growth of parameters, but with piece-linear ones.

It follows that we obtain a lognormal distribution of thermokarst form radius (and also diameter and area) for the every instant of time, i.e.,

$$F_r(x,t) = \Phi\left(\frac{\ln x - a(t)}{\sigma(t)}\right), \text{ ГДе } a(t) = \sum_{k=1}^{n-1} a_k t_k + a_n\left(t - \sum_{k=1}^{n-1} t_k\right),$$
$$\sigma(t) = \sqrt{\sum_{k=1}^{n-1} \sigma_k^2 t_k + \sigma_n^2\left(t - \sum_{k=1}^{n-1} t_k\right)}$$

or for density of distribution

$$f_r(x,t) = \frac{1}{\sqrt{2\pi}\sigma x} e^{-\frac{(\ln x - a)^2}{2\sigma^2}}$$

where

$F_r(x,t)$ – radius distribution
$\Phi(t)$ – Laplace functions
a,σ – distribution parameters, depended from time
t – time from process beginning

It is not difficult to show that the conclusion is preserved, but the expressions change somewhat at gradual changing of climatic parameters of model ($a_0(t)$, $\sigma_0(t)$) for the parameters:

$$a(t) = \int_0^t a_0(u)du, \sigma^2(t) = \int_0^t \sigma_0^2(u)du.$$

For the above constructions, as well as in part 1.1, the circular shape of the lakes is not essential, if we assume that the geometric similarity of the outlines remains in the process of growth. In this case, the lake acts as a cylinder, but not with a circular directrix, but with a directrix of arbitrary shape.

The question, what is the influence of the degree of thermal abrasion process activity on the proposed model, is also important. In other words, what if the activity of the processes has sharply decreased at some time period and the growth of the planned outline of the lakes of the area under consideration is temporarily stopped. The situation described above is also close to the situation, where the growth of lake areas is practically stopped due to climatic changes, and then it continues again, as soon as conditions favorable for thermoabrasion are reached.

The described model can be used for answering these questions. Let us consider the situation of the developing resumption after stopping.

Select the time interval of stopping as t_i. Obviously, because of stopping, the relationships are rightly for the interval

$$a_i = 0, \quad \sigma_i^2 = 0$$

and, accordingly, we have just a special case of the model, namely, for this i:

$$\zeta_i = \zeta_{i-1}.$$

Consequently, the situation is analogous to that of simply shifting the numbers of intervals greater than i by one, and hence, after the time interval under consideration, the lognormality is preserved. The fact that the sizes of lakes water surface can unexpectedly decrease and the situation of the lakes growth stopping does not affect the results of the analysis; it is also true for the models relating to the case of relatively constant climate. The interruption in the lake growth does not affect the lognormal distribution.

The empirical data represented in the previous part show that the presented model is supported by empirical material.

Mathematical Model 3.1 of Morphological Pattern Development of Lacustrine-Thermokarst Plains Under the Climate Change; "Synchronous Start" Case, Quasi-uniform Growth

An alternative was also studied, for completeness of the research, based on the assumption of a quasi-uniform lake growth (model 3.1). At the same time, climatic changes were also assumed.

The model belongs to the case of plot with homogeneous physics-geographical and geocryological conditions, as a base model.

The second assumption was replaced in base of model (model 3.1):

2a. The growth of lake sizes occurs independently of each other due to thermal abrasion effect and has a constant probability distribution.

Thus, the synchronous start variant is also considered when the process of the initial lake appearance is relatively short in comparison with the period of their development, and in future the number of depressions remains constant. The lake is considered as the cylindrical object in this schematization.

There is no assumption that the climatic characteristics are constant over time, but climate is allowed to change in the model, as well as in the model 3.0.

As it was shown in previous part, it follows from the assumptions underlying the model that the distribution of the number (k) of thermokarst depressions (centers) on a randomly chosen plot obeys the Poisson law.

Let us consider the changing of thermokarst depression sizes; the analysis is largely analogous to the analysis carried out for model 3.0. We divide the entire time interval t into n intervals of length t_i, $i = 1, \ldots, n$ where the climate can be considered as approximately constant; consider the i-th interval $\left(\sum_{k=1}^{i-1} t_k, \sum_{k=1}^{i} t_k\right)$

It follows form the results of model 1.0 that the process of thermokarst form radiuses growth can be approximately regarded at every time interval with considerable t as Markov random process with a continuous time with a transition function:

$$f(v,x,t) = \frac{1}{\sqrt{2\pi}\sigma_i\sqrt{t}} e^{-\frac{((x-v)-a_i t)^2}{2\sigma_i^2 t}}, t = r - s, \quad \sum_{k=1}^{i-1} t_k \le s < r \le \sum_{k=1}^{i} t_k$$

where a_i, σ_i – distribution parameters (in given i-th time interval), v – initial size of thermokarst depression, x – depression size in time t (in given i-th time interval)

In other words, if the considering thermokarst depression radius is equal to ζ_{i-1} at the beginning of the analyzed time interval, then the next expression is rightly

$$\zeta_i = \zeta_{i-1} + \eta_i,$$

where η_i is the random value, which is equal to the difference of the thermokarst form radius (ζ_i) at the end of time interval and the radius (ζ_{i-1}) at the beginning of the considered (i-th) time interval. This value, as it follows from the previous expression, has a normal distribution with parameters

$$M[\eta_i] = a_i t_i, D[\eta_i] = \sigma_i^2 t_i, i = 1, \ldots, n-1. \tag{18.5}$$

From the above, conditionally taking the initial thermokarst form size as zero, it is not difficult to obtain the relation:

$$\zeta(t) = \eta_1 + \eta_2 + \ldots + \eta_{n-1} + \eta_n, \sum_{i=1}^{n-1} t_i \le t \le \sum_{i=1}^{n} t_i,$$

where

$\zeta(t)$ – thermokarst form size at the time t

η_i ($i = 1, \ldots, n-1$) – normal distributed independent random values with the above (18.5) parameters

η_n – normal distributed independent random values with parameters

$$M[\eta_n] = a_n\left(t - \sum_{i=1}^{n-1} t_i\right), D[\eta_n] = \sigma_n^2\left(t - \sum_{i=1}^{n-1} t_i\right).$$

Therefore, taking into account the summation of independent normally distributed values, it follows that the size of the thermokarst form at the end of the time interval (0, t) is a normally distributed random variable. Thus, the distribution of thermokarst form radiuses obeys the normal law, even under climatic changing, but not linear, but with piece-linear parameter growth.

It follows that we get normal distribution of thermokarst form radiuses for the every instant of time, i.e.,

$$F_r(x,t) = \Phi\left(\frac{\ln x - a}{\sigma}\right), \text{ ГДе } a(t) = \sum_{k=1}^{n-1} a_k t_k + a_n\left(t - \sum_{k=1}^{n-1} t_k\right),$$

$$\sigma(t) = \sqrt{\sum_{k=1}^{n-1} \sigma_k^2 t_k + \sigma_n^2\left(t - \sum_{k=1}^{n-1} t_k\right)}$$

where

$F_r(x,t)$ – radius distribution
$\Phi(t)$ – Laplace function
a,σ – distribution parameters, depended from the time
t – time from the process beginning

where $F_r(x,t)$ – radius distribution, $\Phi(t)$ – Laplace function, a, σ – distribution parameters, depended from the time, t – time from the process beginning

It is not difficult to show that the conclusion is preserved on condition of gradual changing of climatic parameters and, accordantly, of the gradual changing of model parameters ($a_0(t)$, $\sigma_0(t)$), but the expressions change somewhat for the parameters:

$$a(t) = \int_0^t a_0(u)du, \sigma^2(t) = \int_0^t \sigma_0^2(u)du.$$

The fact that the sizes of lakes water surface can unexpectedly decrease and the situation of the lakes growth stopping does not affect the results of the analysis; it is also true for the models relating to the case of relatively constant climate. The interruption in the lake growth does not affect the normal distribution.

18.5 Empirical Verification

An empirical research was made of the real distribution of thermokarst lake parameters within the lacustrine-thermokarst plains to analyze this theoretical consideration.

There were selected 16 plots in different regions for the research (Fig. 18.2).

We used remote sensing data with resolution on area 5–30 m as starting.

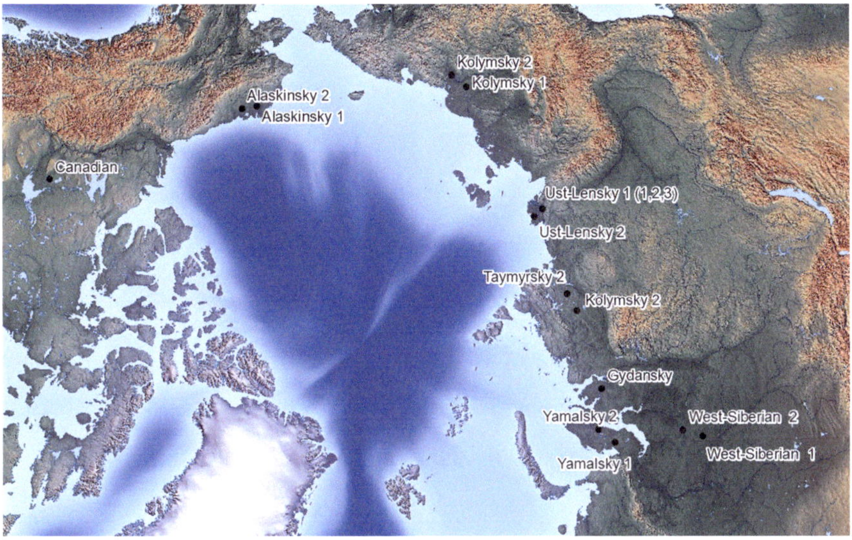

Fig. 18.2 Overview map of key plots. Background map http://www.naturalearthdata.com

The treatment methods included:

- Key plot selection
- Thermokarst lake separation
- "Throwing" randomly the testing sites within every key plot and counting the number of thermokarst lake centers on testing sites
- Determination of lake areas and verification of accordance of lake area distribution within every key plot to (a) lognormal distribution and (b) "logarithmic" distribution
- Counting of the middle lake diameters and verification of accordance of middle diameters distribution within every key plot to (a) normal distribution and (b) uniform distribution

Plots selection was visual by remote sensing data, literary [6–8], and cartographical materials. The main requirement was genetic, landscape, and morphological homogeneity, achieved by analyzing images and additional sources of information. Thus, territories with another origin of lakes, usually expressed in their morphology – the territories of alluvial plains, the territories where the morphology of lakes was subordinate to the grid of polygonal-vein ices, etc. – were excluded from the plots. The model does not include the case of substantial development of drained thermokarst lakes (so-called a khasyreis), another model deals with this process. Thus, we excluded those plots, where lake draining with the subsequent forming of khasyreis was essential. If the plot had single residual lakes, then such plots are not included in the sampling.

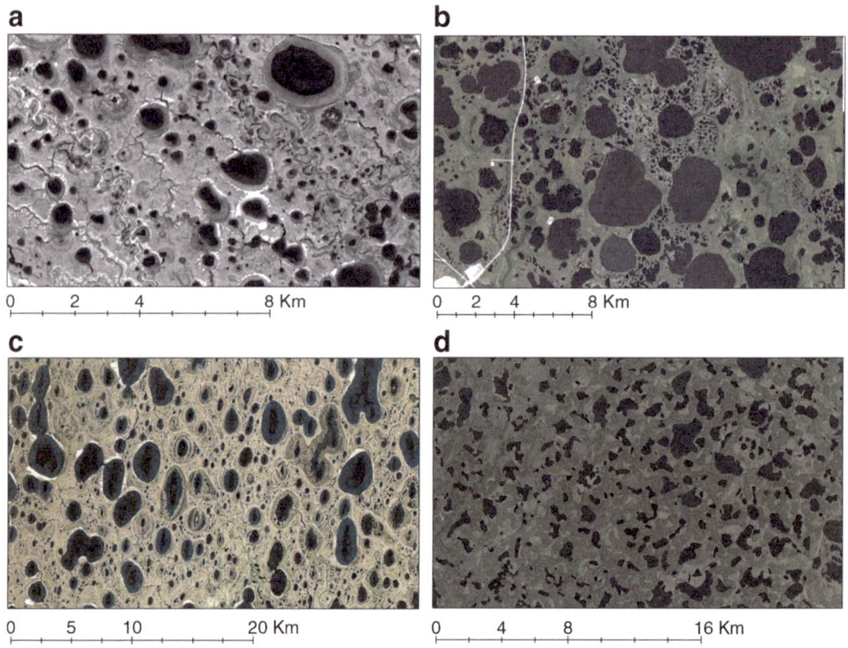

Fig. 18.3 Fragments of the key plot images of the different types of lacustrine-thermokarst plains by the remote sensing data. (**a**) Yamal. (**b**) West Siberian. (**c**) Cat's eye. (**d**) Meandering

As a result, chosen plots are diverse in geomorphological, geocryological, and physics-geographical terms. So, different plots are located within the watershed plains, river, or sea terraces. The plots belong both to the area of continuous permafrost development and to the zones of intermittent and island permafrost development.

The space images of key plots are shown in Fig. 18.3. As can be seen, the plots can be conditionally divided into some morphological types.

- Yamal. Lakes are characterized by a simple rounded shape in key areas of the Yamal Peninsula and the Canadian plot. They are characterized by a large spreading of inter-lake channels, which, however, do not lead to the lake draining with subsequent formation ones.
- West Siberian. There are lakes of complex shape in these plots, besides rounded lakes, as well as lakes, which are the result of merger of two or more lakes during their growth. In particular, such lakes are typical for researching key plots, located on the West Siberian Plain, in the area of the rivers Piakupur, Valoktayagun, and Vynghapur.
- Meandering. There are lakes of irregular, meandering form on these plots, which can be explained by the primary relief warps. Such lakes are typical for plots of Taimyr and Kolyma.

- Oval ("Cat's eye"). Lakes of these plots are characterized by oval shape, as usual, of general orientation. The deeper core is observed in the centers of lakes, which has elongated shape too; often (but not always) it has the same orientation, differing by darker-colored water. They are situated at the Alaska and Lena's estuary.

The determination of thermokarst lakes was carried out both by an automated method and by the involvement of experts. In any case, an additional analysis was made to avoid interpretation errors after the determination, for example, sections of watercourses flowing into the lake were separated, which often joined the lakes during automated contouring of the latter. Several accesses were analyzed for every plot, each one obtained from the data of a separate survey period; this was done to exclude the effect of interyearly variability.

"Throwing" of the testing sites was done using a random number generator (100 throws). Sites had a round shape.

The number of thermokarst lake centers that got into the site was determined for each site using a software package. The procedure was done for the sites of four different sizes. The base size (R) was determined in a way that the site size was equal to the share of the total area, which suited in average on one lake in the plot under consideration. The other three sizes for testing sites were determined as 1, 5R, 2R, 3R. Thus, four accesses of lake center number were formed on the randomly chosen site, each for sites of its own size.

The correspondence between the empirically obtained distributions was verified theoretically by means of a software package for statistical analysis using the Pearson criteria (chi-square), based on a known methods, under the conditions of its application (e.g., [16]). Preliminary, free distribution parameters were determined by the access. Then the criteria value was determined, and obtained value was compared with critical value at a significance level of 0.99. The critical value was determined by the number of degrees of freedom, which depended on the number intervals and on the number of free parameters. The access was divided on these intervals by program during the determination of criteria value.

The access of the average diameter of the contours was formed from the access of areas using the relationship

$$d = 2\sqrt{\frac{s}{\pi}},$$

where s is the lake area.

The analysis showed a significant correspondence between the distributions of the number of thermokarst lake centers at a randomly chosen site to the Poisson distribution. The examples of correspondence of distributions are shown in Fig. 18.4. At the same time, the accordance is not observed in a number of cases, at certain sizes of the site. This allows us to talk about the implicit possible heterogeneity of the reference plots that appear in some scales of consideration.

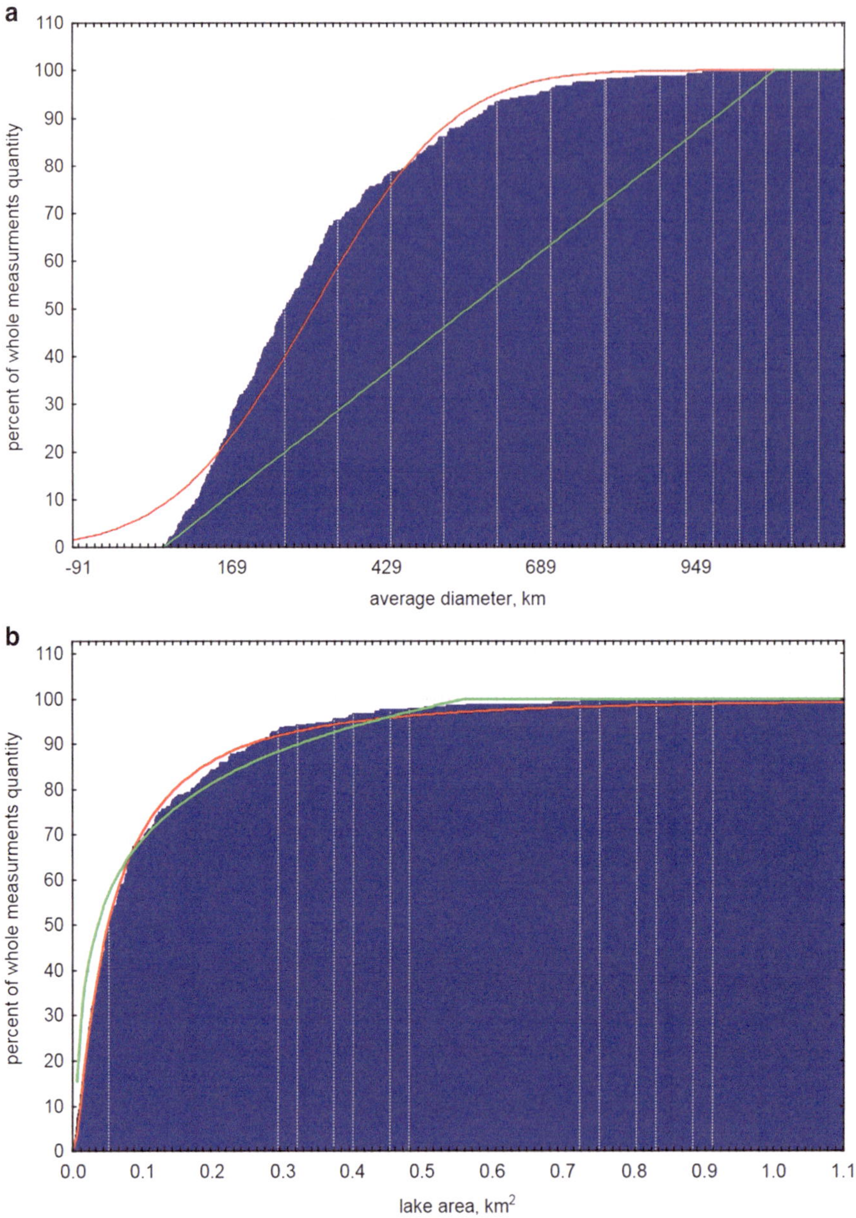

Fig. 18.4 The example of the correspondence of empirical distributions to the different types of theoretical distributions (plot Taymyrsky 1): (**a**) average lake diameter (red line, normal distribution; green line, uniform distribution; boundary of the colored area, empirical distribution function), (**b**) area of lakes (red line, lognormal distribution; green line, "logarithmic" distribution; boundary of the colored area, empirical distribution function)

Table 18.2 The results of comparison of empirical and theoretical distributions of the thermokarst lake sizes

		Area		Average diameter	
Plot name	Access volume	Lognormal	Exponential	Normal	Uniform
Alaskinsky 1	100	0.02284	0.01698	0.00000	0.00000
Alaskinsky 2	108	0.11164	0.00021	0.00000	0.00000
Canadian	154	0.12693	0.00002	0.00000	0.00000
Gydansky	74	0.51741	0.00478	0.00000	0.00000
Kolymsky 1	154	0.21566	0.15734	0.00017	0.00000
Kolymsky 2	576	0.00007	0.00000	0.00000	0.00000
Taymyrsky 1	345	0.11210	0.00088	0.00000	0.00000
Taymyrsky 2	209	0.63060	0.77808	0.00003	0.00000
Ust-Lensky 1–1	145	0.01123	0.00004	0.00000	0.00000
Ust-Lensky 1–2	91	0.15471	0.31057	0.00054	0.00000
Ust-Lensky 1–3	383	0.00001	0.00000	0.00000	0.00000
Ust-Lensky 2	167	0.00638	0.00000	0.00000	0.00000
Yamalsky 1	103	0.30411	0.38489	0.00459	0.00000
Yamalsky 2	176	0.01199	0.00000	0.00000	0.00000
West Siberian 2	84	0.08807	0.00224	0.00000	0.00000
West Siberian 1	78	0.58722	0.12276	0.00000	0.00000

The analysis of the distribution of thermokarst lake sizes included accesses of volumes from 74 to 576 lakes per one reference plot for one survey period. The analysis of the results shows that the overwhelming majority of plots obtained agreement with the lognormal distribution law (14 out of 17 at the significance level of 0.99 Table 18.2). Figure 18.4a, b graphically shows an example of the correspondence of empirical distributions to different types of theoretical distributions for two plots. The general data by all plots by coherence of empirical and theoretical distributions are given in Table 18.2.

18.6 Conclusions

Thus, we can draw the following conclusions:

1. Theoretical statistical distributions of the lake number on the randomly chosen site within lacustrine-thermokarst plains correspond to the Poisson distribution on the plots relatively homogenous by the geomorphological, geocryological, and landscape conditions and under various assumptions about the nature of development of the lacustrine-thermokarst plains; this is largely confirmed by empirical data.
2. Theoretical statistical distributions of the thermokarst lake sizes within the lacustrine-thermokarst plains are different under various assumptions about the

nature of development of the lacustrine-thermokarst plains on the plots relatively homogenous by the geomorphological, geocryological, and landscape conditions.
3. The analysis of the empirical data shows that the observed distributions of the lake number at randomly chosen site correspond to the Poisson distribution and the distributions of the thermokarst lake sizes in the overwhelming majority correspond to the lognormal distribution and do not agree with other types of distributions being investigated.
4. The analysis leads to a conclusion in favor of a model based on the proportionality of the growth rate of the thermokarst lake size to the average heat loss density through the side surface in case of the synchronous start of the processes of thermokarst lake appearance for every plot.

Acknowledgments The research was supported by RFBR grant (project No. № 18-05-00723).

References

1. Andresen, C., & Lougheed, V. (2015). Disappearing Arctic tundra ponds: Fine-scale analysis of surface hydrology in drained thaw lake basins over a 65 year period (1948–2013). *Journal of Geophysical Research: Biogeosciences, 120*, 466–479. https://doi.org/10.1002/2014JG002778.
2. Arp, C. D., Jones, B., Bondurant, A., Grosse, G., Parsekian, A. D., Romanovsky, V., & Hinkel, K. (2016). Extreme sensitivity of shallow lakes and sublake permafrost to Arctic climate change. In F. Günther & A. Morgenstern (Eds.), *XI. International conference on permafrost – book of abstracts* (pp. 20–24 June 2016). Potsdam: Bibliothek Wissenschaftspark Albert Einstein. https://doi.org/10.2312/GFZ.LIS.2016.001.
3. Bondurant, A. C., Arp, C., Jones, B., & Engram, M. (2016). *Rates and mechanisms of expansion in thermokarst lakes with bedfast and floating ice regimes on the Arctic coastal plain of northern Alaska 2015* (pp. 703–705), XI. International Conference on Permafrost – Book of Abstracts, 20–24 June 2016, Potsdam, Germany. Bibliothek Wissenschaftspark Albert Einstein. https://doi.org/10.2312/GFZ.LIS.2016.001.
4. Burn, C. R., & Smith, M. W. (1990). Development of thermokarst lakes during the holocene at sites near Mayo, Yukon territory. *Permafrost and Periglacial Processes, 1*, 161–176.
5. Dneprovskaya, V. P., Bryksina, N. A., & Polishchuk, Y. U. M. (2009). Izucheniye izmeneniy termokarsta v zone preryvistogo rasprostraneniya vechnoy merzloty Zapadnoy Sibiri na osnove kosmicheskikh snimkov (Study of changes in thermokarst in the zone of intermittent propagation of the permafrost of Western Siberia on the basis of satellite imagery), Issledovaniye Zemli iz kosmosa. – No 4. – pp 88–96.
6. Ershov, E. D. (Ed.). (1989). *Geokriologiya SSSR. Srednyaya Sibir' (Geocryology of the USSR. Middle Siberia)*. Moscow: Nedra 414 p.
7. Ershov, E. D. (Ed.). (1988). *Geokriologiya SSSR. Yevropeyskaya territoriya SSSR (Geocryology of the USSR. European territory of the USSR)*. Moscow: Nedra 358 p.
8. Ershov, E. D. (Ed.). (1989). *Geokriologiya SSSR. Zapadnaya Sibir' (Geocryology of the USSR. Western Siberia)*. Moscow: Nedra 454 p.
9. Grechishchev, S. Y., Chistotinov, L. V., & Shur, Y. U. L. (1980). *Kriogennyye fiziko-geologicheskiye protsessy i ikh prognoz. (Cryogenic physico-geological processes and their prognosis)*. Moscow: Nedra 384 p.
10. Grosse, G., Jones, B. M., Nitze, I., Lindgren, P. R., Walter Anthony, K. M., & Romanovsky, V. E. (2016). *Massive thermokarst lake area loss in continuous ice-rich permafrost of the*

northern Seward Peninsula, Northwestern Alaska, 1949–2015 (pp. 739–740), XI. International Conference on Permafrost – Book of Abstracts, 20–24 June 2016, Potsdam, Germany. Bibliothek Wissenschaftspark Albert Einstein. https://doi.org/10.2312/GFZ.LIS.2016.001.
11. Kapralova V.N. (2014) Zakonomernosti razvitiya termokarstovykh protsessov v predelakh ozerno-termokarstovykh ravnin (na osnove podkhodov matematicheskoy morfologii landshafta) (Regularities of the development of thermokarst processes within the lacustrine-thermokarst plains based on the approaches of the mathematical morphology of the landscape), Author's Candidate's summery, Moscow IGE RAN, 24 p.
12. Kapralova, V. N., & Victorov, A. S. (2009). *Modelirovaniye morfologicheskoy struktury ozerno-termokarstovykh ravnin i yego geoekologicheskoye znacheniye. (Modeling of the morphological structure of the lake-thermokarst plains and its geo-ecological significance)*, Modelirovaniye pri reshenii geoekologicheskikh zadach. Sergeyevskiye chteniya. Is. 11. – Moscow, Geos, 174–178.
13. Karlin, S. (1971). *Osnovy teorii sluchaynykh protsessov. (Fundamentals of the theory of random processes)*. Moscow: Mir 536 p.
14. Karlsson, J. M., Lyon, S. W., & Destouni, G. (2014). Temporal behavior of lake size-distribution in a thawing permafrost landscape in Northwestern Siberia. *Remote Sensing, 6*, 621–636.
15. Korolyuk, V. S., Portenko, N. I., Skorokhod, A. V., et al. (1985). *Spravochnik po teorii veroyatnosti i matematicheskoy statistike (A handbook on probability theory and mathematical statistics)*. Moscow: Nauka 640 p.
16. Kramer, G. (1970). *Matematicheskiye metody statistiki. (Mathematical methods of statistics)*. Moscow: Mir 648 p.
17. Kravtsova, V. I., & Bystrova, A. G. (2009). Izucheniye izmeneniy rasprostraneniya termokarstovykh ozer Rossii po raznovremennym kosmicheskim snimkam. *Kriosfera Zemli, 15*(2), 16–26.
18. Kravtsova, V. I., & Tarasenko, T. V. (2009). Izucheniye i kartografirovaniye dinamiki termokarstovykh ozer na territorii Zapadnoy Sibiri po raznovremennym kosmicheskim snimkam, Vos'moye sibirskoye soveshchaniye po klimato-ekologicheskomu monitoringu. Materialy rossiyskoy konferentsii 8–10 oktyabrya 2009 (Study and mapping of the dynamics of thermokarst lakes in the territory of Western Siberia for various space images, The Eighth Siberian Conference on Climate and Environmental Monitoring. Materials of the Russian Conference October 8–10, 2009) (pp. 273–275). Tomsk: Agraf-Press.
19. Kudryavcev, V. A. (1958). O thermokarste,Vopr fis. Geogr. Polar stran Tom 1 (About thermokarst, Questions of physical geography of polar regions, vol 1).
20. Makarycheva, E., Sergeev, D., Kapralova, V., & Jin, H. (2014). *Water level regime of thermokarst lakes in the mountain areas* (p. 206). Book of Abstracts of EUCOP4 – 4th European Conference on Permafrost. 2014.
21. Metodicheskoye rukovodstvo po inzhenerno-geologicheskoy s"yemke masshtaba 1:200 000 (1:100000-1:500000). (Methodical guidelines for engineering-geological survey of a scale of 1: 200 000 (1: 100000-1: 500000)). (1978). Moscow: Nedra. 391 p.
22. Morgenstern, A., Grosse, G., Günther, F., Fedorova, I., & Schirrmeister, L. (2011). Spatial analyses of thermokarst lakes and basins in Yedoma landscapes of the Lena Delta. *The Cryosphere, 5*(4), 849–867. https://doi.org/10.5194/tc-5-849-2011.
23. Morgenstern, A., Mathias, U., Günther, F., Roessler, S., Fedorova, I. V., Rudaya, N., Wetterich, Boike, J., & Schirrmeister, L. (2013). Evolution of thermokarst in East Siberian ice-rich permafrost: A case study. *Geomorphology, 201*, 363–379. https://doi.org/10.1016/j.geomorph.2013.07.011.
24. Muster, S., Roth, K., Langer, M., Lange, S., Cresto Aleina, F., Bartsch, A., Morgenstern, A., Grosse, G., Jones, B., Sannel, A. B. K., Sjöberg, Y., Günther, F., Andresen, C., Veremeeva, A., Lindgren, P. R., Bouchard, F., Lara, M. J., Fortier, D., Charbonneau, S., Virtanen, T. A., Hugelius, G., Palmtag, J., Siewert, M. B., Riley, W. J., Koven, C. D., & Boike, J. (2017). PeRL:

A circum-Arctic permafrost region pond and Lake database. *Earth System Science Data, 9*, 317–348. https://doi.org/10.5194/essd-9-317-2017.
25. Olefeldt D., Goswami, S., Grosse, G., Hayes, D. J., Hugelius, G., Kuhry, P., Sannel, B., Schuur, E. A. G., & Turetsky, M. R. (2016). Arctic circumpolar distribution and soil carbon of thermokarst landscapes. 2015.
26. Perl'shteyn, G. Z., Pavlov, A. V., Levashov, A. V., & Sergeyev, D. O. (2005). Netemperaturnyye faktory teploobmena deyatel'nogo sloya s atmosferoy (Non-temperature factors of heat exchange of the active layer with the atmosphere), Materialy Tret'yey konferentsii geokriologov Rossii, MGU, (pp. 86–91).
27. Polishchuk, V. Y. U., & Polishchuk, Y. U. M. (2013). *Geoimitatsionnoye modelirovaniye poley termokarstovykh ozer v zonakh merzloty (Geoimitational modeling of fields of thermokarst lakes in permafrost zones)*. Khanty-Mansiysk: UIP YUGU 129 p.
28. Polishchuk, Y. U. M., Bryksina, N. A., & Kupriyanov, M. A. (2016). Sravnitel'nyy analiz izmeneniy razmerov ozer v zone merzloty Zapadnoy i Vostochnoy Sibiri po kosmicheskim snimkam (A comparative analysis of changes in the size of lakes in the permafrost zone of Western and Eastern Siberia from satellite imagery). *Geoinformatika*, (1), 64–67.
29. Riordan, B., Verbyla, D., & McGuire, A. D. (2006). Shrinking ponds in subarctic Alaska based on 1950–2002 remotely sensed images. *Journal of Geophysical Research, 111*, G04002. https://doi.org/10.1029/2005JG000150.
30. Sejourne, A., Costard, F., Fedorov, A., Gargani, J., Skorve, J., Masse, M., & Mege, D. (2015). Evolution of the banks of thermokarst lakes in Central Yakutiya (Central Siberia) due to retrogressive thaw slump activity controlled by insolation. *Geomorphology, 241*, 31–40, https://www.sciencedirect.com/science/article/pii/S0169555X15001841.
31. Shur, Y. U. L. (1977). *Termokarst (k teplofizicheskim osnovam ucheniya o zakonomernostyakh razvitiya protsessa) (Thermokarst (to the thermophysical foundations of the doctrine of the laws of the development of the process))*. Moscow: Nedra 80 p.
32. Smith, L. C., Sheng, Y., Macdonald, G. M., & Hinzman, L. D. (2005). Disappearing Arctic lakes. *Science, 308*(3), 14.
33. Streleckaya, I. D., & Turkina, O. S. (1987). *Merzlotnye ploskobugristye torfyaniki Nadym-Purovskogo mezhdurechya Zapadnoy Sibiri, Issledovaniya merzlyh gruntov v rayonah osvoyeniya (Permafrost flathial peatlands of Nadym-Purovsky area of West Siberia, Researches of permafrost at regions of engineering activity)–M* (pp. 41–49). Stroyizdat.
34. Tomirdiaro, S. V. (1972). *Vechnaya merzlota i osvoyeniye gornykh stran i nizmennostey. (Permafrost and development of mountainous countries and lowlands)*. Magadan: Knizhnoye izdatel'stvo 174 p.
35. Tumskoy V. Ye. (2002). *Termokarst i yego rol' v razvitii regiona morya Laptevykh v pozdnem pleystotsene i golotsene. (Thermokarst and its role in the development of the Laptev Sea region in the late Pleistocene and Holocene), Author's Candidate's summery*. Moscow state university: Moscow.
36. Tumskoy V. Ye., Nikol'skiy P. A., Basilyan A. E., Kuznetsova T. V., & Gavrilov A. V. (2000). *Evolyutsiya mnogoletnemerzlykh porod na poberezh'ye proliva Dmitriya Lapteva v pozdnem kaynozoye – Tezisy konferentsii "Ritmy prirodnykh protsessov v kriosfere Zemli" (Evolution of permafrost on the strait of Dmitrii Laptev in the Late Cenozoic – theses of the conference "Rhythms of Natural Processes in the Earth's Cryosphere")* (pp. 123–125). Pushchino.
37. Veremeeva, A., & Gubin, S. (2009). Modern tundra landscapes of the Kolyma Lowland and their evolution in the Holocene. *Permafrost and Periglacial Processes, 20*(4), 399–406.
38. Victorov, A. S. (1995). Matematicheskaya model' termokarstovykh ozernykh ravnin kak odna iz osnov interpretatsii materialov kosmicheskikh s"yemok, *Issledovaniye Zemli iz kosmosa,* (5), 42–50.
39. Victorov, A. S. (1998). *Matematicheskaya morfologiya landshafta (Mathematical morphology of the landscape)*. Moscow: Tratek 191 p.

40. Victorov, A. S. (2002). *Mathematical models of landscape patterns for geological interpretation of remote sensing data, Proceedings of the 8th annual Conference of International Association for Mathematical Geology* (pp. 571–576). Berlin: IAMG 2002.
41. Victorov, A. S. (2006). *Osnovnyye problemy matematicheskoy morfologii landshafta (The main problems of the mathematical morphology of the landscape)*. Moscow: Nauka 252 p.
42. Voskresenskiy, K. S. (1999). Sovremennyye rel'yefoobrazuyushchiye protsessy na ravninakh Severa Rossii, (Contemporary relief-forming processes on the plains of the North of Russia). Author's Doctor's summery.

Chapter 19
Seismic-Ecological Monitoring of Seismic and Special Hazard Object

Oksana Popova, Vladimir Makeev, and Michael Popov

Abstract The paper presents generalized results of seismic-ecological monitoring from natural sources in the Russian Federation regions of different tectonic activities. The method is developed for estimating the stressed state of medium according to the energy characteristics of converted PS waves from distant earthquakes. The values of geodynamic indicators are assessed for the media of different tectonic activities, the main regularities of changes in the geodynamic environment are found, and the cyclicity of geodynamic indicators is revealed on the basis of long-term continuous monitoring. The influence of such external natural factor as distant catastrophic earthquakes on the change of geodynamic indicators of the medium's stress state and the strengthening of seismic activity are revealed.

19.1 Introduction

Earth science has a wide range of methods to study processes occurring in the depths and on the surface of the Earth. The addition of different nature rhythms (e.g., the influence of the Sun, Moon, and planets) can lead to resonances, causing the catastrophes. Because the geological environment is heterogeneous and composed of solid particles, voids, fluids, layers, blocks, etc. with different states of stress, the rhythms and the catastrophes are transmitted in different ways and are not written in the same way in rocks. To study these processes, the geophysical monitoring and, as a special case, the seismic monitoring are necessary.

Seismic monitoring is called continuous long-term or cyclic (periodic) observations for seismic parameters' changes of specific geological environment. It is based on the organization of a network of continuous long-term (or periodic) observations on the explored area, the operator of the processing, and the interpretation of seismic data with access to the rating prediction.

O. Popova (✉) · V. Makeev · M. Popov
Sergeev Institute of Environmental Geoscience, Russian Academy of Sciences, Moscow, Russia

V. Svalova (ed.), *Natural Hazards and Risk Research in Russia*,
Innovation and Discovery in Russian Science and Engineering,
https://doi.org/10.1007/978-3-319-91833-4_19

In recent years, the issue of safety of vital technical facilities such as hydroelectric power plants, nuclear power plants, and subsoil use has become acute. The concept of ecological monitoring appears. First of all, this applies to large and environmentally hazardous facilities, the territories of exploited oil and gas fields, coal basins, trunk pipeline routes, burials of highly toxic radioactive waste, large chemical enterprises, specially protected ecological and resort areas, and urban agglomerations with high population density located in high-altitude regions with seismic risk.

The tasks set for ecological monitoring are largely addressed through local seismic monitoring, since it allows tracking the change in geophysical parameters of the environment in time. Local seismic monitoring, which allows in time to assess the state of the geophysical environment in limited areas in order to assess the degree of environmental hazard and reduce the risk of natural hazards, can be called seismic-ecological monitoring.

19.2 Objects of the Research and Technique of Observation

Regions for analyzing local seismic-ecological monitoring materials can be divided into three categories: (1) seismic zones, (2) urbanized megacities, and (3) critical sites (e.g., nuclear power plants). The duration of observation fluctuated at these sites from 2–3 months to 11 years, depending on monitoring tasks.

The region of the Caucasian Mineral Waters is included in the seismically dangerous regions. As a megapolis, materials of seismic-ecological monitoring of Moscow and Moscow region are considered. Most of the research was carried out on the territory of especially hazardous objects, such as areas of planned and operating nuclear power plants for security. These objects are the Tomsk poligon (Seversk NPP), Nizhny Novgorod poligon (Nizhny Novgorod NPP), Middle Urals poligon (Beloyarsk NPP), Kola poligon (Kola NPP), and Tver poligon (Tver NPP). The regions Tomsk and Middle Urals are classified as seismic areas with weak and moderate seismic activity, while the Nizhny Novgorod region and the Tver region are platform aseismic regions. The Kola region is a shield, which is characterized by activation of seismic activity.

Seismic-Ecological Monitoring Monitoring in the abovementioned poligons was carried out from natural sources (distant earthquakes, local and remote earthquakes, explosions, and microseismic backgrounds were recorded). Three-component digital seismic stations and three-component geophones with one vertical (Z) and two horizontal (X and Y) components were used. The components of the geophone were oriented strictly fixed and identical – the horizontal component X to the north, the horizontal component Y to the east, and the vertical component Z to the zenith.

In local seismic-ecological monitoring in the abovementioned poligons, an area arrangement of observation points (a local network) consisting of three-component digital seismic stations was used. The area of the investigation is usually of the order of (60 * 60–40 * 60 km^2). The optimal number of stations in the area is on the average 10–15, but not less than 5. The distance between the stations was 12 and 20 km, optimally 16 km. During observations, regularly, usually once a day, a calibration pulse of a certain shape and amplitude was recorded; this is necessary to evaluate the amplitude-spectral readings.

19.3 The Technique for Estimation of the Medium's Stress State

The method for estimating the stress state and studying the geodynamics of the medium was developed on the basis of an analysis of the energy of converted PS waves from distant earthquakes based on materials of long-term local seismic-ecological monitoring in the Caucasian Mineral Waters (Cauminwater) region (1995–2006) and described in detail in [7]. Converted PS waves from distant earthquakes with epicentral distances $\Delta > 60^{\circ}$ allow to study the environment directly under the observation point, since the rays along which the oscillations propagate are close to the vertical in this case, unlike explosions, when it is difficult to estimate at which part of the path from source to the receiver there were changes in the environment.

The technique is based on the fact that the energy of the transverse and converted PS waves (and the converted waves also belong to the transverse waves) essentially depends on the degree of fracturing, porosity, fluidization, and anisotropic properties of the medium [4].

The anisotropic properties of the medium and its fluid saturation are determined by the stress state in turn [2, 3]. If the medium is isotropic and the boundaries are subhorizontal, then only SV-type converted waves are formed, which are polarized in the ray plane. A ray plane is a vertical plane passing through a source and a receiver. In an anisotropic medium, three types of volume waves propagate but not two: one longitudinal (*P*) and two transverse (SV and SH); there is a so-called splitting of the transverse wave. Waves of types SV and SH differ in propagation, velocity, and polarization. The splitting of the transverse wave SH is a sign of an anisotropic medium. Analysis of the records of PS waves, which are differently polarized, of the type SV and SH, makes it possible to study the anisotropic properties of the medium, which are to a great extent determined by its stressed state [2, 3].

According to the developed technique, the medium's stress state at some point in space is estimated by the following indicators [8]:

- The anisotropy indicator of the medium γ (dimensionless value) under each observation point ($\gamma = E_R/E_V$), where E_V and E_R are the energy of the radial (v) and of the tangential (r) components of the PS wave from distant earthquakes, respectively. The anisotropy indicator γ has nothing in common with the anisotropy coefficient γ, which is calculated from the velocities taking into account the type of anisotropy [13]. The anisotropy indicator allows us only to ascertain the presence of anisotropy in the medium to a greater or lesser extent.

- Integral indicator S (dimensionless value) of the medium's stress state in the observation region determined by calculation based on the study of the parameter γ distribution over the area where the observational monitoring network was (19.1)

$$S = \int_{X1}^{X2}\int_{Y1}^{Y2}\int_{\delta}^{\infty}\gamma(x,y)dxdyd\gamma \Bigg/ \int_{X1}^{X2}\int_{Y1}^{Y2}\delta\cdot x\cdot y\cdot dxdy,, \tag{19.1}$$

where x and y are coordinates of monitoring poligon's area and δ is the threshold level. Exceeding the values of γ over the quantity δ ($\gamma > \delta$) indicates an increase in the anisotropic properties of the medium and, correspondingly, the stress state in the study area.

The γ and S indicators can be estimated for different depth intervals and different time observation intervals. The time interval varies from 10 to 40 days (an average of 1 month), depending on the quality of the distant earthquakes' records.

19.4 Peculiarities of Changing Dynamic Characteristics of Converted PS Waves, Indicators of Anisotropicity, and Stress State in Different Tectonically Active Regions

The dependences of the medium's indicator anisotropy on the observation time γ (t) are constructed for each observation point. In addition, for different observation time intervals and different depth levels, the distribution of $\gamma = Er/Ev$ over the study area (poligon) is constructed, and the integral indicator of the medium's stress state S is estimated.

Thus, as a result of processing the records of PS wave from distant earthquakes, the time series are obtained: $Ev(t)$, $Er(t)$, and $\gamma(t)$ for each observation point and $S(t)$ for the poligon's station network and a set of schemes for the distribution of the parameter γ for different time intervals and for different depth intervals.

The proposed technique for estimating the stress state of the medium was used in the processing of all the experimental materials listed above.

Research objectives:

- To show how the geodynamic indicators of the medium's stress state vary in time for each of the studied regions, to estimate the range of these changes in seismically active and aseismic environments
- To determine the existence of cyclicity in geodynamic indicators' change
- To determine the influence of external natural factors on stress state's indicators of the environment in regions of different geodynamic activities

The longest continuous seismic-ecological monitoring from all of the abovementioned regions was carried out in the Caucasian Mineral Waters region (11 years), the Moscow megalopolis (2.5 years), the Tomsk region (Seversk NPP) (about 2 years), and the Nizhny Novgorod region (Nizhny Novgorod NPP) (1 year). On these materials it was succeeded to reveal regularities in geodynamic indicators' change for stress state of the medium γ and S in time in different tectonic zones and to carry out generalization.

The Variability of the PS Waves' Energy Characteristics in Time and the Level of the Values for the γ and S Indicators in Various Media

For all the regions, where seismic-ecological monitoring was conducted, it was noted the variability in time and on the area of observation for the energy parameters characterizing the stress state of the medium (*Ev* and *Er*) of the PS waves formed at different depths, as well as for the anisotropy indicator γ and indicator S. Figure 19.1 shows the schemes of the anisotropy indicator γ for the Cauminwater, Beloyarsky, and Nizhegorodsky poligons for different time observation intervals for the studied depth interval (0–30 km). Converted waves from deeper boundaries are not

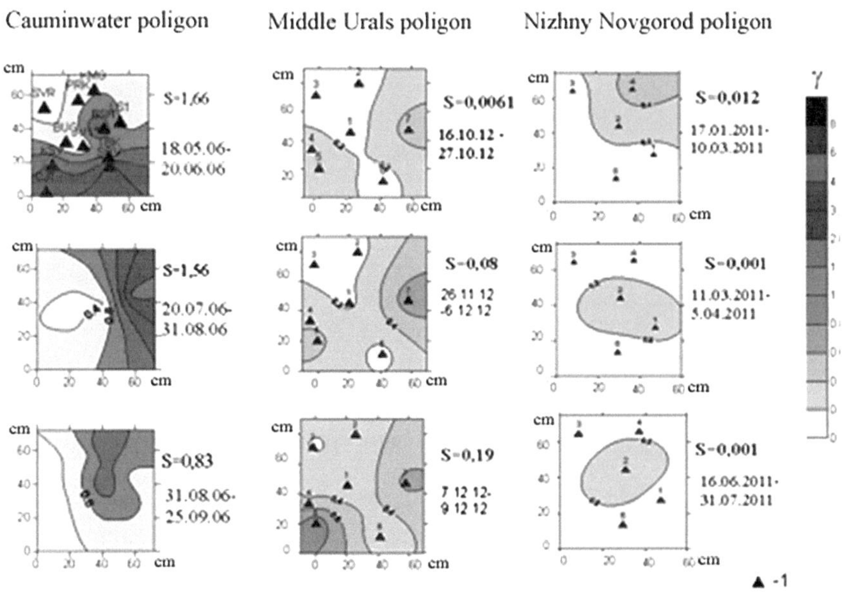

Fig. 19.1 Schemes of anisotropy indicators γ distribution for the Cauminwater, Middle Urals (Beloyarsk NPP), and Nizhny Novgorod poligons at different time intervals. 1- points of observation

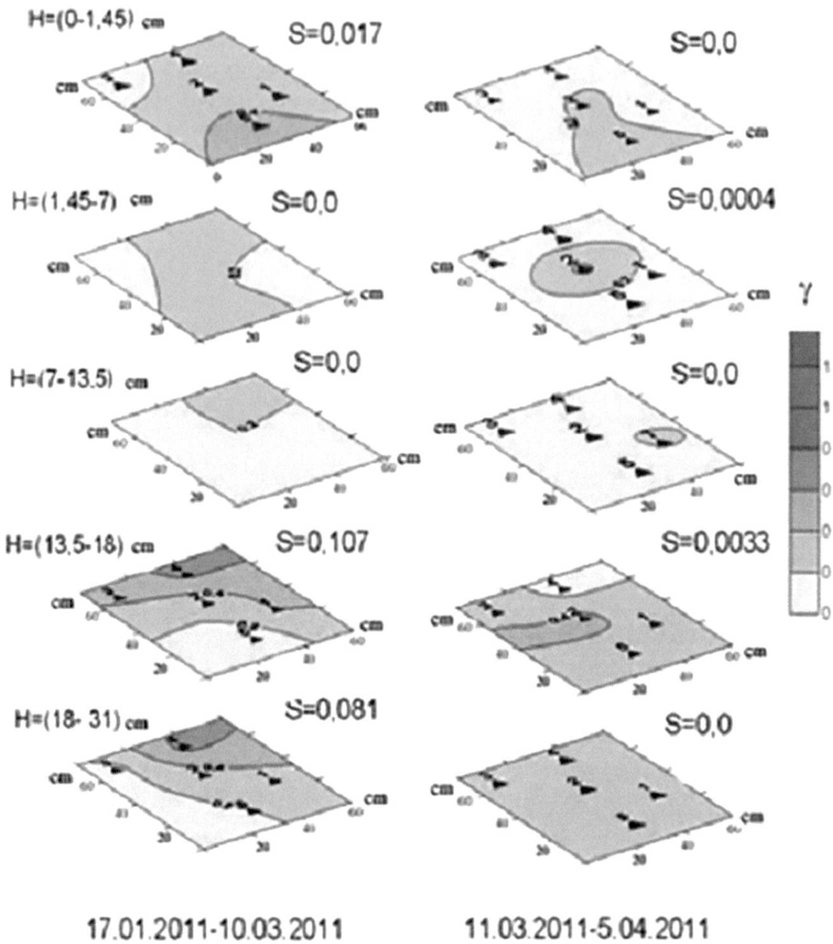

Fig. 19.2 Schemes of anisotropy indicator's distribution for Nizhny Novgorod poligons at two time intervals and for different depth ranges

analyzed, since the rays of these waves will have large drifts and do not characterize the medium below the observation point. Figure 19.2 shows schemes of indicator γ for two time intervals of the Nizhny Novgorod poligon, but for different depth ranges.

The range of changes in the values of γ and S for the regions of the longest monitoring is given in Tables 19.1, 19.2, 19.3, and 19.4. As can be seen from the tables and figures, the values of γ and S in seismically active regions change much more widely than in weak seismic and aseismic (platform) regions. For the Cauminwater poligon's medium during 7 years, the values of γ varied from 0 to 8.0 and the values of S from close to 0 to 5.3 (see Table 19.1). High values of S correspond to the manifestation of seismic activity.

Table 19.1 Geodynamic characteristics of the Cauminwater poligon's medium

Year of research	Time intervals for which the geodynamic parameters were determined	The anisotropy indicator of the medium $\gamma = E_R/E_V$	Integral indicator of the medium's stress state S
2000	20.05.00–15.06.00	0.2–2.4	1.35
2001	23.03.01–03.05.01	0.2–1.2	0.63
	22.09.01–21.10.01	0.2–2.4	1.7
2002	09.05.02–15.06.02	0.2–1.0	0.4
2004	25.07.04–19.09.04	0.2–2.8	2.6
	23.09.04–23.10.04	0.2–2.4	2.1
	24.10.04–24.11.04	0.2–2.0	0.9
2005	01.04.05–16.04.05	0.2–8.0	5.3
2006	18.05.06–20.05.06	0.2–2.0	0.8

Table 19.2 Geodynamic characteristics of the Tomsk poligon's environment

Year of research	Time intervals for which the geodynamic parameters were determined	The anisotropy indicator of the medium $\gamma = E_R/E_V$	Integral indicator of the medium's stress state S
2008	11.07.08–31.07.08	0.0–0.8	0.149
	27.08.08–10.09.08	0.0–0.4	0.001
	11.09.08–10.10.08	0.0–0.8	0.04
	11.10.08–10.11.08	0.0–0.6	0.017
2009	03.01.09–25.01.09	0.0–0.6	0.001
	20.01.09–10.02.09	0.0–0.8	0.074
	04.04.09–17.04.09	0.0–1.0	0.126
	18.04.09–12.05.09	0.0–1.0	0.16
	14.05.09–20.06.09	0.0–0.6	0.017
	05.09.09–29.09.09	0.0–0.4	0.003
2010	**12.12.00–10.01.10**	**0.0 1.4**	**0.18**
	06.02.10–25.02.10	0.0–0.6	0.000
	26.02.10–04.03.10	**0.0–2.8**	**0.542**
	05.03.10–14.03.10	0.0–0.8	0.003

On the territory of a region weak seismic activity (Tomsk poligon), during 2 years of observation, the γ and S indicators changed insignificantly (see Table 19.2) and only twice significantly exceeded the background level, increasing to 1.4 and 2.8 (in bold). In these time intervals, the values of the integral indicator S have lawfully increased to 0.18 and 0.542, reflecting the appearance of sections with the anomalous stress state of rocks within the poligon territory [10]. As a result of the contrast anomaly's appearance, a weak local tectonic earthquake with a magnitude $M = 1.4$ occurred on February 5, 2010, in the area of the Tomsk poligon (the territory of the

Table 19.3 Geodynamic characteristics of the Nizhny Novgorod poligon's environment

Year of research	Time intervals for which the geodynamic parameters were determined	The anisotropy indicator of the medium $\gamma = E_R/E_V$	Integral indicator of the medium's stress state S
2011	17.01.11–31.07.08	0–0.4	0.012
	11.03.11–05.04.11	0–0.2	0
	16.06.11–31.07.11	0–0.2	0.001
	01.09.11–21.10.11	0–0.6	0.061
	21.10.11–07.11.11	0–0.4	0.0015

Table 19.4 Geodynamic characteristics of the Moscow poligon's environment

Year of research	Time intervals for which the geodynamic parameters were determined	The anisotropy indicator of the medium $\gamma = E_R/E_V$	Integral indicator of the medium's stress state S
1997	04.02.97–05.04.97	0–1.2	1.036
	06.04.97–02.06.97	0–0.8	0.38
	26.06.97–27.07.97	0–2.4	0.91
	28.07.97–20.09.97	0.2–1.2	1.42
	01.10.97–05.11.97	0–0.1	0
	06.11.97–02.12.97	0–0.8	0.217
1998	03.12.97–10.01.98	0.2–2.0	2.08
	11.01.98–22.02.98	0.2 2.8	1.79
	23.02.98–23.03.98	0–0.8	0.532
	23.03.98–12.04.98	0–0.4	0.81
	13.04.98–12.05.98	0–0.4	0.218
	14.05.98–23.05.98	0–0.4	0.133
	25.05.98–20.06.98	0–0.4	0.11
	21.06.98–0.07.98	0–0.4	0.225
	21.07.98–5.08.98	0–1.2	1.5
	26.08.98–0.09.98	0–0.8	0.71
	10.09.98–0.11.98	0–0.8	0.276

Seversk NPP), whose epicenter is confined to the marginal part of the isolated anomaly.

For the platform Nizhny Novgorod region [10], the values of the anisotropy indicator γ varied from 0 to 0.6 and the stress state indicator S from 0 to 0.061 (see Table 19.3, Fig. 19.1).

The maximum values of S for the seismically dangerous Cauminwater poligon are more than 10 times higher than the maximum values of this indicator for the weakly seismically active Tomsk poligon and 100 times higher for the aseismic Nizhny Novgorod poligon. Such a ratio of the compared parameters' values obtained for the three poligons suggests a significant difference in the geodynamics and stress state of the medium in these regions.

However, the seismic monitoring results on the territory of the Moscow megapolis (platform region) for 2 years of research (see Table 19.4) and the Tver poligon are strikingly different from the data on the platform Nizhegorodsky Poligon (see Tables 19.3 and 19.4). On the territory of the Moscow megapolis, during the time of observation, contrast sites with high values of the anisotropy indicator γ to 2.0–2.8 appeared, while the S indicator accordingly increased to values of 1.5–2.08. Similar results were obtained for Tver poligon.

Probably, Moscow and Tver region cannot be called a region of passive geodynamics. The geodynamic estimates γ and S for these regions are quite high and are closer to those for the weakly seismically active Tomsk region.

This allows us to assume a sufficiently high degree of geodynamic activity in the areas of the Moscow megapolis and the planned Tver NPP, which should be taken into account in the construction of high-rise buildings and technological and industrial objects [6].

The Cyclicity Observed from the Evaluation of the Energy Parameters of the PS Waves and from the Values of the Geodynamic Indicators γ and S in Time

Analysis of long-term seismic-ecological monitoring results for the region of the Caucasian Mineral Waters and the Moscow region showed that the time series of the energy values of the waves $E_v(t)$ and $E_r(t)$ (respectively, radial and tangential components) revealed a clear periodicity of these parameters' variations for different depth levels at all points of observation (Figs. 19.3a, b).

The period of these variations is approximately 170–200 days, an average of 180–185 days, that is, 0.5 years. For the Cauminwater region, the contrast of the semiannual periodicity is higher at depths 0–15 km and weaker at greater depths (15–30 km) (Fig. 19.3a). In the platform area of the Moscow megapolis, the greatest variation's contrast in the energy of the converted waves corresponds to large depths (15–30 km) (Fig. 19.3b), i.e., these depths are more sensitive to semiannual changes in the stress state. This fact requires further observations and confirmation. Semiannual cyclicity is also manifested in the dependences $\gamma(t)$ and $S(t)$ for these regions (Fig. 19.3d). The nature of the semiannual cyclicity, according to a number of researchers, is probably due to a change in the rotation speed of the Earth [1, 11, 12]. However, the influence of the seasonal factor is also possible [12].

Dependence $S(t)$ for the research poligon of the Cauminwater made it possible to reveal the low-frequency cyclicity of the change in the stress state indicators connected, apparently, with the preparation of strong earthquakes of magnitude $M \geq 4.3$ in this region. The cycle duration is 2–3 years [8]; each cycle is completed by seismicity with $M \geq 4.3$. Semiannual cyclicity is superimposed on the low frequency and complicates it.

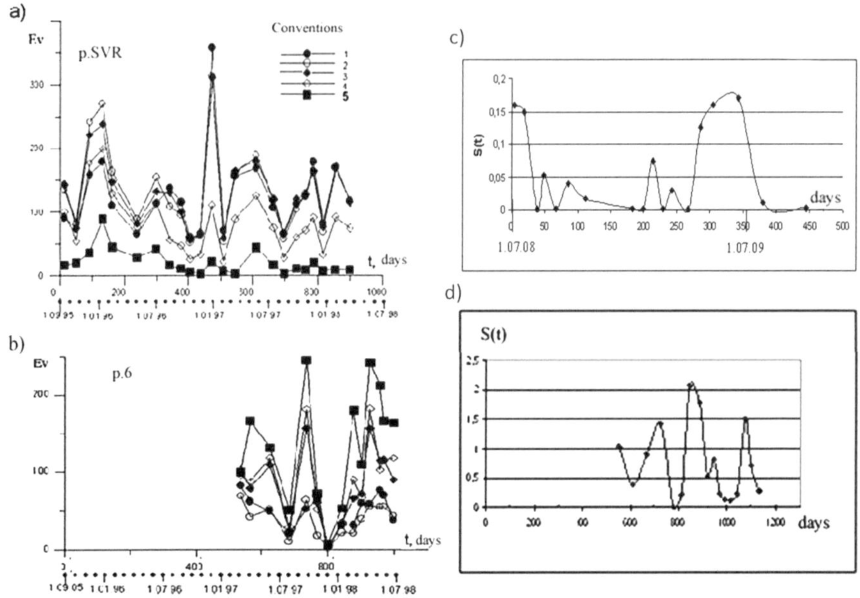

Fig. 19.3 The cyclicity of PS wave energy parameters and the stress state medium's indicators: change in time of the energy PS waves (radial component *Ev*) at one of the observation points during the seismic-ecological monitoring for (**a**) Cauminwater and (**b**) Moscow poligons; change in time of the stress state medium's indicator *S* in the poligon of the Severskaya NPP (Tomsk region) (**c**) and in the Moscow poligon (**d**)

Shorter time series *S*(*t*) for the Tomsk poligon, Moscow poligon, Middle Urals poligon, and Nizhny Novgorod poligons did not allow us to distinguish low-frequency cyclicity associated with the tectonic activity of the regions. Perhaps, this cyclicity for weakly seismically active and aseismic platform regions does not exist. For the *S*(*t*) dependence of the Tomsk poligon, annual cyclicity is revealed (Fig. 19.3c), but this cyclicity is likely to be of the same nature as the semiannual cyclicity of the Cauminwater and Moscow regions.

Thus, the high-frequency cyclicity with periods of 180 days (0.5 years) and 1 year is revealed in the variation value energy of PS waves and the indicators of the medium's stress state in time $\gamma(t)$ and $S(t)$ in both seismic and in the aseismic regions. Low-frequency cyclicity can be detected only for very long continuous observations, if this cyclicity exists.

Influence of External Natural Factors on the Indicators of the Medium's Stress State

One of the key problems is the disclosure of the mechanisms of interaction between different geophysical fields and the manifestation of their total impact on the stress state of the geological environment (synergetics of the process).

In this aspect, the study of external impacts by seismic-ecological monitoring in seismic areas and regions location of especially important objects is extremely important, since these impacts can be a trigger in the preparation of tectonic or man-made disasters. The daily tidal effects of the Sun and the Moon, lunar and solar eclipses, the influence of the Moon in its various phases, solar activity (magnetic storms), changes in the Earth's rotation speed, and the impact of catastrophic distant earthquakes should be attributed to external natural influences on the geological environment.

To estimate the effect of catastrophic distant earthquakes on the geodynamic indicators of the medium's stress state, the technique described above was used, which makes it possible to estimate the medium's stress state from the energy-converted PS waves [7].

The impact of catastrophic distant earthquakes was considered in two aspects:

1. The effect of distant catastrophic earthquakes on the increase of the indicators anisotropy γ and stress state S and, accordingly, on the enhancement of local seismic activity in the region which is remoted from the epicenter of a catastrophic earthquake by a distance of about 7000 km
2. The effect of a distant catastrophic earthquake on the cessation of a local earthquake focus work

The First Aspect As a result of the analysis of seismic-ecological monitoring results in the area of the seismic dangerous environment of the Cauminwater poligon, it was established that after distant catastrophic events with $M > 7.0$ (e.g., such as the Sumatran earthquake 2004 with $M = 9.0$, remote from the monitoring network by about 7000 km), accompanied by intense low-frequency surface waves, the indicators of the anisotropy γ and the medium's stress state S changed sharply. It was fixed the increasing of local seismic activity with magnitudes $M \geq 4.3$ [7, 8] which can be considered induced [5].

In the region of moderate seismic activity, such as the Beloyarsk poligon (Middle Urals), the formation of an area with increased tension was estimated by indicators γ and S, associated with the impact of a catastrophic distant earthquake from the Japanese region (Honshu Island December 7, 2012, $M = 7.3$). There was no increase of local seismic activity in the monitoring area in this case.

The effect of catastrophic distant earthquakes on the geodynamic indicators γ and S of the geological environment in the aseismic regions was analyzed for the platform regions of the Nizhny Novgorod NPP where, during the monitoring, a catastrophic Fukushima earthquake was registered on 11.03.2011 (Japan, $M = 9.0$) and, for the Tver NPP, where a catastrophic distant earthquake from the Alaska region (January 5, 2013, $M = 7.5$) was recorded. Intensive low-frequency surface waves from a catastrophic earthquake were recorded in the poligon of the Nizhny Novgorod NPP for 1 h and in the poligon of the Tver NPP for 30 min. Nevertheless, the increase in the values of the anisotropy indicator γ and the state stress indicator S in the environment of the Nizhny Novgorod and Tver poligons after the registration of catastrophic earthquakes with a magnitude $M > 7.0$ was not observed.

The Second Aspect The impact of distant catastrophic earthquakes in seismic regions was manifested in another aspect, namely, in stopping the work of a local earthquake focus.

On February 6, 2006, 150 km to the south of monitoring station network of the Cauminwater (the Dzirul massif area), a strong event occurred (the North Caucasus earthquake, with a magnitude $M = 5.9$). It was accompanied by a large number of aftershocks with a magnitude of $M > 4.1$, which formed a focal zone. Aftershock activity in the focal zone continued until mid-July 2006, while area of high-value γ clearly moved to the southern part of the study region, toward maximum impact from the focal region located 150 km south of the network of stations [9].

On July 17, 2006 on the island Java, there was a catastrophic earthquake ($M = 7.2$) (remoteness from the Cauminwater poligon is 7800 km). Intensive low-frequency surface waves from this earthquake were recorded by network of stations for 2 h. After registering this earthquake, the region of increased values of γ moved to the eastern part of the poligon – to the side where the earthquake from Java came – while aftershock activity in the focal zone of North Caucasus earthquake immediately stopped.

Thus, it is obvious that the impact of catastrophic ($M > 7.0$) earthquakes dramatically changes the structure of the anisotropic properties of the remote seismic medium, in some cases causing increased seismic activation and in others sharply stopping the activation of a strong local earthquake source. In the platform aseismic environment, the effect of distant catastrophic earthquakes may not appear, since such media are likely to have a different structure and greater stability than the seismic environment.

19.5 Conclusions

Based on the analysis of seismic-ecological monitoring results in regions of different tectonic activities, the following conclusions can be drawn:

1. A technique for estimating the stress state of a medium is developed.
2. It is concluded that the geodynamic indicators of the stress state vary with time in media of different tectonic activities, but the level and contrast of the changes in the estimated geodynamic indicators differ significantly in 10–100 times in seismic and aseismic media.
3. It was revealed that cyclicity of the stress state indicator of the medium changes with a period of 0.5 years and 1.0 year, associated with a change in the rotation speed of the Earth or with a climatic factor for regions of any tectonic activity. For seismically dangerous areas, cyclicity associated with the cycle of tectonic activity in the region is also revealed.

4. The influence of such external natural factor as distant catastrophic earthquakes on the change of geodynamic indicators of the stress state of the environment and the strengthening of seismic activity are revealed, especially in regions of high seismic activity.

All the features of the change in the stress state of the environment in different areas of local seismic-ecological monitoring made it possible to conclude that such monitoring is necessary at all stages of planning and operation of especially dangerous objects, which will allow us to assess the danger of natural and man-made disasters. In seismically dangerous regions, such monitoring should be continuous; in the platform regions, it can be cyclic.

References

1. Barsukov, O. M. (1994). Annual variations of seismicity and rotation speed of the Earth. *Physics of the Earth, 4*, 96–98.
2. Crampin, S. A. (1978). Seismic-wave propagation through a cracked solid: Polarization as possible dilatancy diagnostic. *Geophysical Journal of the Royal Astronomical Society, 53*, 426–467.
3. Crampin, S. A. (1981). Review of wave motion in anisotropic and cracked elastic-media. *Wave Motion, 3*, 343–391.
4. Gik, L. D. (1997). Physical simulation of the seismic waves' propagation of in porous and fractured media. *Geology and Geophysics, 38*(4), 804–815.
5. Nikolaev, A. V. (1994). *Problems of induced seismicity. Induced seismicity* (p. 222). Moscow: Nauka.
6. Pomerantseva, I. V., & Mozzhenko, A. N. (1997). *Seismic research with Earth equipment* (p. 256). Moscow: Nedra.
7. Popova, O. G., Gray, A. V., Konovalov Yu, F., & Nedyadko, V. V. (2006). Influence of catastrophic earthquakes on the stressed state of remote territories. Sb. of geophysical readings to them. In V. V. Fedynsky (Ed.), *Geophysics of the XXI century: 2006* (pp. 200–2004). Moscow: Mir.
8. Popova, O. G., Gray, A. V., & Konovalov Yu, F. (2008). Results of long-term seismic monitoring in the seismic region of the Caucasian Mineral Waters. *Geoecology, 2*, 35–40.
9. Popova, O. G., & Zhigalin, A. D. (2011). *Influences of catastrophic events (M> 6.8–7.0) on seismic activity in remote seismic regions from the epicenter*. Sb. mater. XVII Vseros. Conf. with intern. "Problems of seismotectonics", Moscow-Voronezh September 20–24, 2011, Voronezh: Publishing House of Voronezh State University (pp. 432–436).
10. Popova, O. G., Zhigalin, A. D., Popov, M. G., Arakelyan, F. O., Nedyadko, V. V., & Lavrik, E. V. (2015). Geodynamic monitoring in the design of power engineering objects. *Geoecology, 5*, 460–468.
11. Rulev, B. G. (1991). *Annual frequency in the emission of micro-earthquakes and uneven earth rotation. Earthquakes and the processes of their preparation* (pp. 127–138). Moscow: Nauka.
12. Shabarov, A. P., & Tarasov, B. G. (2003). Geodynamic state of an array of rocks in the cycles of the Earth and the Sun, Tez. In V. V. Fedynsky (Ed.), *Fifth geophysical readings named after V.V. Fedynsky* (pp. 51–52). Moscow: Center Geon Them.
13. Shneerson, M. B. (2006). *Introduction to seismic anisotropy: Theory and practice*. (p. 160). Moscow, GERS.

Chapter 20
Landslide Risk Assessment, Management, and Reduction for City Territories

Valentina Svalova

Abstract Geological and landslide risk management can be viewed as a series of events leading to landslide risk reduction, including risk analysis, risk assessment, risk mapping, vulnerability evaluation, the concept of acceptable risk, monitoring organization, engineering-technical methods, insurance, and others. This chapter describes the methodology for landslide risk assessment and mapping in urban areas. The construction of a landslide risk map in a territory of Moscow is proposed. Based on preliminary expert estimates, areas with high landslide risk are in the vicinity of the Moscow and Yauza Rivers, as well as in areas of contrasting relief along riverbeds of paleochannels in the city center. These areas may be considered "hot spots" on the risk map.

20.1 Introduction

Landslide risk management can be viewed as a series of events leading to landslide risk reduction. Natural risk is a relatively new concept that has not been fully explored. As such, there are many definitions of natural risk. Often, a scientific study or a scientific approach to a problem begins with the presentation of the author's position, including his or her chosen definition of natural risk for that particular problem. However, this individualistic approach is difficult to avoid.

One of the main systematic approaches to hazards research is the classification of hazards. The concept of risk management is a new scientific development, which provides a foundation for systematic hazard investigations. Development of the risk concept requires advances in the methods used for risk assessment and calculation. The theory of risk would need to be approached as a scientific discipline with a strong mathematical aspect. It would be necessary to determine common approaches to risk calculation for different types of natural hazards. For example, the methods

V. Svalova (✉)
IEG RAS, Moscow, Russia
e-mail: inter@geoenv.ru

V. Svalova (ed.), *Natural Hazards and Risk Research in Russia*,
Innovation and Discovery in Russian Science and Engineering,
https://doi.org/10.1007/978-3-319-91833-4_20

used for seismic risk assessment, which are the most commonly used, should be extended to landslides, karst, suffusion, flooding, pollution, and other types of complex and multi-risk natural hazards.

The concept of *risk* arose from everyday scenarios such as gambling, finances, and business. By extending the concept of risk to scientific research, it became the foundation for systematic investigations of natural and man-made hazards and disasters. In common terms, risk is the potential possibility to gain or lose something (e.g., life, health, property, money, environment). A risky situation can occur when one is met with uncertainty resulting from action or inaction. Risk is a consequence of unpredictable outcomes.

In scientific risk analysis, risk is considered to be a measure of the probability of damage to life, health, property, finances, or the environment. Risk is defined as the probability of a natural hazard event multiplied by the damage from possible consequences. Risk analysis uses available information for hazard identification and vulnerability evaluation. Vulnerability is the degree of loss of a given element or set of elements exposed to a natural or manmade hazard; it is expressed on a scale of 0 (no loss) to 1 (total loss).

Risk assessment is the process of determining whether an existing risk is acceptable or nonacceptable, as part of the risk analysis and risk evaluation processes. Sometimes, risk assessment is considered to be a risk calculation based on selected parameters and an established ranking of risk criteria. An acceptable risk is defined by the level of human and property loss that can be tolerated by an individual or community. The probability of acceptable risk is very small. The concept of acceptable risk arises from the understanding that absolute safety is an unachievable purpose.

Risk management is the complete process of risk assessment and risk reduction. Risk reduction uses a variety of methods and measures, including legislative, organizational, economic, engineering, information, and others. In a narrow sense, risk management may be considered as a measure for risk reduction. In this context, landslide risk management can be viewed as a series of events leading to the risk reduction and avoidance of landslides. It includes landslide monitoring, landslide forecasting, engineering works, slope strengthening, insurance and others.

By defining a systematic approach to natural hazards research that is based on the concept of risk, it is possible to determine a process and schema for establishing criteria to classify the risk posed by different types of natural and manmade hazards and disasters. In addition, the impact that a hazardous event or process would have on a population and structures could be quantified, and strategies for risk reduction and avoidance could be improved.

20.2 Risk Management

The main aspects of a systematic approach to risk management in natural hazards and disaster research are as follows (Table 20.1):

Table 20.1 Risk management structure

Hazard Identification + **Vulnerability evaluation**	**Risk analysis** + **Acceptable risk**	**Risk assessment** + **Risk reduction**	**Risk management**

1. Hazard identification
2. Vulnerability evaluation
3. Risk analysis
4. Concept of acceptable risk
5. Risk assessment
6. Risk mapping
7. Measures for risk reduction:

   (a) Legislative
   (b) Organizational and administrative
   (c) Economic, including insurance
   (d) Engineering and technical
   (e) Modeling
   (f) Monitoring
   (g) Information

According to the most common definition, risk is the probability of a natural hazard event multiplied by the possible damage:

$$R = P \times D,$$

where R is the risk, P is the probability, and D is the damage.

For multiple-risk assessment, it is possible to use sum of risks of different hazards:

$$R = \sum R_i$$

To construct a risk map, it is necessary to use maps of natural hazards and their possible damage. These maps can be on a local, regional, federal (subglobal), or global level. Geological risk mapping is an important step in solving the problem of natural risk management [1, 8, 19–25]. Due to the complexity and diversity of the problem, a combination of probabilistic and deterministic approaches and expert estimates is needed. Some main aspects of landslides risk management are considered in the next section.

20.3 Landslide Risk Assessment and Mapping

A landslide is a major geological hazard that poses a serious threat to human population and various infrastructures, such as highways, rail routes, and civil structures (e.g., dams, buildings). Landslides occur very often during other major natural disasters, such as earthquakes, floods, and volcanoes. The word *landslide* represents only a type of movement (e.g., sliding). However, it is generally used as a term that covers all types of land movements, including falls, creeps, spreads, flows, and other complex movements. A more correct term to represent all of these movements may be *mass movement* or *mass wastage*. However, the term *landslide* has been accepted and is being used commonly around the world as a synonym for *mass wastage*.

The likelihood of a landslide process depends on the stability of the landslide slope, trigger mechanisms (e.g., precipitation, earthquakes), and technological factors. The first step in landslide risk assessment is studying the physical and mechanical sliding processes under different conditions. However, landslide process mechanics are still not fully understood, so landslide prediction is not always possible. Even the statistical frequency of landslide activation for a particular area varies very widely.

As an example, the next section examines an approach to the construction of a landslide risk map for a territory of Moscow.

Study Area

Landslide processes in Moscow have been well investigated [2–7, 9–18], and the landslide hazard has been mapped. Landslides cover approximately 3% of the city, where there are 15 deep landslides and many smaller ones. Recently in Moscow, there has been significant activation of landslide processes.

To assess landslide hazard, the height of the slope, the landslide body volume, mass velocity, rock properties, topography of the surrounding area, the range of possible promotion of landslide masses, hydrogeological conditions, and trigger mechanisms have to be taken into account. The selection of taxons (special areas) with varying degrees of landslide hazard in the city is a completely solvable task. Gradation is possible using either three degrees of danger (high, medium, low) or five degrees of danger (very high, high, medium, low, not dangerous), depending on the requirements of the task.

The most expensive land and buildings in Moscow are located in the city center, including the oldest historic buildings. This area is also the most vulnerable to natural hazards. It has the most expensive new above-ground and underground construction, subway lines, complex traffic, and high-density telecommunications, as well as a dense population. This nighttime aerial photograph of Moscow also interestingly demonstrates the density of communications and possible damage

Fig. 20.1 Aerial photograph of Moscow at night

(Fig. 20.1). Thus, we can assume that the closer one gets to the center of Moscow, the greater is the potential damage from possible landslide process.

Hazardous industrial production is concentrated in Moscow's periphery. The protected zones of Moscow in Vorobiovy Hills and Kolomenskoye also have high cultural value, and the potential damage there is highly evaluated. Thus, a first approximation of a landslide risk map for Moscow may be an overlay of landslide hazard maps and population density, building density, land prices, road density, and infrastructure maps. Areas with the highest degree of landslide hazard and the highest damage would be the areas of highest landslide risk in the territory of Moscow.

The methodology for risk evaluation and mapping is proposed in the following section.

Methodology for Landslide Risk Mapping

For an automated analysis of the factual material and the construction of risk maps, we need to determine the intersection of the landslide hazard map and the integrated map of possible damage. That is, for each *i*th fragment *Ri* of the risk map, we need to calculate the product of the probability of a landslide event, *Pi*, for the amount of different *j*th possible damages from landslides, which could be damage to land, buildings, transportation, communications, people, and others:

$$R_i = P_i \sum_j D_{ij}$$

Maps of landslide hazards are calibrated from 0 to 1, which reflects the probability of landslide events ($0 \leq P \leq 1$). Thus, for example, gradation is possible on a scale of (0, 0,25, 0,5, 0,75, 1), where 0 corresponds to no danger of landslides, 0.25

is low danger, 0.5 is average danger, 0.75 is high danger, and 1 is very high danger of a landslide process. This assessment can predict natural events. In principle, it is possible to construct a landslide hazard map as the intersection of maps of factual material, such as maps of relief contrast, rock strength, slope stability, speed of motion of the surface, the density of rainfall, seismicity, etc. Of course, this would require additional research and evaluation.

For a comprehensive assessment of the damage in each region, it is suggested to calibrate the possible damage of each option on a three-point system (0, 1, 2), where 0 represents no damage, 1 is moderate damage, and 2 is high damage. For example, the parameters here would be cost of land, cost of housing, density of buildings, population density, and density of roads and communications. Higher values (the value of land, housing, etc.) predict greater damage in the case of a hazardous event. Thus, the total possible damage for these five parameters could vary from 0 to 10.

The risk for each element also ranges from 0 to 10. This is the risk in relative terms (high-low) on 10-point scale:

$$D_i = \sum_j D_{ij}, j = 1 - 5,\ D_{ij} = (0, 1, 2),\ 0 \leq D_i \leq 10,$$
$$0 \leq R_i \leq 10.$$

After dividing the map of the area into squares and calculating the risk for each square, we can obtain a map of the area at risk on a 10-point scale. On the basis of preliminary expert estimates, the areas at risk are in the vicinity of the Moscow and Yauza Rivers, as well as in the areas of contrasting relief along riverbeds of paleochannels in the city center.

Areas with high landslide risk include the Andronievskaya embankment (Figs. 20.2 and 20.3), the Nikolo-Yamskaya embankment (Fig. 20.4), the Kotelnicheskaya embankment (Fig. 20.5), and Samotechnaya Street (Fig. 20.6) in the center of Moscow.

Areas with the highest landslide risk include the Vorobiovy Mountains (Hills) (Figs. 20.7 and 20.8) and Kremlin Hill (Figs. 20.9 and 20.10). They are shown as white circles in the map of geological danger in Moscow (Fig. 20.11).

Fig. 20.2 Andronievskaya embankment with Svjato-Andronikov monastery

Fig. 20.3 Cracks near Svjato-Andronikov monastery

Fig. 20.4 Nikolo-Yamskaya embankment

Fig. 20.5 Kotelnicheskaya embankment

Fig. 20.6 Samotechnaya Street

Fig. 20.7 Vorobiovy Mountains with Moscow State University, ski jumps and Metro bridge

Fig. 20.8 Vorobiovy Mountains with the Russian Academy of Sciences building, Andreevsky monastery and new homes

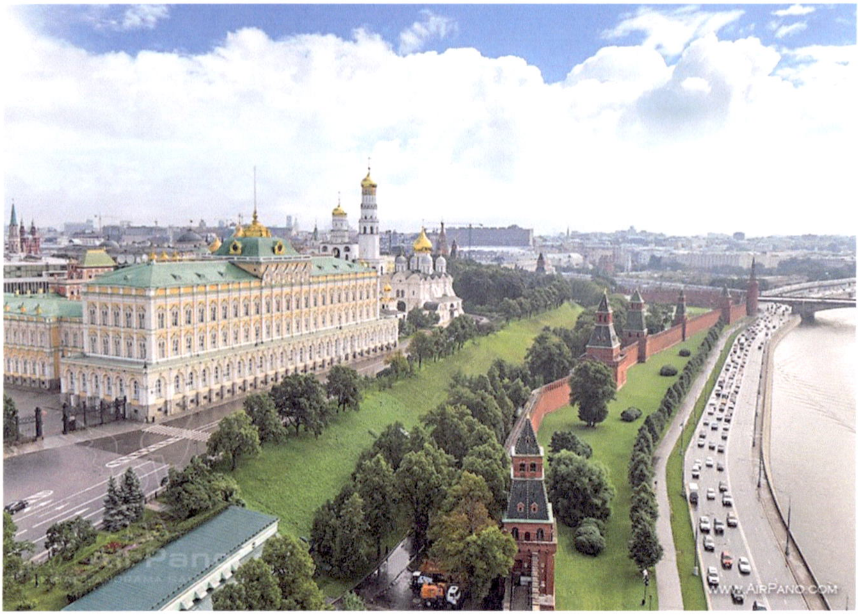

Fig. 20.9 Kremlin embankment

Fig. 20.10 Center of Moscow with Kremlin Hill and Moscow River

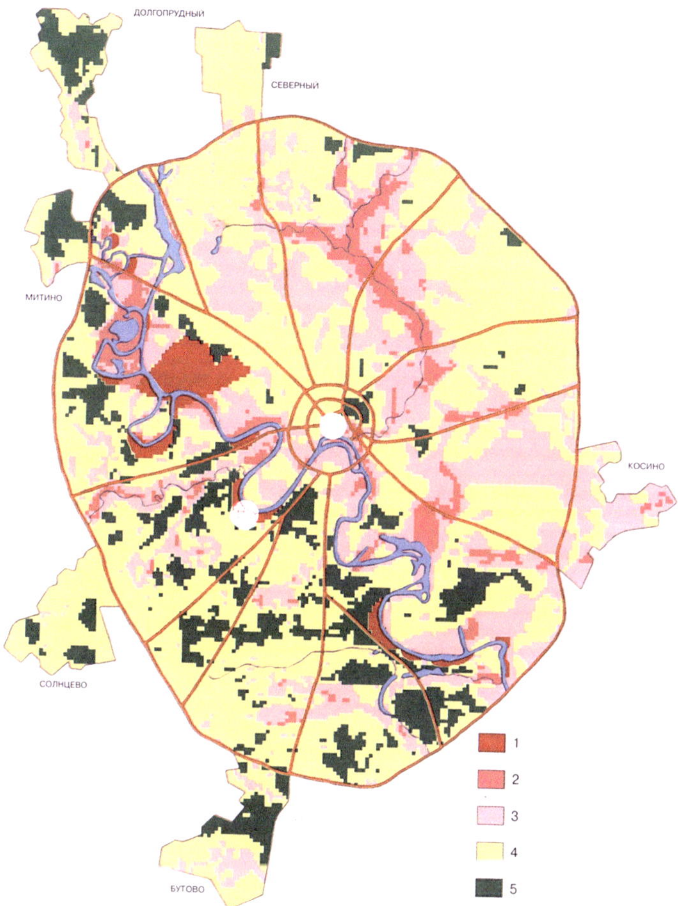

Fig. 20.11 Map of geological danger in Moscow, including landslides, karsts, and underflooding. (Osipov V.I., Kutepon V.M., Mironov O.K.) [3]. Landslides are shown near rivers in red and pink. 1, very high danger; 2, high danger; 3, moderate danger; 4, low danger; 5, no danger. White circles indicate risk "hot spots." Kremlin Hill (center) and Vorobiovy Mountaims (southwest) are shown.

These areas may be considered "hot spots" on the risk map. Even though the population density is not very high in some of these areas, other characteristics (e.g., cost of land, the historical importance of the object, the density of underground utilities) are greater contributors to the high risk assessment. These areas must be prioritized for risk management and reduction, including monitoring organization, strengthening the slope, restricting the construction of new buildings, and limiting activity.

References

1. Corominas, J., van Westen, C., Frattini, P., Cascini, L., Mallet, J.-P., et al. (2014). Recommendations for the quantitative analysis of landslide risk. *Bulletin of Engineering Geology and the Environment, 73*(2), 209–263.
2. Kutepov, V. M., Sheko, A. I., Anisimova, N. G., Burova, V. N., Victorov, A. S., et al. (2002). *Natural hazards in Russia. Exogenous geological hazards*. Moscow: KRUK.
3. Kutepov, V. M., Postoev, G. P., & Svalova, V. B. (2004). Landslide hazards estimation on sites of modern and historical constructions in Moscow. Proceedings of 32 IGC, Italy, 2004.
4. Osipov, V., & Medvedev (1997). Moscow. Geology and town. Moscow textbooks and kartolitografiya, Moscow.
5. Osipov, V. I., Shojgu, S. K., Vladimirov, V. A., Vorobjev, Y. L., Avdod'in, V. P., et al. (2002). *Natural hazards in Russia. Natural hazards and society*. Moscow: KRUK.
6. Postoev, G. P., Erysh, I. F., Salomatin, V. N., et al. (1989). *Artificial activation of landslides*. Russia: Nedra.
7. Postoev, G. P., & Svalova, V. B. (2005). *Landslides risk reduction and monitoring for urban territories in Russia. Proceedings of the First General Assembly of ICL (International Consortium on Landslides), Landslides: Risk analysis and sustainable disaster management* (pp. 297–303). Washington, DC: Springer.
8. Ragozin, A. (Ed.). (2003). *Natural hazards of Russia. Evaluation and management of natural risk*. Moscow: KRUK.
9. Svalova, V. B. (2001). Mechanical-mathematical modeling and monitoring for landslide processes. *Journal of Environmental Science nd Engineering, 5*(10), 1282–1287.
10. Svalova, V. B. (2011). Monitoring and modeling of landslide processes. *Monitoring Science and Technology, 2*(7), 19–27.
11. Svalova, V. B. (2011). Landslide process simulation and monitoring. Proceedings of ENGEOPRO, Moscow.
12. Svalova, V. B. (2012). Mechanical-mathematical modeling and monitoring for landslides. Proceedings of IPL (International Program on Landslides) Symposium, UNESCO, Paris.
13. Svalova, V. B. (2014). Modeling and monitoring for landslide processes. In K. Linwood (Ed.), *Natural disasters – typhoons and landslides – risk prediction, crisis management and environmental impacts* (pp. 177–198). New York: Nova Science Publishers.
14. Svalova, V. B. (2014). *Mechanical-mathematical modeling and monitoring for landslide processes*. IPL 163 Project. Proceedings of the World Landslide Forum 3 (Vol. 4, pp. 24–27). Beijing.
15. Svalova, V. B. (2014). *Modeling and monitoring for landslide processes: Case study of Moscow and Taiwan*. Proceedings of the World Landslide Forum 3 (Vol. 4, pp. 628–632). Beijing.
16. Svalova, V. B. (2015). *Mechanical modeling and geophysical monitoring for landslide processes*. Proceedings of IAEG XII Congress "Engineering geology for society and territory" (Vol. 2, pp. 345–348), Torino-2014, Italy, Springer.
17. Svalova, V. B. (2016). Monitoring and modeling of landslide hazard in Moscow. *Engineering Protection, 1*(12), 34–38.
18. Svalova, V. B., & Postoev, G. P. (2008). *Landslide process activization on sites of cultural heritage in Moscow, Russia*. Proceedings of the First World Landslide Forum 2008, Tokyo, Japan.
19. Svalova, V. B. (2016). Monitoring and reducing the risk of landslides in Taiwan. *Monitoring. Science and Technology, 3*, 13–25.
20. Svalova, V. B. (2016). Analysis of landslide risk in Taiwan. "Commonwealth". *Russia-China Scientific Journal, 4*, 136–141.
21. Svalova, V. B. (2016). Analysis and management of risk of landslides. *Scientia. Physics and Mathematics, 2*, 28–31.

22. Svalova, V. B. (2016). Reducing the rissk of landslides. *Uniform All-Russia Scientific Bulletin, 2*(3), 79–83.
23. Svalova, V. B. (2016). Landslides modeling, monitoring, risk management and reduction. *EESJ (East European Scientific Journal, Poland), 7*(11), 43–52.
24. Svalova, V. B. (2016). Risk analysis, evaluation and management for landslide processes. *Sciences of Europe (Praha, Czech Republic), 4*(6), 15–25.
25. Svalova, V. (2017). *Landslide risk: Assessment, management and reduction*. New York: Nova Science Publishers.

Chapter 21
Mechanical-Mathematical Modeling for Landslide Processes

Valentina Svalova

Abstract Mechanical-mathematical model of high viscous fluid was used for modeling of the matter movement on landslide slopes. Equation of continuity and approximated Navier-Stokes equation for slow motions in a thin layer of the matter were used. The results of modeling give a possibility to define the place of highest velocity on landslide surface, which could be the best place for monitoring post position. A model can be used for comparison of calculated and measured velocities of the matter and gives possibility to investigate some fundamental aspects of the matter movement on landslide slope.

21.1 Introduction

The most popular and maybe important problem in landslide research is landslide slope stability. The beginning of the process is very important for safety problems. It is rather well investigated and widely known [1–5]. There are many criteria for estimation and calculation of the slope stability. Investigations are provided on the basis of elastic model mostly and cannot describe velocities of the process and development of dangerous situation. It is impossible to say that the problem of landslide stability has final decision. Landslides arise suddenly very often. And even if we know that landslides arise after heavy rains or strong earthquakes, it is not always possible to forecast the place and intensity of landslide, to assess landslide risk and vulnerability of objects, territories, and people.

Every type of landslides demands specific approach and method. It is important if there is separation of material from the slope. Different methods are used for rockfalls or debris flow.

Each analytical or numerical method has advantages and disadvantages. It is difficult to solve all deformation process, from initial small deformation to large

V. Svalova (✉)
IEG RAS, Moscow, Russia
e-mail: inter@geoenv.ru

V. Svalova (ed.), *Natural Hazards and Risk Research in Russia*,
Innovation and Discovery in Russian Science and Engineering,
https://doi.org/10.1007/978-3-319-91833-4_21

deformation with high calculation accuracy. It is important for an engineer to select an appropriate mechanical-mathematical and numerical method for each deformation level based on engineering judgment.

One of the methods of studying landslide processes is mechanical-mathematical modeling of gravitational movement of the matter on landslide slope. At different stages of the process development, the landslide movement can be described by various mechanical and rheological models. At the stage of formation of cracks, losses of stability, and break of blocks, the models of the elastic medium and model of destruction are applied. During slow movement of soil on the slope, the model of high viscous incompressible fluid can be used. Such model allows to calculate velocities of movement in the layer of the matter and to compare them with results of velocity monitoring. Boundary conditions of the problem also depend on concrete situation.

So, the condition of sticking is used on the bottom border of the layer in case of slow movement. The condition of sliding or more complex boundary condition is possible on the bottom border, if the process of debris flow, underwater landslide, or snow avalanche is considered. The choice of adequate model of the process and statement of initial and boundary conditions are the special mechanical problems.

21.2 Mechanical-Mathematical Model

Let's consider movement of landslide masses on the slope as movement of high viscous incompressible fluid described by equation of Navier-Stokes and continuity:

$$\frac{d\vec{v}}{dt} = \bar{F} - \frac{1}{\rho}\,\mathrm{grad}p + \frac{\mu}{\rho}\Delta\vec{v}$$

$$\mathrm{div}\ \vec{v} = 0$$

$\vec{v}$, vector of velocity; F, force of gravity; p, pressure; ρ, density, μ, viscosity; t, time.

Such a model can be successfully used for the lithosphere movement simulation [6–9] and for soil movements on the slope [10–32]. The problem is to estimate effective viscosity for real matter. It can be done by comparing calculated and measured velocities of the matter movements.

Let the characteristic horizontal scale of a body of landslip L considerably surpasses its thickness h. We shall count also a landslip extended enough in the plan that allows to consider three-dimensional model as two-dimensional one for sections of landslide bodies. Following works [6–9] and applying the method of decomposition on small parameter, it is possible to get the equation of continuity and an approximated equation of the Navier-Stokes in dimensionless form for slow motions in a thin layer:

$$\begin{cases} \dfrac{\partial P}{\partial X} = \alpha\mu \dfrac{\partial^2 U}{\partial Z^2} \\ \dfrac{\partial P}{\partial Z} = -\rho \end{cases}$$

$$\frac{\partial U}{\partial X} + \frac{\partial W}{\partial Z} = 0$$

$$\alpha = \frac{F}{R\left(\frac{h}{L}\right)^3}$$

$$R = \frac{u_0 L \rho_0}{\mu_0}$$

$$F = \frac{{u_0}^2}{gL}$$

P is dimensionless pressure; U, W, dimensionless velocities; F, Froude number; R, Reynolds number; ρ, density; μ, viscosity; ρ_0, μ_0, u_0, scales of density, viscosity, and velocity.

Then it is possible to get the velocities and pressure in the layer:

$$P = \rho(\varsigma^* - Z)$$

$$U = U_0 + \frac{\rho}{2\alpha\mu}\frac{\partial\varsigma^*}{\partial X}\left[(\varsigma^* - Z)^2 - (\varsigma^* - \varsigma_0)^2\right]$$

$$W = W_0 + \frac{\partial U_0}{\partial X}(\varsigma_0 - Z) +$$
$$\frac{\rho}{\alpha\mu}\frac{\partial^2\varsigma^*}{\partial X^2}\left[\frac{1}{6}(\varsigma^* - Z)^3 + \frac{1}{3}(\varsigma^* - \varsigma_0)^3 -\right.$$
$$\left.\frac{1}{2}(\varsigma^* - Z)(\varsigma^* - \varsigma_0)^2\right] +$$
$$\frac{\rho}{2\alpha\mu}\left(\frac{\partial\varsigma^*}{\partial X}\right)^2(Z - \varsigma_0)^2 + \frac{\rho}{\alpha\mu}\frac{\partial\varsigma^*}{\partial X}\frac{\partial\varsigma_0}{\partial X}\varsigma_0(\varsigma^* - \varsigma_0)$$

ς_0 – the bottom border of the layer;
ς^* – the top border.

Let on the bottom border the condition of sticking is satisfied:

$$U_0 = W_0 = 0$$

The discharge of matter along the layer is:

$$Q = \int_{\varsigma_0}^{\varsigma^*} U dZ = -\frac{\rho}{3\alpha\mu}\frac{\partial \varsigma^*}{\partial X}(\varsigma^* - \varsigma_0)^3$$

Since $Q = \text{const}$ along the X, then:

$$\frac{\partial Q}{\partial X} = 0$$

$$\frac{\partial^2 \varsigma^*}{\partial X^2}(\varsigma^* - \varsigma_0)^7 + \frac{9\alpha\mu Q}{\rho}\left[\frac{3\alpha\mu Q}{\rho} + (\varsigma^* - \varsigma_0)^3 \frac{\partial \varsigma_0}{\partial X}\right] = 0$$

The condition of convexity of upper boundary is:

$$\frac{\partial^2 \varsigma^*}{\partial X^2} < 0 \Rightarrow$$

$$\frac{3\alpha\mu Q}{\rho} > -(\varsigma^* - \varsigma_0)^3 \frac{\partial \varsigma^0}{\partial X}$$

This condition enables to analyze the form of the surface of moving matter (Fig. 21.1).

Structure of clinoforms (convex) can arise, if:

1. Q is large, that is, flux is high.
2. μ is large. It means that matter spreads bad and can support big angle.
3. ρ is small. It means that matter has large specific volume and is friable.

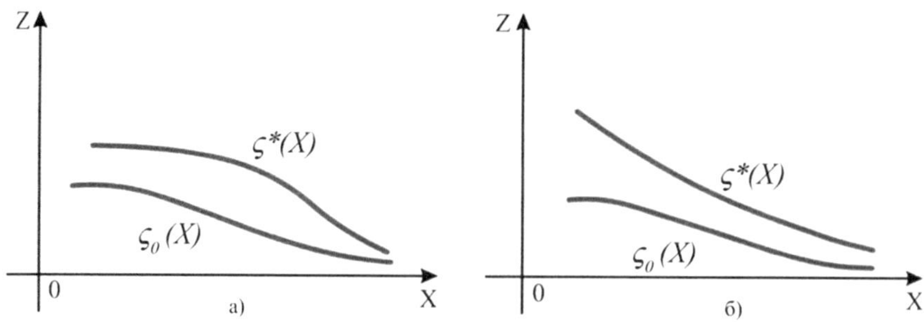

Fig. 21.1 Sketches of dimensionless landslide surface. Various possible forms of landslide ground surfaces: (**a**) convex, (**b**) concave

4. grad ς_0 is small, that is, angle of lower boundary is small.
5. $(\varsigma^* - \varsigma_0)$ is small, that is, thickness of sedimentary layer is small. Under fixed Q it means that velocity of flux is high and formation of clinoforms and even overturning of rocks are possible.

All these conditions seem to be natural enough for the explanation of formation of structures such as inflows and clinoforms of sedimentary cover that confirms correctness of the model.

It is important to define the place of maximal velocity on the slope. An optimum place for location of monitoring post is the point of maximal speeds of movement of landslide masses.

Let's consider the massif of sedimentary rocks with the top border ς^* representing landslide slope. The bottom border ς_0 is compatible with an axis X. The maximum of horizontal speed U is reached on the top border ς^* of the massif according to condition:

$$\frac{\partial U}{\partial Z} = -\frac{\rho}{\alpha\mu}\frac{\partial \varsigma^*}{\partial X}(\varsigma^* - Z) = 0 \Rightarrow Z = \varsigma^*$$

The point of the maximal horizontal speed on the surface ς^* can be found from condition of equality to zero of the first derivative:

$$\frac{\partial U^*}{\partial X} = 0, \text{ where } U^* = -\frac{\rho}{2\alpha\mu}\frac{\partial \varsigma^*}{\partial X}(\varsigma^*)^2$$

From here it is easy to receive the condition:

$$\frac{\partial^2 \varsigma^*}{\partial X^2}\varsigma^* + 2\left(\frac{\partial \varsigma^*}{\partial X}\right)^2 = 0 \tag{21.1}$$

It is necessary to stress that $\varsigma^*(X)$ is a known function – the surface of landslide slope. And the received condition allows to find a point on a slope where speed of movement is maximal.

Let's consider for illustration of the received decision the surface of landslide as (Fig. 21.2):

$$\varsigma^*(X) = -\text{th}X + 1$$

It reflexes roughly the form of landslide surface.
Then the condition (21.1) gives:

$$\text{th}^2X - \text{th}X - 1 = 0$$

Whence we receive th$X = \frac{1-\sqrt{5}}{2}$ and

$$\varsigma^* = \frac{1+\sqrt{5}}{2} \approx 1.62$$

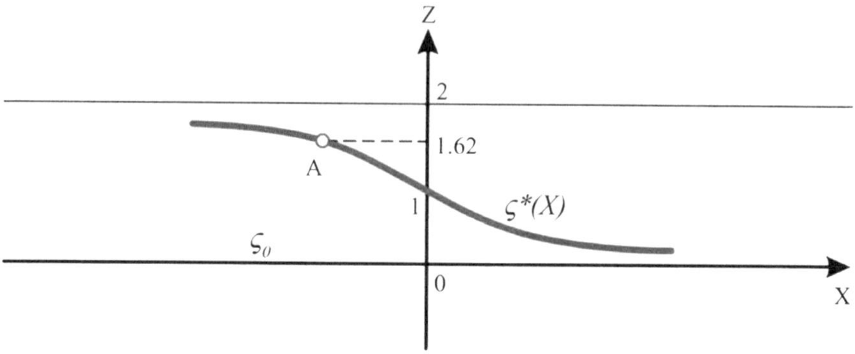

Fig. 21.2 Sketch of dimensionless slope surface. Point A is the point of maximal horizontal speed of movement of masses on the ground surface of the slope

Such position of the point of the maximal horizontal speed seems to be real, and more exact data on the structure of landslide and its surface will enable to define such point on a concrete slope. The point of maximum of speed on a slope defines the place of possible failure of a landslip in case of achievement of limiting pressure in massif of rocks.

There could be several points of local maximum of speed on a slope that characterizes an opportunity of failure of a landslip on each terrace of a slope.

The places of minimum speed can be used for building constructions and oil-gas pipelines. Model velocities of matter can be compared with real velocities that give possibility to define the real mechanical parameters of media.

21.3 Conclusion

Mechanical-mathematical model of high viscous fluid is elaborated for simulation of matter movement on landslide slopes. The results of modeling give possibility to define the place of highest velocity on landslide surface, which could be the best place for monitoring post position. Model can be used for comparison of calculated and measured velocities and estimation of effective viscosity of real matter. The results of modeling give possibility to investigate some fundamental aspects of landslide matter movement.

References

1. Kutepov, V. M., Sheko, A. I., Anisimova, N. G., Burova, V. N., Victorov, A. S., et al. (2002). *Natural hazards in Russia. Exogenous geological hazards*. Moscow: KRUK.
2. Kutepov, V. M., Postoev, G. P., & Svalova, V. B. (2004). *Landslide hazards estimation on sites of modern and historical constructions in Moscow*. Proceedings of 32 IGC, Italy, 2004.

3. Osipov, V., & Medvedev, O. (Eds.). (1997). *Moscow. Geology and town*. Moscow: Moscow Textbooks and Kartolitografiya.
4. Osipov, V. I., Shojgu, S. K., Vladimirov, V. A., Vorobjev, Y. L., Avdod'in, V. P., et al. (2002). *Natural hazards in Russia. Natural hazards and society*. Moscow: KRUK.
5. Postoev, G. P., & Svalova, V. B. (2005). *Landslides risk reduction and monitoring for urban territories in Russia*. Proceedings of the First General Assembly of ICL (International Consortium on Landslides), Landslides: risk analysis and sustainable disaster management, Washington, DC: Springer (pp. 297–303).
6. Svalova, V. B. (1975). *The models of the lithosphere movements*. Ph. D. Thesis at Moscow State University, 130 p.
7. Svalova, V. B. (1992). Mechanical-mathematical models of the formation and evolution of sedimentary basins. *Science de la Terre, Series Information, Nancy, 31*, 201–208.
8. Svalova, V. B. (1993). Mechanical-mathematical simulation of geological structures evolution. *Geoinformatics, 4*(N3), 153–160.
9. Svalova, V. B., & Sharkov, E. V. (1992). Mantle diapirs and surface depression in back-arc areas: A rheological model. *Ofiolity, 17*(1), 165–170.
10. Fathani, T. F., & Nakamura, H. (2005). *Numerical analysis of the movement distance and velocity of landslides*. Proceedings of the International Conference on Landslides, Vancouver, Canada, 10 p.
11. Lang, Y. H., & Nakamura, H. (1998). Characteristics of slip surface of loess landslides and their hazard area prediction. *Journal of Japan Landslide Society, 35*(1), 9–18.
12. Nakamura, H., & Fathani, T. F. (2002). *Hazard area prediction for landslide debris*. Proceedings of the Tenth International Conference and Fieldtrip on Landslide, Krakow (pp. 129–142).
13. Nakamura, H., Tsunaki, R., & Ishihama, S. (1989). *Simulation model for debris movement of landslides*. Proceedings of the Japan-China Symposium on Landslides and Debris Flows, Niigata, Tokyo (pp. 81–86).
14. Suzuki, K. (2001). *Estimation of the travel distance of landslide debris in granite area*. Master Thesis at Tokyo Univ. of Agriculture and Technology. Tokyo (pp. 32–47).
15. Svalova, V. B. (2011). Mechanical-mathematical modeling and monitoring for landslide processes. *Journal of Environmental Science nd Engineering, 5*(10), 1282–1287.
16. Svalova, V. B. (2011). Monitoring and modeling of landslide processes. *Monitoring. Science and Technology, 2*(7), 19–27.
17. Svalova, V. B. (2011). *Landslide process simulation and monitoring*. Proceedings of ENGEOPRO, Moscow.
18. Svalova, V. B. (2012). *Mechanical-mathematical modeling and monitoring for landslides*. Proceedings of IPL (International Program on Landslides) symposium, UNESCO, Paris.
19. Svalova, V. B. (2014). Modeling and monitoring for landslide processes. In K. Linwood (Ed.), *Natural disasters – Typhoons and landslides – Risk prediction, crisis management and environmental impacts* (pp. 177–198). New York: Nova Science Publishers.
20. Svalova, V. B. (2014). *Mechanical-mathematical modeling and monitoring for landslide processes*. IPL 163 Project. Proceedings of the World Landslide Forum 3. Beijing. (Vol. 4, pp. 24–27).
21. Svalova, V. B. (2014). *Modeling and monitoring for landslide processes: Case study of Moscow and Taiwan*. Proceedings of the World Landslide Forum 3. Beijing (Vol. 4, pp. 628–632).
22. Svalova, V. B. (2015). *Mechanical modeling and geophysical monitoring for landslide processes*. Proceedings of IAEG XII Congress "engineering geology for society and territory", 2, Torino-2014, Italy, Springer (pp. 345–348).
23. Svalova, V. B. (2016). Monitoring and modeling of landslide hazard in Moscow. *Engineering Protection, 1*(12), 34–38.
24. Svalova, V. B., & Postoev, G. P. (2008). *Landslide process activization on sites of cultural heritage in Moscow, Russia*. Proceedings of the First World Landslide Forum 2008, Tokyo.

25. Svalova, V. B. (2016). Monitoring and reducing the risk of landslides in Taiwan. *Monitoring. Science and Technology, 3*, 13–25.
26. Svalova, V. B. (2016). Analysis of landslide risk in Taiwan. "Commonwealth". *Russia-China Scientific Journal, 4*, 136–141.
27. Svalova, V. B. (2016). Analysis and management of risk of landslides. *Scientia. Physics and Mathematics, 2*, 28–31.
28. Svalova, V. B. (2016). Reducing the risk of landslides. *Uniform All-Russia Scientific Bulletin, 2* (3), 79–83.
29. Svalova, V. B. (2016). Landslides modeling, monitoring, risk management and reduction. *EESJ (East European Scientific Journal, Poland), 7*(11), 43–52.
30. Svalova, V. B. (2016). Risk analysis, evaluation and management for landslide processes. *Sciences of Europe (Praha, Czech Republic), 4*(6), 15–25.
31. Svalova, V. (2017). *Landslide risk: Assessment, management and reduction* (Vol. 253). New York: Nova Science Publishers.
32. Tsurugaya, K. (2001). *Study of the occurrence and flow mechanism of the large-scale landslides caused by an earthquake*. Master Thesis at Tokyo Univ. of Agriculture and Technology. Tokyo (pp. 80–108).

Chapter 22
Structural Transformations of Permafrost before the Formation of the Yamal Craters

A. N. Khimenkov, D. O. Sergeev, Y. V. Stanilovskaya, A. N. Vlasov, D. B. Volkov-Bogorodsky, V. P. Merzlyakov, and G. S. Tipenko

Abstract This article deals with the genesis of the gas discharge funnel located 30 km south of the Bovanenkovskoye gas condensate field and named the Yamal crater. A new approach contains the theory of the formation of similar funnels due to the structural transformations of the frozen massif under the influence of gases coming from the decomposition of gas hydrates. The leading role of lakes in the formation of local zones of dissociation of gas hydrates was declared. The estimation of the pressure that is necessary for migration of fluids in frozen rocks and the explosion that formed the Yamal crater was carried out.

Legend of Math Symbols

h	Ejection depth, meters
ϕ	Angle of inclination to the conical surface, radians
H	Height of conical body, meters
R_1	Radius of the base of the conical body, meters
R_2	Upper radius of conical body, meters
$[\sigma]$	Maximum permissible tensile stress, mPa
$[\tau]$	Maximum allowable shear stress, mPa
F	Vertical force of adhesion of the lateral surface to the ground, newtons
G	Strength of body weight, newtons
P_0	Atmospheric pressure, newtons
$\rho_{мг}$	Frozen soil density, t/m^3
g	Acceleration of gravity, m/s^2
p_h^*	Critical ejection pressure from the depth, mPa

A. N. Khimenkov (✉) · D. O. Sergeev · Y. V. Stanilovskaya · A. N. Vlasov
D. B. Volkov-Bogorodsky · V. P. Merzlyakov · G. S. Tipenko
Sergeev Institute of Environmental Geoscience RAS (IEG RAS), Moscow, Russia
e-mail: cryo2@yandex.ru

V. Svalova (ed.), *Natural Hazards and Risk Research in Russia*,
Innovation and Discovery in Russian Science and Engineering,
https://doi.org/10.1007/978-3-319-91833-4_22

22.1 Introduction

In 2014 a round funnel was found on Yamal, 30 km south of the Bovanenkovskoye oil and gas condensate field (NGKM). It has about 50 m deep with 30 m diameter in the upper part. From the depth of 8 m, the diameter decreases to 20–24 m. A concentric shaft from the ejected rock is gathered around the funnel. There were no traces of economic activity nearby. Later, in the north of Western Siberia, several similar funnels were found. Most researchers who studied this phenomenon agree that their formation is due to the release of gases that either were formed during the dissociation of gas hydrates or gases coming from great depths [2, 3, 4, 11, 15, 17, 19]. This gave rise to V. I. Bogoyavlensky who interpreted the field data as an evidence of a gas emission [3]. Although this notion is fixed and widely used, it is rather vague and allows a broad interpretation. Emissions, for example, mean the short-term or long-term ejections of any pollutants or heat [6]. The morphology of the funnels and the scatter of rocks around allow an interpretation of the process of their formation as a physical explosion. A physical explosion is associated with the emergence of internal pressure in a cavity filled with a liquid or a gas exceeding the maximum permissible values of the strength characteristics of the enclosing medium. The main reason for their formation is not a chemical reaction but a short-time physical process caused by the transfer of stored potential energy to kinetic energy of the motion of gases and liquids [8]. After that, a closed cavity with a gaseous or liquid substance collapses due to elevated pressure, and then ground ejection occurs with the generation of pressure waves. In this case, the potential energy of the compressed gas or liquid passes into the energy of the shock wave. We will use the established, generally accepted definition of "gas release," as a physical explosion of underground gas.

The purpose of the publication is to consider a new type of geocryological hazard – gas emissions in permafrost following the decomposition of gas hydrates.

22.2 Object of Study

Most amount of information on the structure of funnels of a gas emission was obtained by studying a funnel located 30 km south of the Bovanenkovskoye gas condensate field. The morphology of this phenomenon is the widened conical upper part and the narrowed cylindrical lower part that is very similar to the volcanic crater. As will be shown below, the processes forming the funnel have the analogy with volcanic processes. Therefore, for the considered given object, we use the name Yamal crater offered by Novosibirsk researchers [11].

22.3 Stages of Formation of the Yamal Crater

A study of the structure of deposit in the walls of the Yamal crater made it possible to single out the following sequence of processes that had formed it.

Stage I. Formation of a lake over a perennial frozen ground massif that contains a layer of massive ice. A bowl of thawing is forming under the lake as well as the zone of heightened temperature in lower part of the massif of rocks (Figs. 22.1 and 22.2).

Stage II. The temperature-rising zone in permafrost reaches the gas hydrate layer. The process of dissociation of gas hydrates begins with the release of gas and "supercooled" water. Gas-water fluids under the influence of pressure begin to migrate to the least stable area, which is the zone of high temperature under the lake. They saturate the primary frozen substrate with the gas. On contact with the horizon of hydrate-saturated rocks, a zone of caverns and grottoes measuring tens of centimeters is formed. As thawing, small caverns merge, forming large caves with smooth walls several meters in size. The estimated volume of methane gas hydrate in Yamal crater is about 12.5 m^3 of hydrate [16]. Such insignificant volume completely corresponds to the zone of small caverns at the bottom of the Yamal crater. Similar formations were found in the zones of gas emissions due to decomposition of gas hydrates in the Okhotsk Sea. Here, in the area of the gas fountain, the characteristic forms are found: failure funnels, pits, bottom niches, and caves that are forming a system of interconnected grottoes [10]. Probably the

Fig. 22.1 Aerial view of Yamal crater. (Photo from A. V. Lupachev)

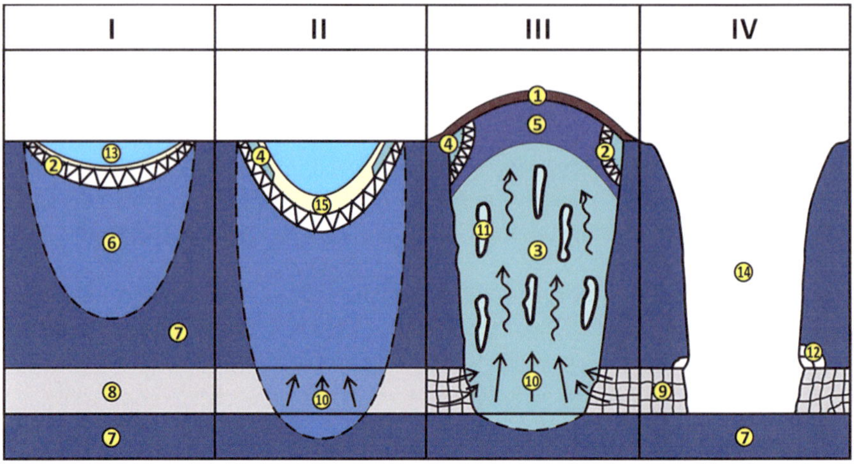

Fig. 22.2 Stages of development of the Yamal crater (stages I, II, III, IV) [13, 14]. Legend: 1, Cover horizon; 2, Layer of the ice between thawed and frozen rocks; 3, Frozen gas-saturated ice-bounded rock with traces of plastic deformations; 4, Infiltration-segregating ice; 5, Gastight roofing of permafrost; 6, Zone of temperature increase in permafrost under the lake; 7, Permafrost outside the heating effect of the lake; 8, Layer of hydrate-containing permafrost; 9, Zone of decompaction in the layer of hydrate-containing permafrost around the crater; 10, Direction of the movement of fluids; 11, Gas fluids; 12, Caves and caverns in the lower part of the crater; 13, Lake; 14, Crater formed after the release of a gas-saturated deposits; 15, Talik. (Picture is drawn by D. Volkov)

development of a system of interconnected cavities is typical for the dissociation of hydrate-saturated rocks (Fig. 22.2).

Stage III. The decomposition of a metastable gas hydrate causes an increase in pressure, which greatly exceeds the reservoir pressure. As a result, the fluids move upward with accompanying plastic and discontinuous deformations. The increase of the pressure in the fluid migration zone leads to distension under the bottom of the thermokarst lake. As the basin is shoaling and the talik freezes, the infiltration-segregated ice is forming due to migration at the lower border of the talik. The transition from subaquatic to subaerial state leads to decreasing of the temperature and increasing of the ground strength. At the same time, a frozen dome-shaped screen is formed. The fragments of this layer are observed during the initial survey of the Yamal crater. A frozen screen is an obstacle to the movement of fluids coming from the bottom. It leads to large mound rising at the Earth surface. The fluid saturation gradually forms a rod in the initial layered ice-bound ground massif. The ground here contains the layered gas-saturated ice with numerous cells of various sizes. The thickness of the formed rod is about 40 m, and the diameter is about 15 m (Fig. 22.2).

Stage IV. When the plastic deformations in the frozen roof reach their limit, the fragile deformation fractures occur in it. The pressure in gas inclusions becomes much greater than the atmospheric pressure. Overcoming the limit of strength of the ground roof leads to sudden drop in pressure. We use the two-stage model of the volcanic explosion to describe the explosion process that formed the Yamal crater [1]. At the stage of preparation, a porous gas-saturated magmatic body is formed in the vole of the volcano. At the stage of development of the explosion, the fragmentation (destruction) of this body occurs because of sharp decompression and releasing of potential energy of compressed gases by eruption of crushing products into the atmosphere. The explosion wave from unloading gradually destroys the porous material of the massif layer by layer [1].

Similar processes are observed in the formation of the Yamal crater. A gas-saturated ice-soil body covered by a gastight deformable roof was formed as a result of the penetration of gas fluids from the dissociation of gas hydrates. The pressure in the gas inclusions is much greater than the atmospheric pressure. After the overcoming of the limit of the roof strength, the pressure dramatically decreases. The pressured gas in the ice-soil gas-saturated rod successively launches the frozen rock layer by layer. As a result a dry crater of gas emission was formed (Fig. 22.2).

22.4 Change in Temperature of Permafrost Under the Influence of Climate and Surface Water

The available permafrost temperatures in Yamal Peninsula show the insufficiency of the climate changes disturbing effects for the possibility of gas hydrates decomposition at depths corresponding to the depth of the funnel (about 60 m). The geotemperatures from the geocryological observatory "Marre-Salé VSEGINGEO" show the fluctuation in the range −5 to −6 °C from 2001 to 2015 [9]. The interval of permanent temperature distribution is observed at depths from 20 to 60 m because the water phase transitions belt in saline deposit layers. Minor fluctuations from climate warming are smoothing out by phase transitions. The data considered allow us to doubt the statements of some authors about the leading role of the sharp increase in summer temperatures during the several years preceding the formation of the Yamal crater [15, 18].

Surface water bodies are the most powerful heating factor in the permafrost zone. Therefore, it is necessary to assessing the permafrost temperature field under the lakes. For the north of Western Siberia, where craters of the gas emission have been found, there are no direct data on the temperatures of subaqueous ground media. The warming effect of the lakes on permafrost can be assessed by analyzing the temperatures in the wells drilled on the "khatsyreys" (large thermokarst depression), where lakes existed earlier. Here, at a depth of 60–80 m, the temperature of the frozen rocks for the third marine terrace is in the range from −3 to −2.5 °C and −1.8 to −1.6 ° C for the floodplain. Temperatures of permafrost outside the lake zones at the depth of 60–80 m as −5.5 to −5.0 °C were found in the third marine terrace and floodplain [7].

G. S. Tipenko performed the calculations of the stationary state of the Yamal permafrost temperature field for different lake dimensions. He showed that under a lake up to 90 m wide, the permafrost rocks do not thaw completely. It corresponds to the diameter of the base of the hillock antecedent the Yamal funnel [14]. According to calculations, the permafrost temperature has a range of −2.0 to −2.5 °C at a depth of 60–80 m. It corresponds to the full-scale measurements (Fig. 22.3).

The zones of warming up to −1 to −3 °C under lakes growing-up even in the low-temperature permafrost conditions. The vertical dimension of these zones can exceed 100 m. When the permafrost temperature is rising, the rock strength decreases, and local defects appear in association with volumetric temperature stresses. The increasing of the temperature under the lake affects the structure and properties of permafrost rocks and ice and forms favorable conditions for the penetration of gas fluids in the massif. It's important that the dissociation of gas hydrates and migration of fluids can occur at a negative but heightened temperature under real or past lake areas.

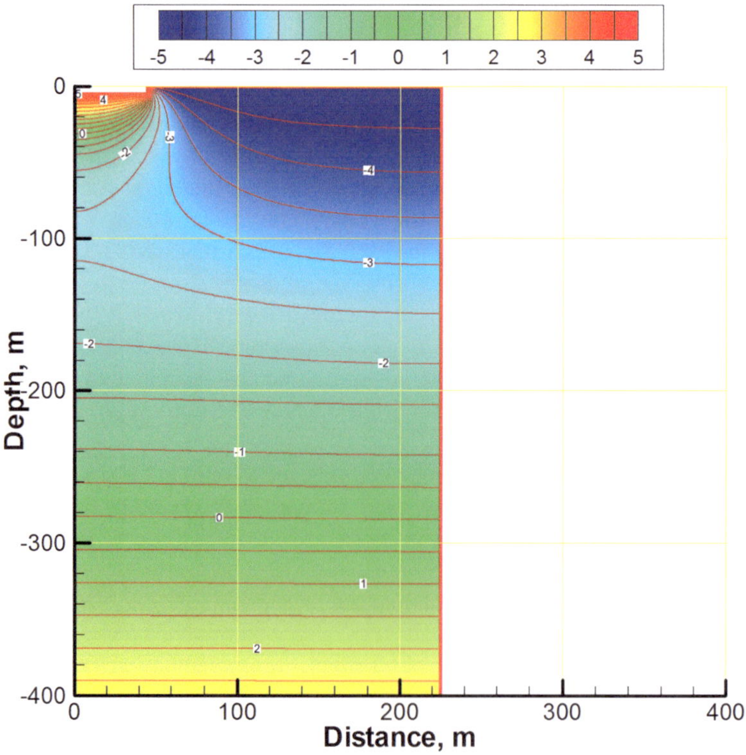

Fig. 22.3 Stationary temperature field under the lake 45 m wide

22.5 The Gas Pressure Triggers the Yamal Crater Formation

The gas hydrate decomposition increases the pressure, which can significantly exceed not only the hydrostatic but also the reservoir pressure [22]. At temperatures in the range of 268.15 K (−5 °C) ÷ 273.15 K (0 °C), the equilibrium pressure is in the range 2.57–2.17 MPa in the system of gas-water (ice)-hydrate [20]. In the area of local permafrost warming under the lakes, soil temperatures are in the range of −1 to −3 °C. A system of pores, defects, and weakened zones is developed in the frozen soils and ice, so pore pressure corresponds to hydrostatic pressure. At a depth of 60–80 m, where a hydrate-containing layer is detected by geophysical survey [19], the hydrostatic pressure is about 0.7 MPa. At this depth in the zone of gas hydrate decomposition, pressures are created higher three times than the hydrostatic pressure. This will cause the filtration of the gas fluids upward. A pressure increase in the upper part of the frozen soils will lead to plastic deformations first, growth of the mound, and subsequent explosion.

During gas hydrate decomposition, a certain balance between pressure and temperature is observed. In the case of incomplete decomposition, when the thermobaric conditions change, the temperature or pressure is returned to the equilibrium value [5]. In the R. Hunt's experiments of modeling the conditions of gas hydrate decomposition in the core, it was shown that after each pressure drop, it returned to values corresponding to the reservoir [12]. In the case of the Yamal crater, the pressure resulting from gas hydrate dissociation will be maintained throughout the formation of gas-saturated ice. In the dissociation zone, the pressure will decrease due to the outflow of gas fluids, but it will be constantly reestablished as long as the thermobaric conditions are preserved or the gas hydrates exist.

We estimate the minimum pressure that is necessary for the explosion, which caused the the Yamal crater formation. It is assumed that the gas under pressure pushes out a body made up of a conical and cylindrical parts (Fig. 22.4). The required pressure value depends on the depth of the ejection.

The conical walls at the upper part of the crater have traces of soil fissure at an angle of 40–60° to the horizontal, while the almost vertical walls of the lower part have smooth surface of frozen soil [4]. Therefore, the upper part of the ejected body is represented in the shape of an inverted truncated cone with lower radius R_1 and upper radius R_2. The height is equal to H.

The detachment occurs under the tensile stresses σ distributed along the lateral surface of the cone. In order to "break" the upper part of the ejected body by pressure from the depth H, it is necessary to overcome the vertical adhesion of the lateral surface with the soil (F), the weight force of the cone (G), and the force of atmospheric pressure (P_0) distributed over the outer surface of the area πR_2^2. Thus, the critical pressure at a depth H is equal to

$$p_H^* = \frac{P}{S}, ГДе P = F + G + P_0, S = \pi {R_1}^2. \qquad (22.1)$$

Fig. 22.4 Scheme of the ratio of the conical and cylindrical parts of the Yamal crater

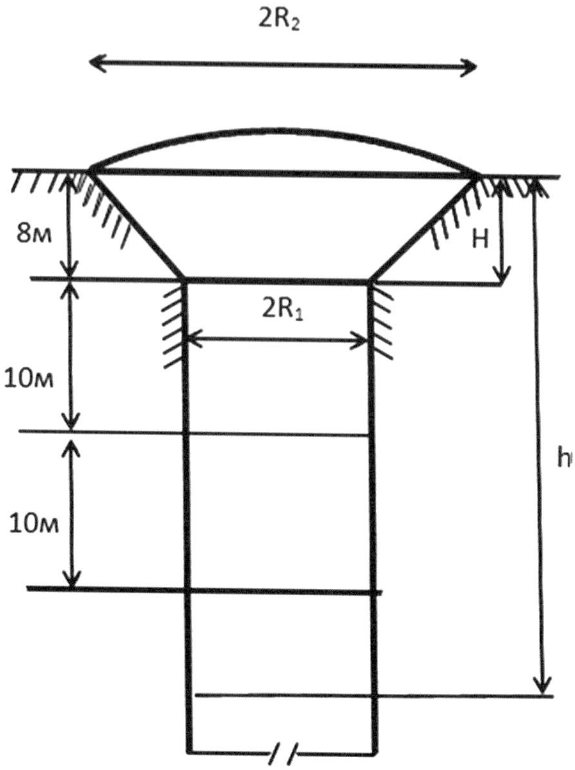

The vertical adhesion $F = \frac{[\sigma]}{\cos\phi}\pi(R_1 + R_2)\sqrt{H^2 + (R_2 - R_1)^2}$,

$$G = g\rho_{Mz}\frac{\pi H}{3}\left(R_1{}^2 + R_2{}^2 + R_1R_2\right), \text{and } P_0 = p_0\pi R_2{}^2,$$

where $[\sigma]$ is the maximum permissible tensile stress and $\phi = \text{arctg}\frac{H}{R_2 - R_1}$.

We set the quantities in the above equations equal to $R_1 = 12.5$ m, $R_2 = 18.5$ m, $H = 8$ m, $\rho = 1.6\text{t/m}^3$, $[\sigma] = 0.4$ MPa,and $g = 9.8$ m/s^2. Then $F = 6.4910$ n, $G = 9.5910$ n, $P_0 = 1.0710$ n, and $P = 8.5310$ n. The critical pressure is $p_H^* = 1.74$ *MPa*. The results are close to the calculations of V. I. Bogoyavlensky and I. A. Garagash, who determined a pressure of even 1.25 MPa for the frozen roof destruction of the Yamal crater [4].

Pressures of the soil outburst at different depths were calculated. At a depth $h > H$, it is necessary to take into account the weight force of the cone (G) and the vertical adhesion of the lateral surface with the soil (F). Assuming a cylindrical body to be circular, its radius is $R_1 = 12.5$ m, and the height is equal to $h - H$. In this case $P = F + G$. Force $G = g\rho_{Mz}\pi R^2(h - H)$ and $F = [\tau] \cdot 2\pi R(h - H)$, where $[\tau]$ is the maximum permissible tangential stress on the cut. Then the critical pressure is equal to

Table 22.1 Dependency of exploding pressure from the depth

Depth, h, m	8	18	28	38	48	58
Pressure, p_h^*, MPa	1.74	3.50	5.26	7.02	8.78	10.54

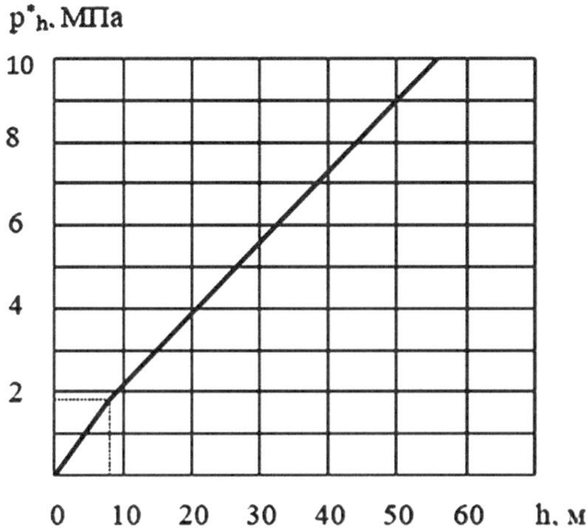

Fig. 22.5 Dependence of the ejection pressure on depth

$$p_{h-H}^{*} = \frac{P}{S}, \tag{22.2}$$

where $P = F + G$, $S = \pi R_1{}^2$,
and

$$p_h^* = p_H^* + p_{h-H}^* \tag{22.3}$$

Setting the value $[\tau] = 1.0$ MPa and the value $h - H = 10$ m, we get $p_{10}^{*} - 1.76\ MPa$. The dependence of the critical pressure on the depth is presented in Table 22.1 and Fig. 22.5.

At a depth of more than 20 m, the pressures created by gas hydrate decomposition in the range of the temperatures mentioned above will not lead to the soil outbreak, especially at a depth of 60 m. The above process of successive layer-by-layer destruction of gas-saturated frozen soils allows to overcome this discrepancy.

22.6 Scenarios of the Possible Development of Explosive Processes in Permafrost

The considered example of Yamal crater formation is only one example for the development of gas emission crater. The difference in natural conditions causes different scenarios for the preparation of explosive processes in permafrost.

With a slight and short-term increase in temperature, gas hydrate dissociation can quickly stop. As a result, a zone of increased porosity will be formed in the permafrost due to small gas cavities.

If the lake is deep enough and the thick talik formed under it, then the gas during the hydrate dissociation will be released from the bottom of the lake. This is observed in many lakes on the Yamal Peninsula [21]. The gas release lasts from several months to several years, but without an explosion since there is a possibility of its free discharge.

If hydrate-bearing permafrost occurs at a shallow depth (the first tens of meters), the pressure of the gases that arise from the gas hydrate decomposition can cause an explosion, bypassing the stage of fluid migration. In this case the mound preceding the explosion will be weakly expressed and even absent.

22.7 Conclusion

The primary reason for the Yamal crater formation is the gas hydrate dissociation in permafrost. This process is caused by local warming of soils under the influence of a surface water body. Under the lake, an area of thermal impact is formed, which has a two-membered structure. Directly below the lake, there is a thaw zone with temperatures above 0 °C, in which below is a zone of high-temperature permafrost (−1… −3 °C). After the zone of high-temperature permafrost reaches the horizon of hydrate-containing permafrost, the mechanism of gas hydrate dissociation is triggered.

The study of the Yamal crater made it possible to reveal the stages of its development. Each stage corresponds to a certain set of transformation processes of permafrost and certain elements of the structure of the crater. As a result of gas hydrate dissociation, a zone of increased pressure exceeding hydrostatic is formed. Then gas fluids penetrate into the overlying layers. The uplift of the gas-saturated icy soil mass leads to the formation of a mound. When the plastic deformations of the frozen roof reach the ultimate strength, it breaks, and the gas-saturated icy soil mass is ejected, and a crater is formed.

The Yamal crater is a special case of a whole group of phenomena, commonly called the gas emission craters. Gas hydrate dissociation can occur in local zones of permafrost temperature rise under surface water bodies, in areas with favorable surface conditions, or in man-caused impacts and occurs at negative temperatures. Depending on the ratio of various factors, scenarios for gas emission crater development and, accordingly, their morphology can differ.

Gas-emission permafrost craters are the geohazards for constructions and operators. As exploration of the Arctic is developed and the thermal impact on the permafrost increases, the risk of explosive processes for engineering structures will increase. This type of geohazards is still poorly understood; therefore, it is necessary to conduct complex field, laboratory, and experimental-methodological work on their study. Nevertheless, this group of processes is not taken into account in the design, calculations, and forecasts of the interaction of the engineering structure with the permafrost, but it is not even included in the classification of dangerous geological processes. The study was carried out at various stages of investigation with partial support from the grant of the Russian Academic State Task #AAAA-A18-118022190065-1 and the RFBR grant #17-05-00294.

References

1. Alidibirov, M. A. (1998). *Mechanism of fragmentation of highly viscous magma in volcanic explosions (experimental study)*: Thesis of Dr. Phys.-Math. sciences. M. NDG RAS, 272 p. (Rus.).
2. Bogoyavlensky, V. I. (2015). Emissions of gas and oil onshore and offshore in the Arctic and the World Ocean. *Drilling and Oil*, vol.6, 4–10 (Rus.).
3. Bogoyavlensky, V. I. (2014). The threat of catastrophic gas emissions from the Arctic cryolithozone. Craters of Yamal and Taimyr. *Drilling and Oil*, vol. 9, 12–17 (Rus.).
4. Bogoyavlensky, V. I., & Garagash, I. A. (2015). Justification of the gas emission craters formation process in the Arctic by mathematical modeling. *Arctic: Ecology and Economics, 3* (19), 12–17 (Rus.).
5. Chuvilin, E. M., Bukhanov, B. A., Grebenkin, S. I., Doroshin, V. V., & Iospa, A. V. (2016). *Experimental study of the strength of frozen hydrate-bearing soils under conditions of self-preservation of pore hydrate*. Fifth Conference of Russian Geocryologists, M. University Book Publ., Vol. 3: (pp. 180–186) (Rus.).
6. Civil protection. Encyclopedia. (2006). In S. K. Shoigu (Ed.), *EMERCOM of Russia* (Vol. No. 2. 1, pp. F–I). Moscow: Moscow Printing Plant (Rus.).
7. Cryosphere of oil and gas condensate fields of the Yamal Peninsula. (2013). Vol. 2: Cryosphere of the Bovanenkovo oil and gas condensate field. Yu. V. Badu, N. A. Gafarova, & E. E. Podborny (Eds.), M.: Gazprom Expo Publ., 424 p. (Rus.).
8. Devisilov. V. A., Drozdova, T. I., & Timofeeva, S. S. (2012). *Theory of combustion and explosion: Workshop: A training manual*. M: FORUM Publ., 352 p. (Rus.).
9. Dubrovin, V. A., Kritsuk, L. N., & Polyakova, E. I. (2015). Temperature, composition and age of the shelf sediments of the Kara Sea in the area of the geocryological observation marre sale. *Cryosphere of the Earth, XIX*(4), 3–16 (Rus.).
10. Dyadin, Y. A., & Gushchin, A. L. (1998). Gas hydrates. *Sorosovsky Educational Magazine*. vol. 3, 55–64 (Rus.).
11. Epov, M. I., Eltsov, I. N., Olenchenko, V. V., Potapov, V. V., Kushnarenko, O. N., Plotnikov, A. E., & Sinitsky, A. I. (2014). Bermuda triangle of Yamal. *Science First-Hand, 5*(59), 14–23 (Rus.).
12. Hunt, J. (1982). *Geochemistry and geology of oil and gas*. Moscow: Mir (Rus.).
13. Khimenkov, A. N., Sergeev, D. O., Stanilovskaya, Y. V., Vlasov, A. N., & Volkov-Bogorodsky, D. B. (2017a). Gas emissions in the cryolithozone, as a new type of geocryological hazards. *Georisk, 3*, 58–65 (Rus.).
14. Khimenkov, A. N., Sergeev, D. O., & Tipenko, G. S. (2017b). Local warming of permafrost massive as the one of causes of gas emission funnel in cryolithozone/theses of conferences "Natural processes in Polar Regions of Earth in Global Warming Epoch, October 9–11, Sochi, Russia:43 (Rus.).
15. Kizyakov, A. I., Sonyushkin, A. V., Leibman, M. O., Zimin, M. V., & Khomutov, A. V. (2015). Geomorphological conditions of gas-emission crater and its dynamics in Central Yamal. *Cryosphere of the Earth, XIX*(2), 15–25 (Rus.).
16. Kozhina, L. Y., Miklyaeva, E. S., Perlova, E. V., Sinitsky, A. I., Tkacheva, E. V., & Cherkasov, V. A. (2015). Dangerous contemporary manifestations of cryoactivity – the main results of the

Yamal crater study. Scientific Herald of the Yamalo-Nenets Autonomous District, № 2/87, Salekhard:19–28 (Rus.).
17. Leibman, M. O., Kizyakov, A. I., Plekhanov, A. V., & Streletskaya, I. D. (2014). New permafrost feature – Deep crater in central Yamal (West Siberia, Russia) as a response to local climate fluctuations. *Geography Environment, 7*(4), 68–80 (Rus.).
18. Leibman, M. O., & Plekhanov, A. V. (2014). Yamal gas emission crater. *Kholod'OK, 2*(12), 9–15 (Rus.).
19. Olenchenko, V. V., Sinitsky, A. I., Antonov, E. Y., Yeltsov, I. N., Kushnarenko, O. N., Plotnikov, A. E., Potapov, V. V., & Epov, M. I. (2015). Results of geophysical researches of the area of new geological formation "Yamal crater". *Cryosphere of the Earth, vol. XIX, #4*, 94–106 (Rus.).
20. Shostak, N. A. (2015). *Modeling the formation and dissociation of hydrates in the development and operation of oil and gas fields*: PhD thesis of tech. sciences. Krasnodar (Rus.).
21. Sizov, O. V. (2015). Remote analysis of the surface gas manifestations consequences in the north of Western Siberia. *Geomatics, 1*, 53–68 (Rus.).
22. Vasilyeva, Z. A., Djafarov, D. S., & Ametova, T. A. (2011). Indirect technogenic signs of gas hydrates indication in the cryolithozone. *Cryosphere of the Earth, XV*(1), 61–67 (Rus.).

Chapter 23
Landslide Activity and Landslide Hazard in Geyser Valley (Kamchatka Peninsula, Russia)

Oleg V. Zerkal and Irina P. Gvozdeva

Abstract Geyser Valley (valley of the Geysernaya River) is located on the territory of Kronotsky State Natural Biosphere Reserve (Kamchatka, Russia) and is one of the most unique nature objects in Russia. The valley is known by numerous geysers and other hydrothermal manifestations and is a tourist attraction. A big landslide formed on June 3, 2007, in Geyser Valley. It dammed up the river affecting more than half of all thermal objects. Some of them were destroyed; some changed their regime.

Formation and the displacement of landslide in the left side of the Geyser Valley occurred in the upper part of Vodopadniy creek. According to witnesses the main displacement looked flow-like movement of ground masses and lasted for 2.5 min. As a result of slope deformations, a landslide tongue was formed with length up to 3.4 km and width 800–900 m in the upper part and 0.2–0.4 km in the lower part. The mechanism of displacement of landslide was complex. Comparative analysis of models of relief after and before sliding has shown that the whole volume of landslide deposits is 22.6 million m^3 (including secondary block slide).

Different-scale landslides in Geyser Valley occurred before. Many slope failures induced by cyclone Elsa in 1981 were noted. The mapping of the area under consideration made it possible to identify over 650 landslides, rockfalls, and other slope phenomena. On January 4, 2014, large-scale rock avalanche also formed in the left side of the Geysernaya River valley.

The Geysernaya River cuts massif composed by lacustrine-volcanic hydrothermally altered deposits. These rocks contain a large number of clay minerals that are broken by tectonic and sedimentary fractures and unstable to humidifying. Geological and geomorphologic factors such as high seismicity, contrast relief, steep slopes, rainfalls, intense snowmelt and inflow of thermal water also control formation of different slope processes – landslides, rockfalls, taluses, debris flows, mudflows, etc.

Slope displacement in the region is very intense that can be hazardous for tourists and also causes damage to geysers and hot spring regime.

O. V. Zerkal (✉) · I. P. Gvozdeva
Laboratory of Engineering Geodynamics and Substantiation of Engineering Protection of Territories, Department of Engineering and Ecological Geology, Geology Faculty, Lomonosov Moscow State University, Moscow, Russia

V. Svalova (ed.), *Natural Hazards and Risk Research in Russia*, Innovation and Discovery in Russian Science and Engineering,
https://doi.org/10.1007/978-3-319-91833-4_23

23.1 Introduction

The unique area of distribution of geysers in Kamchatka – the Geyser Valley – was discovered by Tatyana Ustinova in April 1941. At the present time, it is part of Kronotsky State Natural Biosphere Reserve and included in UNESCO World Natural Heritage Complex Object "Volcanoes of Kamchatka."

A large landslide occurred in the Geyser Valley due to intensive snow melting. The failure took place on June 3, 2007, in the lower reaches of the Geysernaya River, in its left side [1]. The landslide changed strongly valley landscape: the Geysernaya River was dammed up by landslide deposits. The dam thickness was 20–40 m and its length 300 m. The dammed reservoir was formed with a depth of 20 m. Several geysers lost forever, while some of them changed their hydrothermal regime. This failure was the largest to have occurred in the Valley since its discovery.

This event gave rise to intensive research into the development of slope deformations in the valley and clarification of the causes of their activation. In this paper, the results of studies performed by specialists of the Geological Faculty of Lomonosov Moscow State University are presented.

23.2 Nature and Climate of Geyser Valley

The natural and climatic features of the territory are determined by its location near the Pacific coast. It is the active cyclone area characterized by stormy often destructive weather. In accordance with the climatic zoning of Kamchatka, the area of Geyser Valley is located within the eastern mountain volcanic climatic subregion [2]. Removing the Geyser Valley from the coast by almost 40 km and exceeding elevations above sea level by 600–1400 m result in more contrasting air temperatures and less clouds than on the coast. The average annual air temperature in the territory is of the order of +1.9 °C with an annual temperature amplitude of 20–25 °C [2]. The average annual rainfall in the study area is about 1240 mm reaching 2000 mm in some years. Most of the precipitation falls in the form of snow in the cold period of the year and a steady snow cover of up to 2 m in height and in relief depressions – up to 3 m lasts from November to May. The peak precipitation is typical for the autumn, when the daily values of precipitation can reach 60–80 mm. For example, during the passage of the cyclone "Elsa" (October 4–6, 1981), the amount of precipitation was 92 mm, and in the previous cyclone month, the amount of precipitation in Geyser Valley reached 200 mm [3].

Altitudinal climatic zoning determines the distribution of vegetation. Siberian draft pine and draft-alder elfin wood covers middle parts of the slopes up to a height of 600 m. There are groves of stone birch near the river. The watershed areas of the volcanic highlands are characterized by sparse vegetation of mountain tundra with extensive areas with lichen-moss cover, with creeping undersized thickets of elfin and individual bushes.

23.3 Geomorphologic Setting

Valley of the Geysernaya River is located in the eastern part of the Uzon-Geysernaya volcano-tectonic depression formed according to radiocarbon dating about 40 Ka ago and represented isolated reservoir where intense sedimentation occurred. The formation of the Uzon-Geysernaya volcano-tectonic depression took place within the vast volcanic highland formed within the East Kamchatka graben-syncline as a result of modern volcanic activity (Fig. 23.1). Absolute elevation marks within the volcanic plateau in the study area are 900–1100 m. Depression's sides are practically vertical (>70° in the upper part) scarp in the modern relief, having a height of up to 400–500 m in the southeastern part of the depression – in the canyon of the Shumnaya River and the left side of the lower and middle parts of the Geysernaya River declining in the northern part of the depression (the upper reaches of the Geysernaya River, the northern side of the Uzon caldera) to 200–300 m. The maximum amplitude of the relief difference in the depression board, reaching 700–750 m, is noted in the western part of the depression near the Uzon volcano (Fig. 23.2).

The valleys of the main rivers of the study area (the Shumnaya River and its left tributary – the Geysernaya River) are confined to the zone of the ring fault that limits the Uzon-Geysernaya volcano-tectonic depression from the east and southeast. The valley of the Shumnaya River before its merger with the Geysernaya River is a canyon with a width of 1–1.1 km (between the benches of the volcanic plateau) and a depth of 300–350 m. The upper and middle parts of the valley of the Geysernaya River also have the form of a canyon with a depth of 120–150 m in the upper reaches and 300–350 m in the middle part with a width of up to 900–1000 m. The lower part

Fig. 23.1 Orographic diagram of the Uzon-Geysernaya volcano-tectonic depression. (Compiled using STRM data and published materials)

Fig. 23.2 General view on the landslide on June 3, 2007, in the Geysernaya River valley. (Photo of I.P.Gvozdeva)

of the valley of the Geysernaya River where its width increases to 3.5 km is characterized by asymmetry. The steep right slope (40–50°, sometimes up to >60°) from 120 to 200 m in height is a slope of a volcanic plateau (plateau Krugloe) (Fig. 23.3). The left side of the valley, characterized by significantly smaller slopes of 15–30°, is developed in the lacustrine-volcanic sediments of the southeastern depression.

The total catchment area of the Geysernaya River is about 31.5 km^2. The width of the channel part of the rivers is up to 20 m at a depth of the watercourse to 1.5 m. The height of the level of flood waters above the low water is usually 1.5–2 m. Precipitation in the autumn can cause a rise in the level of rivers by 1–1.5 m, reaching 2.5 m and higher during the passage of cyclones.

The average river flow rates vary from 0.5–1 to 2–4 m/s. The lowest water flow in the river is typical for February–March (in low-water years – no more than 1.0–1.2 m^3/s), and the greatest expenditure is observed in June–July, during intensive snowmelt, exceeding winter water consumption of five to seven [4]. During the passage of cyclones during the autumn period, accompanied by heavy precipitation of rainfall, water in the river can reach anomalously large values, amounting to 20–25 m^3/s, and induce activation of erosion and development of mudflows and landslides.

23.4 Geological Setting

Structural-Tectonic Position

Hydrothermal systems of the Geyser Valley and the Uzon caldera are associated with the Uzon-Geysernaya volcano-tectonic depression located in the central part of the East [5]. The formation of the Uzon-Geysernaya volcano-tectonic depression took place at the intersection of two major faults: the volcanic spreading zone and the Uzon-Valaginsky fault [6–8]. The zone of volcanic spreading represents in the study region a system of faults of north–northeast strike, reflecting the large area of the crust stretching beyond the region, about 20 km wide. The Uzon-Valaginsky fault is a regional fault of the sub-latitude orientation.

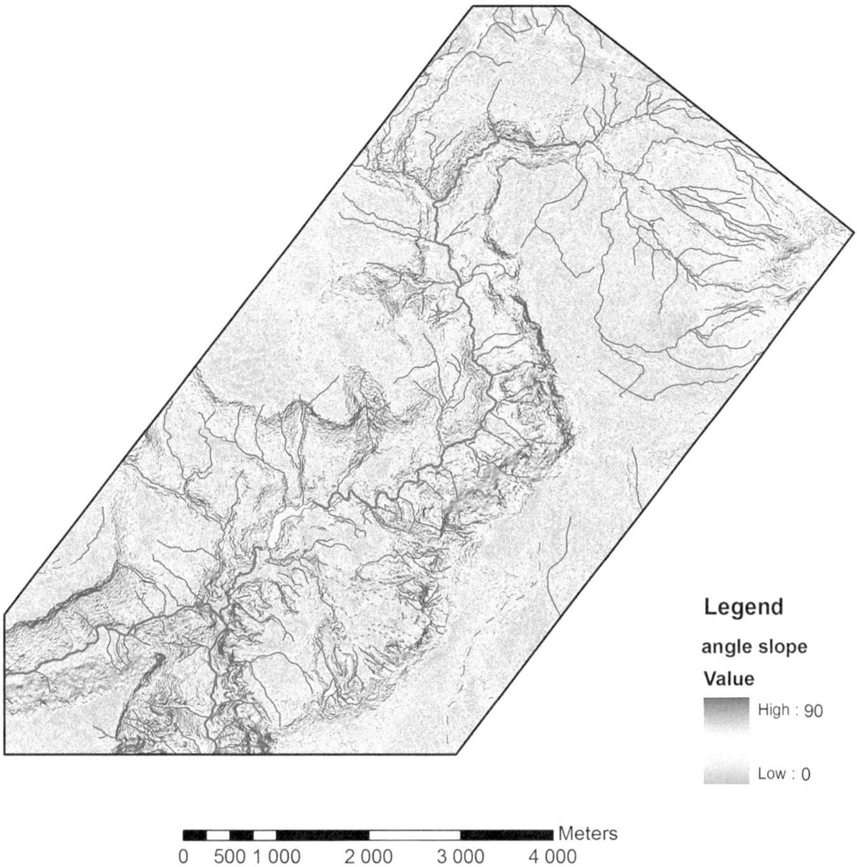

Fig. 23.3 Map of slope angles of the Geysernaya River valley and adjacent areas. (Compiled using topographic basis NIIGAIK 1976)

The Uzon-Geysernaya volcanic-tectonic depression is a typical collapse caldera with a complex structure. It has oval outlines and is stretched in a latitudinal direction. The general dimensions of depression along the edge of the confining ridges are 9–18 km (Fig. 23.1). The structure of the Uzon-Geysernaya depression is complicated by consistently developing calderas of a smaller size – the Uzon caldera and the Geyser caldera. Caldera Uzon occupies the northwestern part of the volcano-tectonic depression. Caldera Geysernaya, inheriting the position of the precaldera center of volcanism, is located in the southeastern part of the depression and has dimensions of about 10 km by 8 km. The unity of the current structural plan for the Uzon-Geysernaya volcano-tectonic depression is determined by the annular fault, expressed in relief by a ledge 300–500 m high.

In tectonic terms, the Uzon-Geysernaya volcano-tectonic depression is confined to the large deflection of the Cretaceous-Paleogene foundation (East Kamchatka or Tyushevskiy), which is filled mainly with Neogene volcano-sedimentary deposits. In the middle-upper Neopleistocene (180–35,000 years ago), numerous calderas were formed on this site, and extensive sheets of pumice tuffs and ignimbrites were formed. The total volume of magma emitted from these eruptions is estimated at 150 km^3 [9] to 220 km^3 [10]. During the formation of the Uzon-Geysernaya depression, about 46 km^3 of magma arrived on the surface. The eruptions that formed the caldera of depression were not one-act ones.

The existence of a long-standing (from the Pliocene) crustal magma chamber, which predetermines the formation of the hydrothermal systems of the Geyser Valley for the region under investigation, is characteristic. Previous studies revealed three cycles of magmatic activity. The first cycle, referring to the beginning of the Middle Neopleistocene, was accompanied by effusions of the andesite and liparite-dacite lavas (in the final stage) of the composition. Streams and extrusions of these lavas form watershed surfaces along the southeastern border of the Uzon-Geysernaya depression (mountain plateau, the base of the Yellow hills, extrusion of Bortovaya, and others (Fig. 23.1)) [5]. The second cycle of magmatic activity is characterized by the outflow of lavas and the formation of extrusions of the dacite-liparite composition occurring in the inner parts of the Uzon-Geysernaya volcano-tectonic depression (lava plateau Krugloe, Ostanets extrusions, Sestrenka, Geysernaya, etc.) (Fig. 23.1) [5]. The third cycle of magmatic activity is a stage of acidic volcanism and has manifested itself in the past 30–40,000 years [11]. The current volcanic activity of the area is confined to the Kikhpinych volcanic center, on the southwestern spurs of which the headwater of the Geysernaya River is located.

Tectonic Dislocations

Study area is characterized by a very complex system of volcano-tectonic faulting associated with the formation Uzon-Geysernaya caldera complex and subsequent volcanic activity. As part of the studies, an analysis was made of the development of tectonic disturbances, in which the previously obtained data were synthesized, the remote materials were deciphered, and a field verification of the results were obtained. As a result of the analysis, fault maps were compiled for the valley of the Geysernaya River and adjacent territories (Fig. 23.4). In the study area, the following systems of major tectonic disturbances were identified:

- Zone of faulting submeridional north–northeast strike (NNE 15–20°)
- A system of sublatitudinal discontinuities (strike WNW 285–290°)
- Ring fault system
- System of diagonal faults west–southwest–east–northeast strike (WSW-ENE 65–70°) (the first)

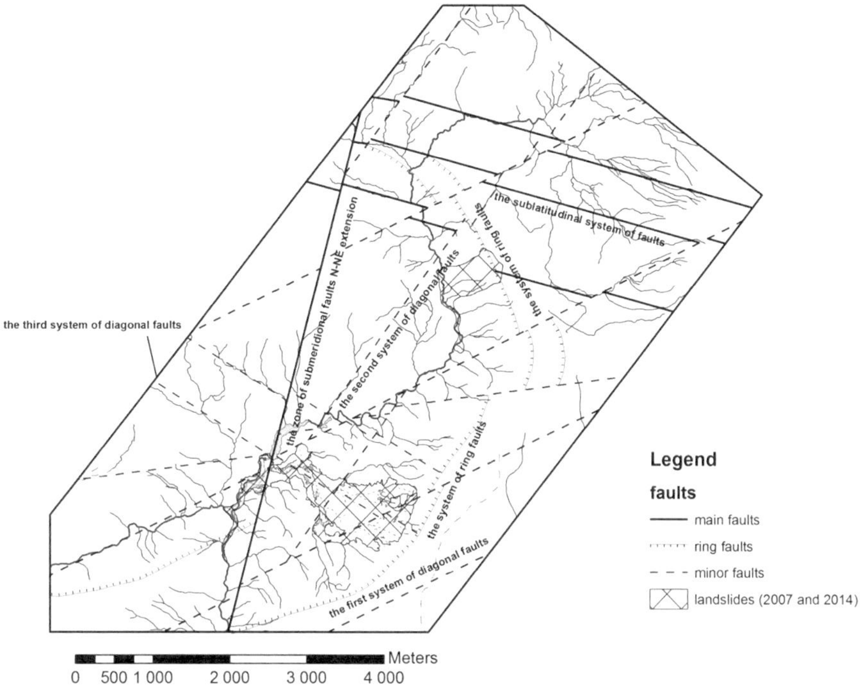

Fig. 23.4 Map of faults in the valley of the Geysernaya River and adjacent areas

- System of diagonal faults southwest–northeast strike (SW-NE 35–40°) (the second)
- System of diagonal faults northwest–southeast strike (the third)

North–northeast-trending strike faults have the character of strike-slip fault with southeast downthrown side. Sometimes there are normal faults with northwest downthrown side. The amplitude of vertical displacements is from several meters to 50 m, and the amplitude of horizontal displacements is up to 100 m. Many thermal manifestations in Geyser Valley are associated with faults of the north–northeastern strike. The orientation of the system of faults of the submeridional strike coincides with the orientation of the previously described zone of the volcanic spreading [10]. With the discontinuities in this area, the newest manifestations of volcanism, such as a series of dikes and chains of extrusive cones, are also associated in the study area.

The sublatitudinal faults and large tectonic cracks are most clearly manifested in the northern part of the region, lying in the bandwidth of 5–6 km. This direction is close to the strike of the long axis of the Uzon-Geysernaya depression. The length of individual dislocations reaches 4 km or more. The amplitude of vertical displacements varies from the first meters to 50–70 m. In addition, along the system of sublatitudinal faults along which the Uzon-Valaginsky regional sublatitudinal fault traces (according to A.E.Shantzer [8]), the northern side of the Uzon-

Geysernaya volcano-tectonic depression is formed. The total amplitude of descent along the system of discharges of the internal block of depression is estimated at 200 m [7].

In addition to regional disruptions (northeast and sublatitudinal strike), ring faults (or arc faults according to V.L.Leonov) are developed in the region under consideration centered on the Uzon-Geysernaya volcano-tectonic depression [7]. The most clearly systems of ring faults are manifested on the northern and southern sides of the depression. They are characterized by the presence of downthrown sides located in the inner part of the depression with vertical displacement amplitudes of up to 5–10 m. Near the southeastern borders of depression, a system of concentric dikes is described which inherit, apparently, a system of arc cracks. The system of arc dislocations is also fixed at the periphery of the Uzon-Geysernaya depression. The initial stage of arc fault (as well as of all depression) dates from the end of the Middle Neopleistocene to the beginning of the Upper Neopleistocene.

The systems (the first and the second) of the diagonal faults of the southwest–northeasterly strike are apparently joints and confined to the zone of the volcanic spreading. The system of diagonal disturbances of the northwest–southeast strike (the third) appears to belong to the zone of feathering faults of the regional sublatitudinal Uzon-Valaginsky fault.

It should be noted that the head scarps of the landslides of 2007 and 2014 are similar in their configuration to the configuration of the ring faults, and the head of the landslide 2007 is located at the intersection of the southern branch of the diagonal faults southwest–northeasterly strike (NE 65–70°) and the third system of diagonal discontinuous disturbances (northwest–southeast strike). This indicates a significant role of tectonic fracturing of the rock massifs in the formation of landslide processes. With that, the main scarp of the landslide on June 3, 2007, does not coincide with the ring rupture of the caldera side, but was formed in the thickness of lacustrine-volcanic sediments filling the caldera. It can be assumed that the landslide separation zone on June 3, 2007, relates to the zone of spreading of groundwater flows which arrives along the zones of tectonic crushing in the thickness of volcanic rocks that form the caldera board and lacustrine-volcanic deposits filling the caldera.

Geothermal Conditions

Geothermal system in Geyser Valley exists, thanks to the shallow long-life partially melted magma body. There are three main discharge areas: Lower Geysers Field (in the lower reaches of the Geysernaya River), Upper Geysers Field (in the middle reaches of the Geysernaya River), and Death Valley (in the source of the Geysernaya River). These discharge areas are connected with faulting zones NNW-SSE and WSW-ENE directions and appear as geysers, boiling springs, steam, and gas jets. Fracturing zones are permeable to fluid upflow. In its turn, tuffites of Geysernaya unit are confining layer. When thermal fluid upflow reaches this caprock layer

Table 23.1 Chemical composition (in mg/L) of springs of main scarp zone of landslide 2007 and neighboring area

Specification	*M*	*Z*, m	pH	CO_3^{2-}	HCO_3^-	Cl^-	SO_4^{2-}	Ca^{2+}	Mg^{2+}	Na^+
Head scarp	88	665	6.38	–	18.91	4.26	44.16	3.60	0.48	25.99
Head scarp	76	662	6.21	–	18.91	5.68	34.56	2.20	0.48	23.92
Medvezhy creek, upper reaches	73	699	6.57	–	21.96	4.26	31.68	1.40	0.48	23.69
Medvezhy creek (discharge of hot spring)	144	674	7.32	–	56.12	4.26	60.00	1.40	0.48	50.14
Thermal field in the upper course of the Medvezhy creek (a-80 °C, b-40 °C)	268	(a) 652	3.77	–	–	0.19	3.57	0.70	0.18	2.88
	242	(b) 652	3.65	–	–	0.39	3.14	0.70	0.18	2.65

dipping toward the river, it separates into steam phase and liquid phase. Hot water flows down along the layer's bedding and discharges as geysers mudpots and hot springs near the Geysernaya River. Steam component moves in opposite direction and rises through the fractured zones and mixes in the upper horizons with meteoric waters. As a result, discharge takes place in the form of steam jets and springs along the side of the caldera [12, 13].These phenomena can be observed at the base of the main scarp of the landslide 2007. Chemical composition of water taken from some sources in the head part of the landslide and in the neighboring Medvezhy creek is given in Table 23.1. The water samples taken from the head scarp of landslide 2007 (absolute height of the sampling points is 660–665 m) is hydrocarbonate – sulfate – calcium – sodium, with a pH of 6.21–6.38 and total dissolved solids of 0.08–0.09 g/l. By mineralization, pH, and composition, they correspond to water from the upper reaches of the nearby Medvezhy creek. These are meteoric waters that are not mixed with the rising flow of steam. Water samples from several sources located to the north of the landslide 2007 at hypsometric marks of 650–670 m (at the level of the base of the main scarp) were investigated. Water from the source located in the upper reaches of the Medvezhy creek (t45q0.5 by Kiryukhin) has slightly alkaline reaction with pH of 7.32, a total dissolved solid of 0.14 g/l, and a temperature of 40 °C. On the thermal field located in the right tributary, the water has an acid reaction (pH 3.65–3.77), with a total dissolved solid of 0.25 g/l. The water temperature in different parts of this field ranges from 40 to 80 °C (Table 23.1).

Characteristic of Lacustrine-Volcanic Strata

In addition to the volcanic activity described above, there was intense sedimentation within the Uzon-Geysernaya depression, which represented an isolated reservoir. According to the spore-pollen analysis, the age of the volcanic-sedimentary strata is 35–45,000 years [14]. The material came from several sources such as pyroclastic flows and ash falls and was brought by rivers. As a result volcanic-sedimentary deposits were accumulated with the thickness reaching several hundred meters. The reservoir was broken through about 9000 years ago, and later conditions for sedimentation did not occur [10]. Afterward, lacustrine-volcanic strata were eroded by the Geysernaya River. At present time the sides of the river are composed of Quaternary tuffites of Ustyevaya (Q_3^3ust), Geysernaya, Pemzovaya, and "Yellow tuff" units (Q_3^4gr, pmz, js), put into the ignimbrite strata and cut by andesitic dykes and rhyolite-dacite extrusions.

The most ancient of the Ustyevaya unit Q_3^3ust is the base of the strata. They are exposed in the lower reaches of the Geysernaya River directly at the water's edge, and their thickness is about 150 m (according to the marks of the roof). It consists of dense coarse-grained lapilli lithic and crystal-rich andesite-dacite tuffs altered by propilitization. These rocks are painted mainly in green colors [15]. The age of the unit is estimated by the boundary between the Lower and Middle Neopleistocene [14]. Thus, these deposits represent the precaldera rock complex. They are not involved in the landslide displacement in the study area.

Lacustrine-volcanic tuffites of the Geysernaya unit (Q_3^4grn) overlap with discordant the deposits of the Ustyevaya unit and ancient volcanic and volcanic-sedimentary rocks. By the first approximation, all tuffites can be subdivided into two large groups: pelitic and coarse-grained. Pelitic tuffites consist of hydrothermally altered volcanic glass coarse-grained tuffites containing pumice and lithoclasts located in fine-grained matrix [16]. The most intensive secondary transformations are characteristic for the lower and middle parts of the strata. The secondary minerals are represented by high-silica zeolites and smectites. Alterations in the upper part are insignificant, except for a high fracture permeability zones where tuffites turn into hydrothermal clays (Fig. 23.5).

Sometimes there are lenses of breccia consisting of igneous and volcanic-clastic rocks in the deposits of a Geysernaya unit. Apparently, these bodies can be interpreted as sediments of ancient mudflows and landslides (Fig. 23.6). The total thickness of the deposits of the Geysernaya unit is estimated at 310 m [5]. Rocks have monoclinal bedding with up to 20–25° toward the river. These deposits form the slopes of the Geysernaya River in the lower and middle reaches, and they are involved in landslide processes. On the other hand, it should be noted that sedimentary deposits are complicated by andesitic dykes and conduits of rhyolite-dacite extrusions. This significantly increases the stability of the slope.

The Pemsovaya unit overlaps with a washout the Geysernaya unit and is widely distributed in the central part of the depression. In the study area, the Pemsovaya unit is found on the left side of the Geysernaya River in the upper part of the strata. They are represented by gray pumice tuffites practically unaffected by secondary hydrothermal processes. Their total thickness is 60–70 m.

Fig. 23.5 Modified to clay tuffites of the Geysernaya unit in fracture zone, exposed by the main scarp (landslide 2007). (Photo of J.V.Frolova)

Fig. 23.6 The lens of the breccia in the tuffites of the geyser unit opened by the landslide 2007. (Photo of I.P.Gvozdeva)

The Yellow tuff unit is represented by practically unchanged light-cream and light-brown pumice lapilli tuffs with tuff-conglomerates and tuff-breccia lenses. These deposits lie with discordant on the eroded surface of the deposits of the Geysernaya and Pemsovaya units. Their total thickness is 210 m [5]. In the study area, they occur in the upper parts of the slopes in the right bank of the Geysernaya River.

23.5 Rock Formations and Their Geotechnical Properties

The upper part of the geological section within the Uzon-Geysernaya volcano-tectonic depression and adjacent territories is composed of volcanogenic and volcanogenic-sedimentary rocks of Eopleistocene-Neopleistocene age [5, 17]. Widespread also are Neopleistocene and modern rockfall, rock avalanche, landslide, and mudflow formations near the sides of the caldera and the slopes of the river valleys [18, 19]. In general, there are three types of geological structure typically for the study area (in terms of consideration of slope processes).

The first type is characteristic of high surfaces of the volcanic plateau. The geological stratum of these sites is represented by stratified volcanic rocks (dacites, andesites), including the thickness of the ignimbrites. The layers of volcanic rocks of the initial cycle of quaternary volcanism form the upper surfaces of the relief. They broke through at later stages of volcanic activity, with the formation of young lava sheets and extrusive domes [5]. This type is distributed in the left bank of the Geysernaya River and forms the ledges of the caldera side. Volcanic rocks characterizing this type of section are mainly represented by andesites and dacites; there are also coverings of ignimbrites. Among the studied rocks, mainly dense differences with a porphyry or cryptocrystalline structure were encountered. On the surface, they are not affected by secondary changes. Their specific density is 2.61–2.65 g/cm^3 which corresponds to medium-acid rocks. Geotechnical properties of volcanic rocks on the left side of the Geysernaya caldera are given in Table 23.2 (№1–3).

Table 23.2 Average values of geotechnical properties of volcanic rocks of the Geysernaya caldera

N	Type of rock	Bulk density g/cm^3	Porosity total (open) %	E_d GPa	Poisson coefficient	UCS MPa	
						Dry	Saturated
1	Dacite	2.26	14 (6)	34	0.30	51	33
2	Andesite	2.19	17 (2)	12	0.31	32	15
3	Ignimbrite	2.22	15 (9)	24	0.28	37	24
4	Liparite	2.03	29 (4)	11	0.32	5	4
5	Volcanic glass	2.27	10 (1)	3	0.22	7	4

After I.P. Gvozdeva [16] with additions

As can be seen from Table 23.2, all samples of studied rocks are characterized by quite high total porosity: 14–17%; the open porosity does not exceed 9%. Apparently, this ratio corresponds to the ratio of isolated pores and micro-cracks in the rock. The presence of micro-cracks and secondary changes along them explains the decrease in strength characteristics in water-saturated rocks by 1.5 times with respect to air-dry samples. Such a sharp drop in strength characteristics promotes the development of landslide and rockfalls within steep (>60° and more) and high (200 m and more) ledges of the caldera sides. The collapse which occurred on January 4, 2014, is one of such manifestations of slope processes.

The second and third types of geological strata are characteristic for internal, tectonically downdip parts of the Uzon-Geysernaya volcano-tectonic depression.

The second type of geological strata is inherent in the zones of internal parts of the caldera, unaffected by volcanic activity in the late stages including the southeastern part of the Geysernaya caldera, where the landslide on June 3, 2007, occurred. In areas with the second type of section, the upper part of the geological strata is represented by sedimentary and volcanic-sedimentary deposits that are accumulated in the lake stage of the geological history of the Uzon-Geysernaya volcano-tectonic depression. Volcanic-sedimentary deposits of Geysernaya (Q_3^4grn) and Pemsovaya (Q_3^4pmz) units were affected by landslide deformations in 2007.

The rocks of these units are represented by tuffites of light colors from fine-grained ("porcelain") to coarse-grained (lithoclastic and pyroclastic) with low and medium densities, high porosity, and low strength. Secondary hydrothermal alterations have different effects on the physical and geotechnical properties of tuffites. Bulk density, modulus of elasticity, and the strength of the rock increase by replacing the primary material with siliceous minerals. In case when the secondary minerals are represented by smectites and zeolites, deformation and strength properties decrease. Argillized zeolitic tuffites are mostly porous and light among all studied rocks [15, 20]. Strength properties of volcanic-sedimentary formations strongly depend on their moistening and drop sharply with water saturation (Table 23.3).

Table 23.3 Geotechnical properties of lacustrine-volcanic rocks of the Geysernaya caldera

N	Type of rock	Bulk density g/cm^3	Porosity %	E_d GPa	Poisson coefficient	UCS MPa	
						Dry	Saturated
1	Lapilli tuffite (Geysernaya and Pemsovaya unit)	1.13–1.28	49–60	2–5	0.26	3–6	1–3
2	Coarse-grained tuffite (Geysernaya and Pemsovaya unit)	1.22–1.36	50–54		0.28–0.30	6–13	4–7
3	Fine-grained tuffite (Geysernaya and Pemsovaya unit	1.12–1.21	51–57		0.26–0.28	12–18	3–5
4	Lapilli pumice tuffite (Yellow tuff unit)	1.15	59	2	0.29	1	–

The third type of geological strata is observed in the central parts of the caldera, including the right side of the Geysernaya River valley. The right side of the river is composed of the volcanic-sedimentary deposits of the Geysernaya and the Yellow tuff units that break through and overlap with young lava flows of andesite, andesite-dacite, and dacite, extrusive domes of the lyparite composition, and ingimbrite cover formed at late Quaternary and modern stages of volcanic activity. Geotechnical properties are shown in Tables 23.2 and 23.3.

Modern slopes of Geyser Valley are complicated by numerous landslide and landslide-rockfall bodies, the toe of which reaches the bottom of the valley and causes the observed tortuosity of the river. As a result of river erosion, landslide formations are exposed in the right-hand side, characterized by substantial fragmentation and tilted bedding formed during displacement. The lower part of the valley of the Geysernaya River is substantially expanded due to the area of development of pre-escaped hills on the left side of the valley. According to the results of the work carried out, it can be assumed that the leading role in the formation of hilly relief in the pre-escarp zone was played by gravitational processes similar to those observed in 2007.

23.6 Landslide Activity in Geyser Valley

Landslides in the Valley of the Geysernaya River

The problem of the development of dangerous geological processes in canyon of the Geysernaya River located in the volcanic and tectonically active area attracted attention since the 1980s of the twentieth century. The main attention in the framework of earlier studies was given to the study of the effect of individual landslides and mudflows on the regime of hydrothermal fields without a regional assessment of landslide activity [2, 3, 12, 18].

To assess the regional development of rockfalls, landslides, and mudflows in the valley of the Geysernaya River, the mapping including thematic interpretation of remote sensing data with subsequent field verification of obtained results was performed, as well as subsequent data analysis. Features of geological and geomorphological settings (volcanic and tectonically active region, the presence of deep slopes, the presence of volcanic and volcanic-sedimentary deposits in the geological strata with a significant decrease in their strength during hydrothermal alternation and/or by humidification) predetermined the widespread development of slope phenomena. The mapping of the slope phenomena in the region showed that within the study area, several groups of landslide forms can be identified (Fig. 23.7):

The first group – large-scale (with the volumes of rocks involved in the displacement from the first million cubic meters to more than 10 million m^3) landslide bodies

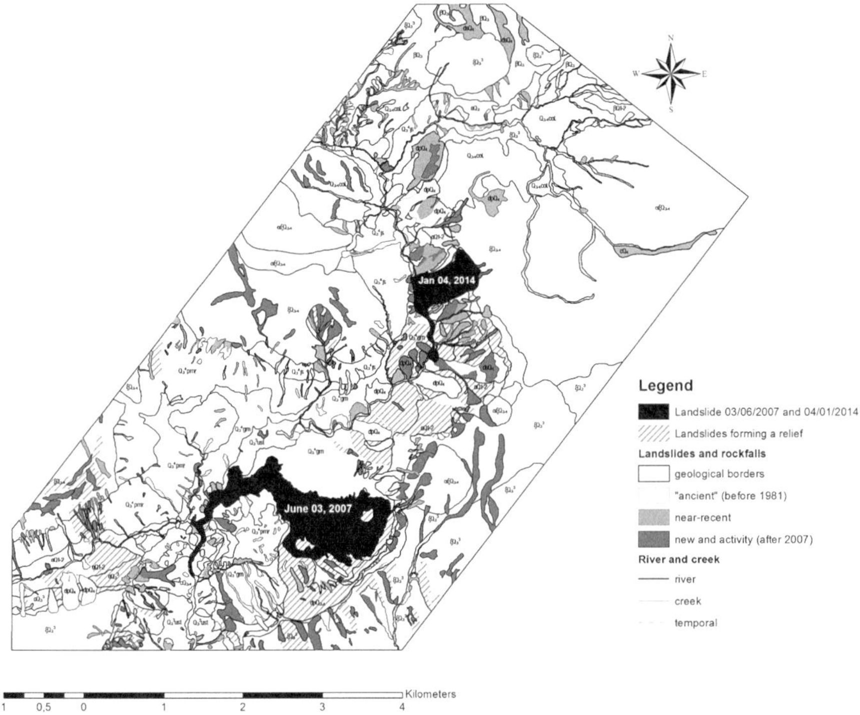

Fig. 23.7 Inventory map of landslides in Geyser Valley. Note to the legend: Volcanogenic-sedimentary rocks: Q_3^3ust – Ustyevaya unit; Q_3^4grn – Geysernaya unit; Q_3^4pmz – Pemsovaya unit; Q_3^4js – "Yellow tuff" units (Based on field research and interpretation of remote sensing data using materials from V.L. Leonov)

that have a complex (combined) displacement mechanism. These landslides form the modern terrain of the territory.

The second group – modern (2007, 2014 years) large-scale (with the volumes of rocks involved in the displacement more than 10 million m^3) landslides and rock avalanches with a complex (combined) displacement mechanism. They create earthen dams flooding the valley and form landslide accumulation in the channel of the Geysernaya River.

The third group – different-scale (with the volumes of rocks involved in the displacement from several tens of thousands to hundreds of thousands cubic meters) landslide bodies, by the mechanism of displacement: landslides, block landslides, and earthflows that complicate the modern relief.

The fourth group – different-scale (with the volumes of rocks involved in the displacement from several thousand to tens of thousands cubic meters) landslide and avalanche bodies, mainly near-surface ones. By the displacement mechanism, landslides, block landslides, and mudflows were formed within the previously formed landslides (secondary landslides) and complicating the modern relief of the territory.

In total, over 650 rockfalls, landslides, and other slope phenomena were identified in the area under consideration, and more than 360 landslides occurred within the limits of development of lacustrine-volcanic sediments. The revealed manifestations of slope phenomena have a different displacement mechanism, including 45 rockfalls, more than 40 rock avalanches; 38 block landslides, 48 complex landslides; and more than 360 earthflows, more than 20 of which were transformed in the process of displacement in the mudflow; and more than 70 development sites of talus were mapped. Among the mapped slope manifestations, the prevailing type is the earthflow (>55%). The formation of the most different-scale landslides apparently is associated with abnormal climatic phenomena, such as the typhoon "Elsa" in 1981 and also the anomalous precipitation in the winter 2007.

It should be noted that the largest slope phenomena (with volumes of more than 10 million m^3), which belong to groups I and II of landslides, are located in the middle part of the left side of the Geysernaya River. Most of them are (by the mechanism of development) to deep block or complex rockfalls and to landslides. Landslides formed on the left side of the valley at least four times overlapped the riverbed of the Geysernaya River. The presented map shows that the landslide formed on June 3, 2007, developed within the preexisting ancient landslide massif (Fig. 23.7). The wide development of large-scale landslides in the area under consideration is due, on the one hand, to the contrasting relief of the territory and, on the other hand, to the features of the structure of the upper part of the geological section, in which soils with high variability of strength properties under humidification are presented.

The Landslide on June 3, 2007

The development of the landslide on June 3, 2007, in Geyser Valley began in the upper reaches of the Vodopadniy creek, left tributary of the Geysernaya River. The main scarp of the landslide is located in the zone of development of volcanogenic-sedimentary formations in the middle part of the main ledge of the Geysernaya caldera in a section with slopes (before landslide) of 35–50° and a local elevation drop (in the riverbank, before landslide) of 80–120 m. The total excess of landslide crown above its toe reaches 420–440 m. It should be noted that the zone of initial failure is not confined to the steepest part of the slope – the bench crest of the Geysernaya caldera, which has slopes of 65–70° with a relative elevation drop of about 50–70 m, located 120–250 m from the main scarp of the landslide. The observed location of the head of the landslide on June 3, 2007, indicates that the slope deformations are not connected with the zone of the main ring fault bounding the caldera but refer to the inside caldera formations. The total width of the main slump zone (across the direction of displacement) was 780–800 m and taking into account secondary deformations along the flanks reached 930–940 m.

According to observations of employees of the Kronotsky State Natural Biosphere Reserve and other eyewitnesses, landslide deformations began on June 3, 2007, at 14:20 local time with flow-like movement of ground masses and lasted for 2.5 min [21, 22]. Given the length of the landslide tongue from the head to the rocky narrowing "Triumphal Gates," where, obviously, the deceleration of the main displacements was about 1.2 km, the average displacement speed of the landslide masses can be estimated at 7 m/s (or about 29 km/h).

As a result of displacements on the left side of the Geysernaya River, in the valley of the Vodopadniy creek was formed a landslide body with a total length along the displacement axis up to 3.4 km (from the main scarp to the end of the landslide). The width of the displacement was up to 0.4–0.6 km in the upper part, up to 250–270 m in the middle part (near the scientific station of Kronotsky State Natural Biosphere Reserve), and no more than 40–50 m in the toe reaching the Shumnaya River. The general trajectory of the landslide movement has a complicated configuration due to the relief features of the left side valley of the Geysernaya River riddled with winding valleys of streams. At the initial interval, the displacement of landslide masses occurred from the southeast to the northwest. In the vicinity of the scientific station of Kronotsky Reserve, the moving collapsing landslide mass hit the slope of the local elevation on which the station is located, turned around, and continued their movement in the southwest direction, inheriting the hollow of the Vodopadniy creek. There are landslide deposits at an altitude of 16–18 m on the slope of local elevation. On the right flank of the landslide tongue, the landslide masses in the turning zone moved up the local hollows in the north direction for a distance of up to 160 m from the main direction of the displacements, coming very close to the buildings of the observation station.

The bulk of the landslide masses, reaching the channel of the Geysernaya River, changed the trajectory of the movement, and their displacements acquired the character of along-the-river migration down the valley. The Geysernaya River was dammed up by landslide deposits. The dam thickness was up to 40 m, its width up to 200 m (at the top), and length up to 300 m. The dammed reservoir was formed with the depth of 20 m and length up to 2 km. Four days later, a partial breakthrough of the dam occurred, and lake level decreased sharp by 9 m [22]. As a result of the breakthrough, the Geysernaya River developed a new V-shaped channel with a width (along the upper edge of the ledges) up to 45–50 m (Fig. 23.15).

At present, the landslide dam of 2007 is almost completely broken through by the mudflow that accompanied the rock avalanche on January 4, 2014, and the lake basin is filled with mud deposits.

Initially, the volume of landslide on June 3, 2007, in the Geyser Valley was estimated at 3–5 million m^3 [22]. Later, based on the analysis of the relief before and after the landslide formation (as of August 23, 1993, and July 12, 2007), V.N. Dvigalo and I.V. Melekestev estimated the total volume of sediment involved in the displacement of 21 million m^3 [23]. According to experts from the Geological Faculty of Lomonosov Moscow State University (also carried out on the basis of the analysis of the relief before and after the landslide formation), the total volume of the landslide on June 3, 2007, amounted to 16.3 million m^3 [18, 19]. Later (probably in

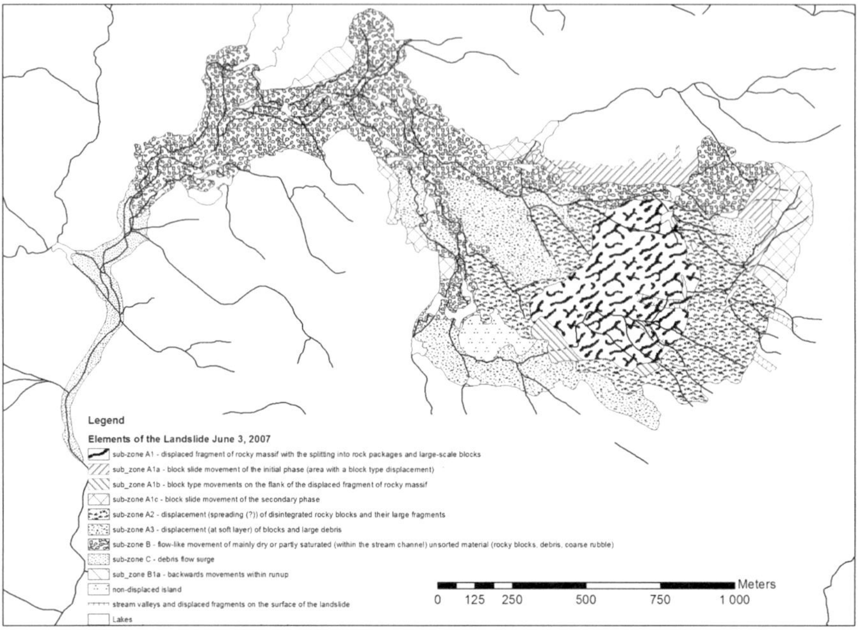

Fig. 23.8 The schematic map of the landslide on June 3, 2007

2008), a secondary landslide block was displaced in northeastern part of the landslide circus with a volume of up to 6.3 million m^3.

The study of the structure of the landslide massif and of the dam showed that the development of the landslide on June 3, 2007, was multistage and the mechanism of its displacement is complex (composite) (Fig. 23.8). The initial phase of the deformations was apparently associated with transformation of the water-saturated tuffites of the Geysernaya unit laid at depths of about 50–70 m into a plastic (?) state and the subsequent dissection and displacement of a large block of the slope massif. Displacement was a kind of spreading, when rock, packs, blocks, and boulders are transported down the slope with flow-like movement in the horizon of water-saturated soils at the base of the landslide body. D.J. Varnes [24] characterized this type of displacement as the rock spread landslide [24]. At the same time, the landslide on June 3, 2007, is characterized by several features. First, the peculiarity of the landslide development on June 3, 2007, was that the development of deformations occurred on a steep slope and was actually localized in the valley of the Vodopadniy creek. Another peculiarity is the possibility of transformation of hydrothermally altered rocks to a fluid-like state with water saturation, which contributed to high velocity of displacement. This landslide on June 3, 2007, differs from the "classic" rock spread landslide where plastic clays, deforming at a low speed, are located at the base of the displacing plates, packages, and blocks of strong rocks. As a result, during the formation of the landslide on June 3, 2007, the spreading was not lateral, but an extended highly mobile landslide tongue was formed which is more typical for earthflows. The surface of formed landslide tongue is uneven; the slope dips do not exceed, as a rule, 10°. There are several parts in the landslide massif which are determined by differences in the mechanism of development of displacements at different stages of the slope process.

The upper part of the landslide on June 3, 2007 (zone A), for up to 1.2 km along the displacement axis (on the left flank – up to 1.34 km) is formed (in the area of major displacements) by volcanic-sedimentary rocks, similar to those observed in head scarp. In this part of the landslide, several subzones can be allocated depending on the degree of fragmentation of the displaced material. The subzone A1 is represented by a large displaced rock massif (in the direction of displacement) with a length of up to 370–380 m and a width of up to 570 m, composed of tuffites, which, in general, retain the primary orientation and the sequence of deposits. It should be noted that the extent (in the displacement direction) of the displaced rocky massif is comparable (about 400 m) with the extent of the previously existing (before the landslide formation) slope from the main scarp to the thalweg of the local diagonal (NE–SW) creek valley.

During landslide displacements, the rock massif was shattered by subvertical cracks and trenches into individual packets and blocks ranging in size from 8–12 to 40 m which experienced slight vertical displacements relative to each other. On the surface of the landslide packets, vegetation (trees and shrubs) and fragments of the stream valleys located on the natural slope above the main scarp have been preserved (Fig. 23.9). Based on the location of the displaced fragments of streambeds and their

Fig. 23.9 The disintegrated surface of the displaced rocky massif in subzone A1. (Photo of O.V. Zerkal)

orientation, the displacement of the array forming the subzone A1 is about 290–300 m with some turning counterclockwise. The total volume of displaced rock massif reached 12 million m^3. On the left (southwestern) flank of the displaced array, two sections of secondary block displacements were formed, allocated to subzones A1a and A1b.

The displaced rocky massif of subzone A1 is separated from the main scarp by a trench with a width of up to 300 m and a depth of up to 20 m. Angular large (from 0.8 to 0.9 m to several meters) boulders of tuffites with fine gravel matrix filled this trench. Numerous seepage of groundwater along the fractures of the main scarp formed small lakes in the most lowered parts. In the northern part, the width of the moat is reduced by 100–110 m and the depth to 10 m due to its filling with debris formed after a later displacement of the secondary block landslide (Fig. 23.10).

The frontal part (in the direction of displacement) of the displaced rocky massif is represented by two subzones – A2 and A3. Subzone A2 has a length (in the form of tongues stretching in the direction of displacement) of up to 250–260 m and a total width of up to 520 m. Crushed by displacement in large boulders and packets, tuffites from coarse-grained to fine-grained ("porcelain") form this zone. Boulders and packets have sizes from the first meters of up to 10–12 m and are divided by gaping cracks up to 2–3 m wide (Fig. 23.11). The surface on this site looks like a debris of rocks with separate tors and rampart rocks. In many cases, packets and boulders were turned around randomly by displacement. In contrast to subzone A1, in subzone A2, areas with fragments of primary vegetation are practically not preserved, and trees, as a rule, are collapsed.

Fig. 23.10 The view of the graben separating the displaced rocky massif of subzone A1 from the main scarp. (Photo of O.V. Zerkal)

Fig. 23.11 The view of surface of the landslide on June 3, 2007, with fallen trees at the subzone A2. (Photo of O.V. Zerkal)

Subzone A3 is a debris field, composed of angular blocks (up to 0.8–0.9 m) and large rubble (Fig. 23.12). Subzone A3 forms a tongue in the central part of the landslide body with a length (in the direction of displacement) of up to 500 m and a total width of 270–280 m, as well as individual sections on the left flank of the landslide. Subzone A3, as it were, forms the external front of the entire boulder zone A.

The middle part of the landslide on June 3, 2007, (zone B) is an extended tongue up to 230 m wide, completely occupying the pre-landslide valley of the Vodopadniy creek and coming out (at the bottom) into the valley of the Geysernaya River. The displacement in this zone occurred in the form of flow-like movement of ground mass transporting large blocks of rocks that were observed by eyewitnesses. The direction of displacement of the landslide masses coincides with the general orientation of the pre-landslide valley of the Vodopadniy creek forming in the right side (in the areas of its turn) run-up (subzones B1a) and then along the river Geysernaya (Fig. 23.13). Zone B has a complex shape in plan. On the right (with respect to the direction of displacement) part of the landslide, the beginning of zone B is located near the crown zone occupying the local stream channel – the right tributary of the Vodopadniy creek. In this part of the landslide, zone B separates zone A from the right flank of the head part of the landslide. On the left part of the landslide, zone B begins at the foot of the relic with a pre-landslide relief, and its configuration inherits the configuration of the central low part of the pre-landslide valley of the

Fig. 23.12 The view of debris plane on surface of the landslide on June 3, 2007, at subzone A3. In the background is undulating-ridge surface (zone B). (Photo of O.V. Zerkal)

Vodopadniy creek. On the left flank of the landslide, zone B also separates zone A from the flank ledge.

The landslide surface in zone B has undulating-ridge appearance. Ridges on the landslide surface are extended in the direction of displacement of the landslide. The width of the ridges in the middle part of the landslide is up to 20–25 m, and the height reaches 4–6 m and more. In the lower part of zone B after the output of the landslide tongue into the valley of the Geysernaya River, the ridges on the surface of the landslide become narrower (the width decreases to 7–10 m) and lower (the height is reduced to 2–4 m). The ridges are composed of large rubble and blocks (from 0.8–0.9 to 2–3 m) of tuffites. In some large blocks, the primary foliation was traced which acquired a multidirectional orientation with a landslide displacement (from subhorizontal (close to initial) to subvertical). The variability in the orientation of the primary foliation indicates that the blocks and debris in the landslide displacement not only moved during flow but also experienced rotational displacements. The hollows between the ridges are filled with gravel, crushed stone, debris (fragments), and also separate blocks (Fig. 23.14).

The thickness of landslide deposits in zone B ranges from 3 to 5 m in the marginal parts and reaches 40 m in the axial part of the landslide tongue in the region of the buried pre-landslide valley of the Vodopadniy creek. In the zone of deceleration of landslide displacements and accumulation of masses in the Geysernaya River valley, the formation of a dam has occurred. The thickness of landslide deposits is up to

Fig. 23.13 Run-up on the right flank in the middle part of the landslide on June 3, 2007 (subzones B1a). (Photo of O.V. Zerkal)

Fig. 23.14 Debris field with undulating-ridge surface of zone B in the middle part of the landslide on June 3, 2007. (Photo of O.V. Zerkal)

Fig. 23.15 The new canal of the Geysernaya River cutting through the landslide dam. (Photo of I.P. Gvozdeva)

10 m in the marginal parts and reaches 50–55 m in the central part. In the areas where the dam was eroded, it is evident that the landslide deposits in zone B have a double-compound structure (Fig. 23.15). The lower part of the landslide tongue is composed of an unstructured, gravelly crushed-grained mass with separate large fragments (up to 10%). In some cases the fragments are flattened with the long-axis orientation in the direction of the landslide movement. The fragments do not form continuous horizons but are located in the layer mainly chaotic. The upper part of the landslide language is the collapse of blocks and boulders. It can be assumed that the lower part of the landslide body is a zone of a clastic substrate where the displacement occurred as flow-like movement of mainly dry or partly saturated material. The upper part of the landslide body is obviously composed of displaced rock blocks mechanically disintegrated due displacement.

The lower part of the landslide on June 3, 2007 (zone C), is represented by deposition of the final phase of displacement, the formation of which occurred after the deceleration of the main landslide mass in the region of the rocky narrowing "Triumphal Gates" in the valley of the Geysernaya River. Downstream from rocky narrowing, only debris flow surge continued to move which reached the valley of the Shumnaya River and formed deposits of zone C. The length of zone C is up to 1 km with a width of 80 m and a thickness of up to 6–8 m in the upper part to 20–30 m and a thickness of 2–3 m in the lower part.

Rock Avalanche on January 04, 2014

The collapse on January, 4, 2014, occurred in the eastern part of the rhyodacite volcanic dome Zheltaya (in the upper reaches of the Goryachaya Rechka creek), whose lavas broke through and covered the volcanic formations of the precaldera stage. The total height of the slope in the area of formation of slope deformations is about 300–350 m, and the length of the slope (from the crest of slope to the riverbed of the Geysernaya River) is 650–680 m. The slope dipping in its upper steepest part reached 70 with an elevation difference of about 100 m. The length of the collapse zone (across the direction of displacement) was about 250–270 m, with the width of the massif (along the slope) up to 80–100 m. Thus, the volume of the primary collapse was 2.2–2.7 million m^3. Earlier, since 1975–1976, there was a large number of tension cracks on this site [21].

A talus slope formed due to collapse with a total width of 380–410 m, reaching the Geysernaya River. It can be assumed that at the initial stage of the displacement, the deformations developed by the mechanism of toppling. This is indicated by the presence of deposits of the rock avalanche overran through the local watershed on the slope up to 30 m high and that overlapped the neighboring valley of the Tepliy creek. Such bedding could not have formed if the displacements that developed in the upper reaches of the Goryachaya Rechka creek at the initial stage were due to a shear mechanism.

It should be noted that the bottom third of the displaced deposits is separated from the upper part by a strip 40–70 m wide with minimal thickness of sediments. The observed situation is typical for rock avalanches, the motion of which forms a transit zone between the collapse and the accumulation areas with virtually no collapse deposits due to the high energy of the motion of the collapsing masses. The slight width of the transit zone (several tens of meters) is explained by the small total length of the slope. At the same time, the collapse energy was sufficient to ensure that the caving masses, reaching the riverbed and overlapping it, partly “climbed” to the opposite side of the valley at 15–18 m and partly changed the trajectory of the traffic, transformed into along-the-channel avalanche. Passing along the Geysernaya River 680–700 m, the displaced masses formed a debris cone with a length of 140–160 m (along the valley) and a width (across the valley) of 120 m and a height of 40–50 m. The volume of collapsing masses in the accumulation zone is estimated at 800–850 thousand m^3. Above the dam, a lake was formed with a depth of up to 15–18 m, a length of up to 400 m, and an average width of up to 80–90 m (maximum width – 180 m).

The displacement of the rock avalanche which reached the Geysernaya River passed along its channel induced a mudflow along the riverbed, which passed down the valley more than 3 km. The mudflow reached the Lake Geysernaya, formed by the landslide on June 3, 2007, where it accumulated a cone with a capacity of 3–4 m (according to other estimates – from 8–9 to 10–15 m [21]). The prerequisites for the formation of a rock avalanche on January 4, 2014, should be considered geological and geomorphological settings on the site of collapse. Among the factors that also contributed to the preparation of the destabilization of the slope are climate influence (cyclone September 17–19, 2013) and seismic activity in the study area.

23.7 Conclusion

Geyser Valley on the territory of Kronotsky State Natural Biosphere Reserve (Kamchatka, Russia) is a unique natural territory, formed as a result of a special combination of natural factors. The current climatic conditions of the region, located near the Pacific coast of Kamchatka, are characterized by active cyclonic activity. The passage of cyclones in the autumn period is accompanied by a peak precipitation, when the daily precipitation amounts can reach 80–90 mm. The features of the Geysernaya River valley (the eastern part of the Uzon-Geysernaya volcano-tectonic depression) are a contrasting relief formed under influence of modern tectonic movements and volcanic activity and the widespread occurrence of low-strength hydrothermal altered volcanic-sedimentary deposits. High rather steep sides of the caldera are composed of high-strength volcanic rocks. At the same time, the formation of the caldera occurred as a result of tectonic subsidence along the zones of ring faults, which predetermines the widespread distribution of tectonic crushing zones oriented subparallel to the ledges of the caldera sides. The noted features of the study area – the probability of anomalous precipitation, contrast relief, tectonic fragmentation, and the presence of low-strength hydrothermal altered lacustrine-volcanic deposits – predetermine the high intensity of slope (landslide, avalanche, mudflow) processes in the valley of the Geyser Valley.

The mapping of the area under consideration made it possible to identify over 650 landslides, rockfalls, and other slope phenomena. Four groups of slope phenomena were identified within the study area: large scale (with volumes of 1–10 million m^3 and more) forming a modern relief; modern (2007 and 2014) large scale (with volumes of 1–10 m^3 and more) transforming the modern relief; scale (with volumes up to hundreds of thousands of cubic meters), complicating the modern relief; and various secondary landslides (with volumes up to 10 thousand m^3), mainly shallow landslides.

Modern (2007 and 2014) large-scale landslides formed in the valley of Geysernaya River are characterized by complex displacement mechanisms. The landslide on June 3, 2007, with a total volume (including secondary offsets) of over 22.6 million m^3 was formed on the left side of the valley within the territory of the development of intracalderal volcanic-sedimentary sequences. It can be attributed to "rock spreads" (according to D.J. Varnes classification), in which the displacement of plates, packets, and blocks of relatively strong rocks occurs along the horizon of soft layer. A feature of the landslide on June 3, 2007, is that in the head part spreading was not lateral, because landslide body is formed within a wide valley. In the middle and lower parts of the landslide body, as a result of flow-like movements, an extended landslide tongue of unsorted coarse material is formed at the base of which is a clastic substratum.

Landslide on January 4, 2014, also is formed on the left side of the Geysernaya River valley. At the same time, the crown zone of the landslide covers the caldera ledge, 300–350 m high, composed of strong volcanogenic rocks. Slope phenomenon on January 4, 2014, is a rock avalanche by type of displacement (according to the classification of D.J. Varnes), with a total length (without taking into account mudflow below the dam) of up to 1.6 km and with a complex trajectory (upper part – down the slope; the lower part – along the riverbed).

During the formation of modern (2007 and 2014) large-scale landslides in the Geyser Valley landslide dams were formed. At the same time, the displacement energy was very high, and landslide movements on the lower part transformed to along-the-river displacement.

Thus, the Geyser Valley is a region of wide development of landslide processes of various types and scales, which are predetermined by the geological and geomorphological settings. Trigger factors of activation of slope deformations are climatic anomalies and high seismicity of the region.

References

1. Gordeev, E. I., Pinegina, T. K., Droznin, V. A., Dvigalo, V. N., & Melekestsev, I. V. (2007). June 03, 2007 natural disaster in the Valley of Geysers in Kamchatka. *EosTrans. AGU*, *88*(52). – Fall Meet. Suppl., Abstract T51A-0297.
2. Kondratiuk, V. I. (1974). *Klimat Kamchatky [Climate of Kamchatka]*. Gidrometeoizdat, Moscow. 204 p. (in Russian).
3. Sugrobova, N. G., & Sugrobov, V. M. (1985). Izmeneniya rejima thermoproyavleniy Doliny Geyserov pod vliyaniem tsyklona "El'sa" [Change in the regime of thermal manifestations of the Geyser Valley under the influence of the cyclone "Elsa"]. *Voprosy Geografii Kamchatki [Questions of geography of Kamchatka]*, (9), 88–94 (in Russian).
4. Muravyev, Y. D., Droznin, V. A., & Bataeva, O. P. (2003). Ob intensivnosty sovremennikh exogennikh protsessov v doline reky Geysernoy [Intensity of contemporary exogenous processes in the Geysernaya river valley]. *Vestnik KRAUNTS. Nauky o Zemle,* (1), 86–88 (in Russian).
5. Belousov, V. I., Grib, E. N., & Leonov, V. L. (1983). Geologicheskie pozitsii gidrotermal'nikh system Doliny Geyserov i kaldery Uzon [Geological positions of hydrothermal systems of the Valley of Geysers and caldera Uzon]. *Volcanology and Seismology,* (1), 65–79 (in Russian).
6. Legler, V. A., & Parfenov, L. M. (1979). Systemi razlomov ostrovnikh dug [Systems of faults of island arches]. In *Tektonicheskoe rayonirovanie i snrukturno-veshestvennaya evolitsiya severo-vostoka Asii [Tectonic zoning and structural-material evolution of northeast Asia]* (pp. 134–155). Nauka, Moscow. (in Russian).
7. Leonov, V. L. (1982a). Razrivnie narushenia raiona Uzonsko-Geysernjy depressii [Faults of the area of Uzonsko-Geysernaya depression]. *Volcanology and Seismology*. (4), 78–84.
8. Shanzer, A. E. (1979). Nekotorye osobennosti evolyutsii tektono-magmaticheskikh struktur Kamchatki v zavisimosti ot eye blokovogo stroyeniya i dvizheniya blokov v pozdnem kaynozoye [Some features of the evolution of tectono-magmatic structures of Kamchatka, depending on its block structure and movement of the blocks during the late Cenozoic]. *Bulletin of the volcanological station*, (57), 53–65 (in Russian).
9. Melekestsev, I. V. (1980). *Vulkanism i reliefoobrazovanie [Volcanism and relief formation]*. Nauka, Moscow. 212 p. (in Russian).
10. Leonov, V. L., & Grib, E. N. (2004). *Strukturnie positsii i vulkanism chetvertichnikh kal'der Kamchtky [The structural position and volcanism of the quaternary calderas, Kamchatka, Russia]*. Dalnauka, Vladivostok. 189 p. (in Russian).
11. Leonov, V. L. (2001). Regional'nie strukturnie positsii visokotemperaturnikh gydrotermal'nikh system na Kamchatke [Regional structural settings of high-temperature hydrothermal systems in Kamchatka]. *Volcanology and Seismology*, (5), 32–47 (in Russian).

12. Kiryukhin, A. V., Rychkova, T. V., Droznin, V. A., Chernykh, E. V., Puzankov, M. Y., & Vergasova, L. P. (2010). Geysers Valley hydrothermal system (Kamchatka): Recent changes related to landslide of June 3, 2007. In *Proc. World Geothermal Congress 2010* (pp. 25–29). Bali.
13. Kiryukhin, A. V., Rychkova, T. V., & Dubrovskaya, I. K. (2012). Formation of the hydrothermal system in Geyser Valley (Kronotsky Nature Reserve, Kamchatka) and triggers of the giant landslide. *Applied Geochemistry, 27*(9), 1753–1766.
14. Egorova, I. A. (1993). Vozrast I paleogeograficheskie usloviya formirovaniya vulkanogenno-osadochnykh otlogeniy Uson-Geysernoy caldernoy depressii na Kamchatke [Age and paleogeographic conditions of formation of volcanic-sedimentary deposits Uzon-Geysernaya caldera in Kamchatka (the results of palynological studies]. *Volcanology and Seismology,* (2), 27–43 (in Russian).
15. Frolova, J. V., Gvozdeva, I. P., & Kuznetsov, N. P. (2015b). Effects of hydrothermal alterations on physical and mechanical properties of rocks in the Geysers Valley (Kamchatka Peninsula) in connection with landslide development. In *Proc. World Geothermal Congress 2015* (pp. 19–25). Melbourn, April 2015.
16. Gvozdeva, I. P., Frolova, J. V., & Zerkal, O. V. (2015). Slope processes hazards in geothermal areas: A case study of the Geysers Valley, Kamchatka. In *Proc. World Geothermal Congress 2015* (pp. 19–25). Melbourn, April 2015.
17. Leonov, V. L. (1982b). Geologicheskoe stroenie kan'ona reky Shumnoy I Uzon-Geysernaya depressiya na Kamchatke [The geological structure of the Shumnaya River canyon and the Uzon-Geysernaya depression in Kamchatka]. *Volcanology and Seysmology*, (2), 100–103 (in Russian).
18. Zerkal, O. V., & Gvozdeva, I. P. (2012). Investigating and mapping of landslides activity in Geyser valley (Kamchatka). In *Problems of decrease in natural hazards and risks: Proc. of the VIII Internat. Conf. "GeoRisk-2012"* (pp. 79–84). RUDN, Moscow. 1. (in Russian).
19. Zerkal, O. V., & Gvozdeva, I. P. (2013). Slope stability analysis in modern hydrothermal active areas. In *Global view of engineering geology and environment*: *Proc. of the Internat. Symp. and 9th Asian Regional Conf. of IAEG* (pp. 225–228). Taylor&Francis.
20. Frolova, J. V., Gvozdeva, I. P., Chernov, M. S., & Kuznetsov, N. P. (2015a). Ingenerno-geologicheskie aspekty gydrotermalnykh preobrasovanij tufogennykh porod Doliny Geyserov (poluostrov Kamchatka) [Engineering aspects of hydrothermal alternations of tuffaceous rocks in the Geyser Valley (Kamchatka peninsula)]. *Ingenernaya geologiya [Engineering geology]*, (6), 30–42 (in Russian).
21. Leonov, A. V., & Leonov, V. L. (2014). Valley of Geysers, Kronotsky Reserve, Kamchatka: Features seen by the 1991 GOSA expedition and changed by the 2007 landslide, 2013 cyclone and 2014 landslide. *The Geyser Gazer Sput. The Newsletter of The Geyser Observation and Study Association, 28*(2), 5–26.
22. Pinegina, T. K., Delemen, I. F., Droznin, V. A., Kalacheva, E. G., Chirkov, S. A., Melekestsev, I. V., Dvigalo, V. N., Leonov, V. L., & Seliverstov, N. I. (2008). Kamchatka Valley of Geysers after the catastrophe on 3 June 2007. *Vestnik DVO RAN [Bulletin of the Far East Branch of the Russian Academy of Sciences].* (1), 33–44 (in Russian).
23. Dvigalo, V. N., & Melekestsev, I. V. (2009). Geologo-geomorfologicheskie posledstvya katasroficheskikh obvalnikh i obval'no-opolznevikh protsessov v kamchtskoy Doline Geyserov (po dannim aerofotogrammetrii) [The geological and geomorphic impact of catastrophic landslides in the Geyser Valley, Kamchatka: Air-photogrammetry]. *Volcanology and Seismology,* (5), 1–14 (in Russian).
24. Varnes, D. J. (1978). Slope movement types and processes. *Landslides, Analysis and Control.* Transportation Research Board, Washington, DC., Special Rep. 176. 11–33.

Chapter 24
The History of Slope Evolution – Primary Cause of its Modern Instability (by Example of the "Vorobyovy Gory" Landslide, Moscow)

Olga S. Barykina, Oleg V. Zerkal, Eugene N. Samarin, and Irina P. Gvozdeva

Abstract Landslide processes on the Moscow city territory have been studied more than a hundred years. The study area is located in the central part of the Vorobyovy Gory, covering the area from the Moscow observation deck to the Moscow metro bridge. The length of the visually defined landslides is up to several hundred meters, and the visible width along the axial part of the landslide is more than 300 m. The volume of soil involved in the landslide deformation is estimated as 2 million m^3. The surface slope, the elevation of which reaches 60–70 m, has a typical landslide terrain. The deposits of Carboniferous, Jurassic, Cretaceous, and Quaternary systems are presented in structure section.

This paper deals with the nature, mechanism, and spatial distribution of landslide processes on the Vorobyovy Gory in Moscow, based on new factual data with the consideration of the regional geological history.

This material made it possible to draw the following conclusions. Firstly, the area involved in landslide processes on the Vorobyovy Gory is characterized by larger values in terms of area and depth than it previously assumed. In the head part, where the displacement zone is located at the depths of 80–100 m, the deformations, confined to the lower part of the Jurassic deposits, have a block character. Secondly, we can talk about a combined mechanism of the development of large-scale landslide massif "Vorobyovy Gory", including plastic flow with the forming of shaft bulging, crash with throwing, block offsets, and other types of deformations. Both primary and secondary displacements can be distinguished in this landslide massif.

O. S. Barykina (✉) · O. V. Zerkal · E. N. Samarin · I. P. Gvozdeva
Laboratory of Engineering Geodynamics and Substantiation of Engineering Protection of Territories, Department of Engineering and Ecological Geology, Geology Faculty, Lomonosov Moscow State University, Moscow, Russia
e-mail: barykina@geol.msu.ru

V. Svalova (ed.), *Natural Hazards and Risk Research in Russia*,
Innovation and Discovery in Russian Science and Engineering,
https://doi.org/10.1007/978-3-319-91833-4_24

24.1 Introduction

Landslide processes have been studied for more than a 150 years at the territory of Moscow [1–3, 9, 8, 11, 14, 15, 18]. At present, more than 200 landslide sites are known on the territory of the city, including 16 sites where large-scale slope deformations develop. One of the largest landslide areas covers the high slope of the starboard side of the Moskva River in the Vorobyovy Gory area (Vorobyovy Gory is the name of own of gently rolling hills in the southwest of Moscow).

Slope deformations on the Vorobyovy Gory have been documented since the beginning of the nineteenth century. For example, in 1817, it was on the slope of the Vorobyovy Gory that a temple was built to commemorate the victory over Napoleon in the Patriotic War of 1812. However, 5 years later, the construction of this grand structure was stopped, largely due to the destruction of the foundations laid by landslips. In the middle of the nineteenth century, the presence of landslides in the Vorobyovy Gory region was reflected at the map of the Moscow Province [21]. For the same reason, in the middle of the twentieth century, the high-rise building, which later became the main building of the Moscow Lomonosov State University, was moved from the edge of the slope into the watershed. In 1959, on the slopes of Vorobyovy Gory near the metro station “Leninskie Gory” (now “Vorobyovy Gory”), an escalator lift gallery was built, closed 30 years later due to damage caused by landslide deformations (Fig. 24.1).

Fig. 24.1 Deformed building of the lower station of escalator gallery. (Photo by Olga Barykina)

Fig. 24.2 Groundmark on the slope of Vorobyovy Gory. (Photo by Olga Barykina)

The landslide on Vorobyovy Gory has been studied by various researchers for more than 150 years (A.V. Pavlov, B.M. Danshin, V.V. Kuntzel, etc.), but only from the middle of the twentieth century, instrumental observations were performed until 1994 [3–6, 8–10, 12, 13, 16, 17, 20]. At the beginning of the twenty-first century, a new monitoring network was set up on the Vorobyovy Gory site from the Trinity Church of the Life-Giving to the metro bridge. Points of observation are located along the cross sections, which are situated in the direction of incidence of the slope, from its edge to the edge of the Moskva River, and include ground-based benchmarks (Fig. 24.2) and inclinometer wells. However, despite so much attention paid to the study of this slope, there are still various opinions, and there are discussions about the mechanism of formation and history of landslide processes on the Vorobyovy Gory.

24.2 Engineering Geological Setting

The Characteristic of Research Area

In tectonics, the territory is located on the southwestern margin of the Moscow syneclise within the central part of the Russian plate as part of the ancient, Precambrian, East European platform.

The city of Moscow is located on a naturally complex territory, which is characterized by a long history of development and a variety of landscapes [10]. The valley complex of the river occupies the most part of the city (river floodplain and its three

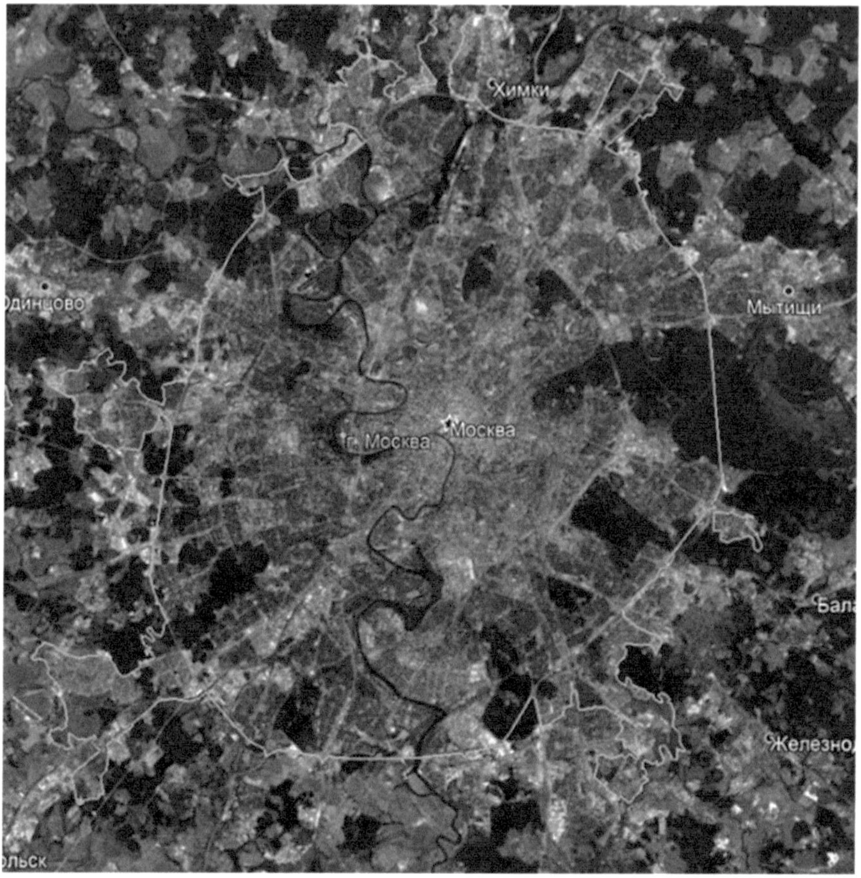

Fig. 24.3 General view of the city of Moscow from Google. The study area is marked by a rectangle

terraces); the southwestern part (where the research area is located) lies within the Teplostan upland; the eastern part is the marginal part of the Mescher lowland – a flat, weakly dissected swampy plain with low absolute marks. Thus, the valley of Moskva River is the main geomorphological object of the territory, occupying a significant part and crossing the city diagonally from the northeast to southwest (Fig. 24.3).

The relief of Moscow inherited preglacial features and was formed because of Quaternary period glaciations, as well as erosion. In the right-sided bends, the Moskva River cuts into the valley side, forming steep landslide slopes, one of which ("Vorobyovy Gory") is the study territory.

In the geological structure of the bedrock slopes affected by landslides, the deposits of Quaternary, Cretaceous, Jurassic, and Carboniferous systems occur at the slight angles in the northeastern direction (Fig. 24.4), within the middle part of the Moskva River.

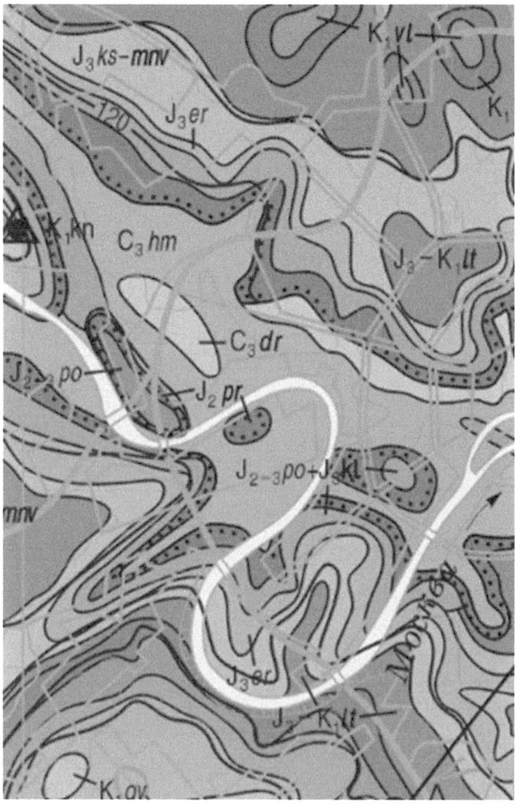

Fig. 24.4 Fragment of State Geological Map (scale 1:200000) with study area. Index: C_3hm, Khamovniki suite; C_3dr, Dorogomilovskaya suite; J_2pr, Pronskaya suite;$J_{2\text{-}3}$po +J_3kl, Podosinovskaya, Podmoskovnaya, and Kolomenskaya suites, combined; J_3er, Ermolinskaya suite; J_3ks-mnv, Kostromskaya, Lytkarinskaya, and Mnevnikovskaya suites, combined; J_3-K_1lt, Lytkarinskaya suite; K_1kn, Kuntsevskaya suite; K_1vl, Vladimir series

The completeness of the Jurassic section is largely determined by the relief of the overlapping surface of the Carboniferous rocks. The most ancient of Jurassic strata smooth out its irregularities, filling paleovalleys. The complex nature of the pre-Jurassic (more accurately, pre-middle-Jurassic) relief on the territory has already been noted by a number of researchers. B.M. Danshin [4] identified the major sub-latitudinal paleovalley as the main element of this relief, which he named "The Main Moscow clough" [4]. Exceeding the paleowatershed above the paleovalley thalweg is 40–60 m.

Geological Setting

The study territory is located in the central part of Vorobyovy Gory, covering the area from the Moscow observation deck, located on a high surface, to the Moscow metro bridge (Fig. 24.5).

Vorobyovy Gory is located on the right side of the valley of the Moskva River and represents a steep, sometimes forested, slope (up to 70 m high) with a peculiar

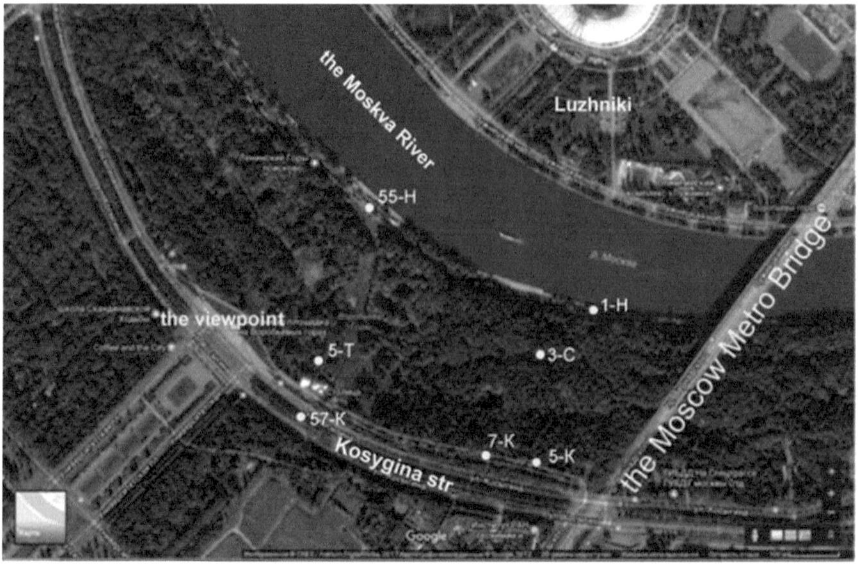

Fig. 24.5 Detailed view of the research area (after Google). The bores are marked by points

Fig. 24.6 The surface of the landslide slope of the Vorobyovy Gory (on the left in the background is the Luzhniki sports complex). (Photo by Olga Barykina)

ridge-landslide relief (Fig. 24.6), stretching along the river. The maximum elevation marks are fixed near Moscow observation deck, from which the slope gradually decreases to the west – to the estuary of the Setun River – and to the east, to the metro bridge. The main scarp is well expressed in relief – its height varies from 12 to 30 m

a

b

Fig. 24.7 Minor scarp of shallow landslide formed in Quaternary deposits in 2006: (**a**) 2006, (**b**) 2016. (Photo by Irina Gvozdeva)

and steepness from 25 to 40° (Fig. 24.6). Erosion forms, such as ruts, gullies, ravines, etc., are developed within the slope also (Fig. 24.7). The lower part of the slope adjoining the Moskva River embankment is significantly technologically altered by anti-landslide measures.

Rocks of the Middle Carboniferous, Bathonian-Tithonian Jurassic, Berriassian-Aptian stages of the Lower Cretaceous and Quaternary formations represented by morainic and aquatic-glacial accumulations take part in the near-surface structure of the watershed part of the Vorobyovy Gory [1, 18]. According to the studied profile,

Fig. 24.8 Carboniferous limestone (C_2). (Photo by Olga Barykina)

extending from the observation deck perpendicular down to the opposite bank of the Moskva River, the structure of the watershed part is presented as follows (according to drilling data, 2015–2017).

At the base of the slope, on the eroded surface of organogenic detrital limestones of the Myachkovskian age (the Moscow level of the Middle Carboniferous) (Fig. 24.8), whose roof was exposed at depths of 110–113 m, lies thin bed seam (up to 2 m) of the deposits of the Callovian stage of the Middle Jurassic, presented by dark brown carbon-bearing clays and gray-brown clays with oolites and large pebbles of limestone.

In fact, the roof of Carboniferous deposits forms a complexly organized pre-Jurassic paleorelief. It should be emphasized that the area of the Vorobyovy Gory from the observation deck to the metro bridge is located within the "Main pre-Jurassic paleovalley." The relative depth of its downcutting, based on the difference in marks of the base of the Jurassic formations (in comparison with the adjacent territories), reaches 40–45 m. According to most researchers, medium-carbon formations in landslide displacement are not involved.

Four suites represent the section of the Oxfordian stage [19], the deposits of which overlap the Callovian formations, successively from the bottom to the top (Fig. 24.9):

- Podosinkovskaya suite composed of gray clays with fauna of belemnites and ammonites, up to 2.7 m thick
- Ratkovskaya suite, represented by a bundle of gray clays and oolitic sands with phosphorites and ammonites, up to 1 m thick
- Podmoskovnaya suite, composed of a pack of dark gray and black, dense, micaceous, fissile shale clays up to 7.0 m thick

Fig. 24.9 Oxford clay. (Photo by Irina Gvozdeva)

Fig. 24.10 Tithonian sands. (Photo by Olga Barykina)

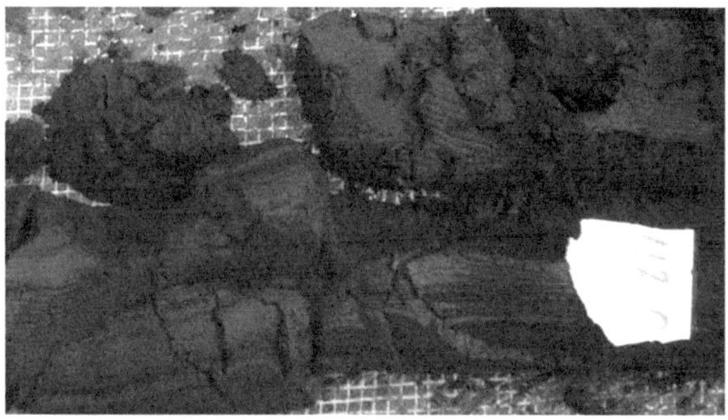

- Kolomenskaya suite composed of brownish-gray and light gray clay, heavily silty, up to 6.0 m

Depth of the roof of Oxford clays within a high, not involved in landslide deformations, of a watershed surface is, of the order, 90–95 m.

The Yarmolinskaya suite, which lies higher in the section, overlays the Jurassic deposits and, in terms of age, covers the upper part of the Oxford stage and the bottoms of Kimmeridgian stage. The Yarmolinskaya suite is composed of black, dense, layered clays having a thickness of up to 5–5.5 m.

Above the geological section, overlapping the Upper Jurassic clays, there is a pack of sands accumulating from the Tithonian to the Aptian age. The deposits of the Tithonian stage are represented by greenish-gray and green quartz-glauconite sands (Fig. 24.10) with numerous phosphorites, which are characterized by the presence of numerous remains of the fauna of belemnites and ammonites. Dark gray clayey aleurites occur at the base of the Tithonian section. The Berriasian stage of the Lower Cretaceous is presented by gray-green, green fine-grained sand deposits. Overlapping their pack of Hauterivian deposits is composed of green fine-grained glauconite sands. Above the section, a thin interlayering of fine- and medium-

Fig. 24.11 Cretaceous sands. (Photo by Olga Barykina)

Fig. 24.12 Quaternary moraine loams. (Photo by Olga Barykina)

grained sands (Fig. 24.11), light and to various degrees of ferruginous, reddish brown with interlayers of lilac-gray and beige clayey aleurites and lilac and dark gray clays represents deposits of the Barremian stage. Deposits of the Aptian stage, which crowns the thickness of the Lower Cretaceous deposits, include brownish-red fine-grained sands, with a thickness of about 15.0 m, a sandy silt-clay packet of fine- and medium-grained micaceous sands, 5.7 m in thickness and light gray, sands, with a thickness of 8.6 m. The total thickness of the sandy layer reaches (in areas with undisturbed bedding) 69–71 m. The roof of the Lower Cretaceous sands lies within a high, not involved in landslide deformations, of a watershed surface at depths of 19–20 m.

The thickness of Quaternary sediments is composed of two horizons of moraine loam (Fig. 24.12) with crushed stone and gravel, separated by fluvioglacial medium- and fine-grained sands formed in the interglacial epoch (Mindel-Riss). The lower horizon of loams has a thickness of up to 7.2 m, and the thickness of the upper horizon (the Moscow stage of glaciation) is up to 6.1 m.

The Moskva River has a depth of about 5–7 m at the site adjacent to Vorobyovy Gory. The thickness of alluvial deposits in the channel part of the valley ranges from 7 to 10 m.

Most researchers (V.V. Kyuntzel, M.N. Paretskaya, etc.), who studied the landslides of the Vorobyovy Gory, consider the clay deposits of the Oxfordian stage as the main deforming horizon of landslide massifs [9, 14]. In this regard, within the slope, overlapping Oxfordian formations, sandy-clayey Upper Jurassic deposits of the Tithonian stage, Lower Cretaceous mainly sandy sediments, and Quaternary glacial, fluvioglacial formations are in landslide occurrence.

The length of visually determined landslide areas is up to several hundred meters, and the visible width along the axial part of the landslide is more than 300 m. The volume of soils involved in landslide deformations in the study area is estimated as 2 million m^3.

24.3 Results and Discussion

The study of structure of the slope in the section was carried out along several parallel profiles corresponding to the axial parts of the landslide bodies, from the high plateau to the Moscow River perpendicular to the slope. Initially, it was considered that the wells (Fig. 24.5), located on the high surface of the plateau, are in the undisturbed landslide deformations of the part of the massif and reveal the bedrock of the rocks.

A detailed study of the core of the wells (Fig. 24.13), located on the surface of the watershed "plateau" near the edge of the slope, as it was considered earlier, showed another picture of the development of geological processes. In the clayey sediments of the Jurassic system, lying in the lower part of the Mesozoic-Cenozoic formations, landslide slip planes were identified. Thus, in the borehole 7-K in-depth intervals 85–86 m (clay deposits of the Oxfordian age of the Jurassic system) and 97.3–99 m (clay deposits of the Oxfordian age of the Jurassic system), steeply dipping

Fig. 24.13 Slickenside (depth interval 85–86 м (borehole 7-K). (Photo by Olga Barykina)

Fig. 24.14 Slickensides from the depth 81,3 м (**a**), 82,5 м (Б), 92 м (**b**) и 100, 5 м (Г). Borehole 5-K. (Photo by Olga Barykina and Irina Gvozdeva)

slickensides were identified (Fig. 24.13). In the borehole 5-K, also located on the edge of the slope in four depth intervals (81.3 m, 82.5 m, 92 m, and 100.5 m), slickensides were detected (Fig. 24.14). The allocated intervals are also confined to Oxford clays.

The obtained data uniquely indicate the presence in natural (not involved in the technogenic activity) occurrence, within the depth of 80–100 m, of explicit zones of landslide deformations (in the form of slip planes). This allows us to state that the deep zone of displacements extends beyond the previously accepted boundary of landslide processes, which was revealed visually by the wide development of cracks, ruptures, etc. on the slope of the Vorobyovy Gory (Fig. 24.7). Based on the above information, the conclusion about the necessity to transfer the boundaries of landslide processes in the study area deep into the watershed "plateau" was done. In this case, all previously described landslide bodies should be considered as secondary, developing within a very large landslide massif, the rupture zone which is located within the modern watershed plateau. Perhaps the surface occurrence of these deep deformations is now retouched by anthropogenic replanning of the territory during its development.

The Mechanism of Formation of a Landslide Massif

The development of landslide in the area of the Vorobyovy Gory is probably related to the location of this region within the deep pre-Jurassic paleovalley of the Moskva River (Fig. 24.4), where the upper part of the Carboniferous deposits (limestones, marls) was eroded and was the accumulation of a thick series of clays during the Jurassic time. Such areas are characterized by a significant, both horizontal and vertical, variability of the geological section, active water exchange between aquifers, which in turn leads to an intensification of the development of geological processes. The data obtained during drilling confirmed the uneven boundary of the roof of Carboniferous deposits. Drilling operations have revealed a difference in the marks of the roof of the Carboniferous deposits, reaching 4–5 m. The surface of Carboniferous formations buried beneath the overlying sediments is eroded and has a complex character and is probably one of the factors of the development of landslide processes in the Vorobyovy Gory region.

Based on newly obtained data, it is possible to clarify the mechanism of landslide processes of the study territory, which is based on the features of the motion of individual elements of the landslide. In our case, we can talk about the simultaneous action of several mechanisms of deformation of soils in different parts of the slope.

The formation of the head part of the landslide massif occurred because of the displacement of large blocks composed of Mesozoic-Cenozoic deposits. This is confirmed by the presence of a series of slip planes encountered during drilling in the thickness of Jurassic clays. The thickness of such landslide blocks in the study territory, based on the drilling data, reaches 80–100 m. It should be noted that the identified slip zones are in a section at absolute elevations substantially below the current level of the river, indicating that the basis for landslide displacements was a lower erosion level, indicating the duration (in geological time) of the development of landslide deformations in the study area. Thus, the head of the modern large-scale landslide massif "Vorobyovy Gory" is a fragment of the relict landslide (according to the terminology proposed by International Geotechnical Society [7]).

Three levels of landslide blocks, the surfaces of which rise above the edge of the river on 15–16 m, 29–33 m, and 47–49 m, respectively, form the main part of the modern large-scale landslide massif "Vorobyovy Gory." The nature of landslide displacement is reflected in the peculiarities of the variability of the structure of the geological section within each of the blocks.

Within the lower block, a normal stratigraphic sequence is established only for the lower part of the open geological section (depth 40 m and below). In this depth interval, near the riverbed part of the valley, there are clays of the Oxfordian stage (Ratkovskaya, Podmoskovnaya, and Kolomna suites). Nevertheless, a certain reduction (up to 0.5–1.0 m) of thickness (as compared to the structure of undisturbed areas) was established for these deposits and an increase in the altitude of the boundaries by approximately 2 m as well. The clay of the Ermolinskaya suite the tops of the Oxfordian stage – the bottom of Kimmeridgian stage – has a thickness of 12.3 m, which is more than twice higher than their thickness in undisturbed occurrence within the limits of the high watershed areas. Above the clays of the

Fig. 24.15 Fragment of landslide folded zone. (Photo by Olga Barykina)

Ermolinskaya suite, a zone of landslide folded rock was revealed, represented by separate fragments (pieces) of black clays of the Yermolinskaya suite, up to several centimeters in diameter, located in a matrix formed by grayish-green sands of Berriasian (Fig. 24.15). The thickness of the folded zone is up to 1 m. The zone of landslide folded is overlaying by the displaced deposits having a reverse (!) in sequence of occurrence. At the base of the stratum, a horizon of fine-grained silty sands of the Berriasian age with a thickness of up to 2.5 m was found, above which lie Tithonian glauconite fine-grained sands (up to 1 m thick). They are covered with dark gray clayey siltstones with a thickness of less than 1 m, lying in an undisturbed state at the base of Tithonian formations. The total thickness of the pack, characterized by the reverse occurrence, is up to 5 m. Above, a zone of interlayering (0.6–0.9 m) of sand and silts of the Tithonian age and of the Yermolin clay was discovered. The upper part of the section of the lower block is formed by alluvial formations (thickness up to 7 m), at the base of which there are floodplain clays.

The geological structure of landslide blocks of the middle part of the slope is not the same than the structure of the blocks in the lower part of the landslide massif. The section of the second landslide block is characterized by a certain elevation (relatively to the position within the unbroken sections of watershed plateau) of the boundaries of all the suites by 2–3 m. In the lower part of the landslide block, the thickness of the clays of the Oxfordian stage (Ratkovskaya, Podmoskovnaya, and Kolomenskaya suites) does not change. Above the section, for the clays of the Ermolinskaya suite (the tops of the Oxfordian stage – the bottoms of Kimmeridgian stage), the aleurite sand formations of the Tithonian-Berriasian stages, a reduction of each of the suites on 1–2 m is typical. A deluvial cover of redeposited Lower Cretaceous sands forms the upper part of the section of the second landslide block.

The geological structure of landslide blocks of the upper tier differs from the structure of the blocks in the middle and lower parts of the landslide massif "Vorobyovy Gory." In contrast to the structure of the middle tier of landslide blocks, where relative elevation of stratigraphic boundaries is observed, the opposite picture is observed for the upper part of the landslide massif. All stratigraphic boundaries within the upper block are reduced by approximately 1.5–2 m. However, for landslide blocks of the upper tier, as well as the structure of the middle tier of landslide blocks, the thickness of all the suites is reduced by 1–2 m, increasing in the upper part of the Lower Cretaceous formations up to 4 m.

Thereby, based on the peculiarities of the variability of the geological structure of landslide blocks, the mechanism of displacement of the modern large-scale landslide massif "Vorobyovy Gory" can be characterized as follows. At the initial stage, because of moistening of the upper part of the Oxford clay, their transition to a plastic state occurred, which led to extrusion of clays in the direction of lowering the relief. At the same time, with the landslide movements in the lower part of the slope and the adjacent part of the river bed, a ridge compression was formed, as evidenced by an increase in the thickness of the clay of the Ermolinskaya suite (the upper Oxfordian stage – the bottom Kimmeridgian stage) by more than twice. In some stages of landslide displacement, because of a change in the orientation of the overlying sediments, a landslide block collapsed, accompanied by tipping (by toppling mechanism), disintegration, and the subsequent formation of a landslide body in the form of a sand-clay avalanche. As the result of deformations in the lower part of the slope, a regressive development of landslide displacements with the separation of blocks has occurred in its upper part. The landslide blocks that were broken during the displacement tested tipping to the side of the slope, as evidenced by the somewhat elevated (relatively undisturbed massif) position of the stratigraphic boundaries in the middle of the slope and their reduced position in the upper part of the slope.

Separately, it should be noted that the development of the modern large-scale landslide massif "Vorobyovy Gory" occurred within the relict landslide massif.

The third type of deformations developed on the slope can be attributed to relatively shallow landslides, which only capture the upper part of the geological section. With the development of these displacements, the deformation zone does not leave the Lower Cretaceous sandy-argillaceous deposits. The zone of shear deformations is located at depths of up to 20 m. This is confirmed by the drilling data, which showed, in particular, the doubling (in relation to unbiased parts) of the Apt sand deposits.

It should be noted that all described landslide processes are of different age.

24.4 Conclusion

Thus, based on the newly obtained evidence, the following conclusions can be drawn. Firstly, the territory involved in landslide processes on the Vorobyovy Gory is characterized by much larger values, both in area and in depth, than it was previously assumed, which is confirmed by the drilling data. In the head part, where the displacement zone is located at depths of 80–100 m, the deformations, confined to the lower part of the Jurassic deposits, have a block character. Secondly, one can speak of a combined mechanism for the development of a large-scale landslide massif "Vorobyovy Gory" (according to the terminology proposed by (International, 1993)), which includes plastic flow with the formation of a ridge compression, collapse with tipping, block displacement, and other types of deformations. As a part of a landslide massif, it is possible to distinguish both primary and secondary displacements.

References

1. Barykina O. S., Zerkal O. V., Samarin E. N., & Gvozdeva I. P. (2017). K voprosu o razvitii opolznevykh protsessov na Vorobyevykh gorakh (Moskva) [On the development of landslide processes on Vorobyovy gory hills (Moscow)]. Inzhenerno-geologicheskiye zadachi sovremennosti I metody ikh resheniya [Engineering-geological problems of the present and methods for their solution]: Proc. conf. – Geomarketing. Moscow. 111–117 (in Russian).
2. Churinov M. V. (1957). Kharakteristika opolzney pravogo berega r.Moskvy na uchastke Leninskikh gor i vozmozhnost stroitelnogo osvoyeniya etoy territorii [Description of landslides of the right bank of the Moskva River on the site of the Lenin Mountains and the possibility of building development of this territory]. Voprosy gidrogeologii i inzhenernoy geologii [Questions of hydrogeology and engineering geology]: Proc. VSEGINGEO. – Gosgeoltekhizdat. Moscow. 15. 62–78 (in Russian).
3. Danshin, B. M. (1937). Geologicheskoye stroyeniye Leninskikh gor v svyazi s nekotorymi voprosami stratigrafii otlozheniy melovoy sistemyi opolznevymi yavleniyami po beregu r. Moskvy [The geological structure of the Leninskie Hills in connection with some issues of the stratigraphy of deposits of the Cretaceous system and landslide phenomena along the banks of the Moskva River]. Izvestiya Moskovskogo geologicheskogo tresta [News of the Moscow Geological Trust]. (4). 3–23 (in Russian).
4. Danshin, B.M. (1947). Geologicheskoe stronye i poleznye iskopaemye Moskvy i eyo okrestnostey [Geological aspects and mineral deposits of Moscow and its outskirts]. Mosc. Soc. Nat. Moscow. 307 (in Russian).
5. Golodkovskaya, G. A., & Lebedeva, N. I. (1984). Inzhenerno-geologicheskoye rayonirovaniye territorii Moskvy [Engineering-geological zoning of the territory of Moscow]. Inzhenernaya geologiya [Engineeringgeology]. (3). 87–102 (in Russian).
6. Gulakyan, K. A., & Kuntzel, V. V. (1970). Klassifikatsiya opolzney po mekhanizmu ikh razvitiya [Classification of landslides by the mechanism of their development]. Proc. VSEGINGEO. 29. 58–64 (in Russian).
7. International Geotechnical Society's UNESCO Working Party on World Landslide Inventory (WP/WLI). (1993). *A multi-lingual landslide glossary* (Vol. 59). Vancouver: Bitech Publ.
8. Korcheobokov, N. A., Romanov, A. V., & Yakovlev, S. Y. (1938). *Opolzni Leninskikh gor. Geologiya v rekonstruktsii goroda Moskvy [Landslides of the Leninskie Gory hills. Geology in*

the reconstruction of Moscow] (pp. 377–390). Moscow-Leningrad: Academy of Sciences of the USSR (in Russian).
9. Kuntzel, V. V. (1965). O vozraste glubokikh opolzney Moskvy i Podmoskovia. svyazannykh s yurskimi glinistymi otlozheniyami [On the age of deep landslides in Moscow and the Moscow region, associated with Jurassic clay deposits]. Bull. Moscow. of the Isp. nature. Subdiv. geol., XL (8), 93–100 (in Russian).
10. Moskva. Gorod i geologiya [Moscow. Geology and the city]. (1997). Editors Medvedev O.P., Osipov V.I. Moscow textbooks and Kartolitographia. Moscow. 399 (in Russian).
11. Nikitin, S. N. (1897a). *Geologicheskaya karta okrestnostey Moskvy [Geological map of the environs of Moscow]*. St. Petersburg: Geological Committee (in Russian).
12. Nikitin, S. N. (1897b). Les environs de Moscou. Guide des excursions du VII Congres Geologique International. St-Petersbourg. 1–16.
13. Olferiev, A. G. (2012). Stratigraficheskye podrazdelenija urskikh otlojeniy Podmoscovija [Jurassic stratigraphic subdivisions of Moscow basin]. Bull. Moscow. of the Isp. nature. Subdiv. geol., 87(4) 32–55 (in Russian).
14. Paretskaya, M. N. (1975). Zavisimost morfologii opolzney vydavlivaniya Podmoskovia ot prochnostiy urskikh glin [Dependence of the morphology of landslides extrusion of the Moscow suburbs on the strength of Jurassic clays]. Proc. VSEGINGEO. 81. 94–97 (in Russian).
15. Pavlov, A. P. (1890). Novyye dannyye po geologii Vorobyevykh gor [New data on the geology of the Vorobyovy Hills]. Bulletin of Natural Science. (7). 301–304 (in Russian).
16. Pavlov, A. P. (1910). Zametka ob obrazovanii opolzney v glinistykh i glinisto-peschanykh porodakh [A note on the formation of landslides in clayey and clayey-sandy rocks]. Bull. Soc. Nat. Moscou. 4. 29–30 (in Russian).
17. Pavlov A. V. (1911). Dokladnaya zapiska Moskovskoy Gorodskoy uprave o stroyenii mestnosti po linii napornyy rezervuar – Yakht-klub – Moskva-reka i o prichinakh spolzaniya nagornogo otkosa mezhdu napornym rezervuarom i vostochnym krayem s. Vorobyeva [Memorandum to the Moscow City council on the structure of the area along the pressure tank line – yacht Club - Moscow River and the reasons for the slope of the uphill slope between the pressure tank and the eastern edge of the village Vorobyov]. City Printing House. Moscow (in Russian).
18. Shkolin A. A., & Malenkina S. Y. (2015). Sravneniye tipov razrezov verkhneyy ury (Volzhskiy yarus) – nizhnego mela yugo-vostoka Moskovskogo regiona [Comparison of the types of sections of the Upper Jurassic (Volga level) -a lower chalk of the southeast of the Moscow region]. Jurassic system of Russia: problems of stratigraphy and paleogeography: Proc. VI Russian Workshop. ALEF. Makhachkala. 304–308 (in Russian).
19. Mitta, V. V., Alexeev, A. S., & Shik, S. M. (Eds.). (2012). *Unified regional stratigraphic scheme of the Jurassic sediments of the East European Platform. Explanatory letter* (p. 64). Moscow: GINRAS - FGUPVNIGNI (in Russian).
20. Zerkal, O., Barykina, O., Samarin, E., Gvozdeva, I. (2017). The influence of paleo-landslide activity on the modern slope stability. Proc. of 2017 IPL Symposium, UNESCO-ICL, Paris. 89–92.
21. Shubert F. F. (1860). Topograficheskaya karta Moskovskoj gubernii, gravirovannaya v Voenno-Topograficheskom depo v 1860 godu na 40 listah. Masshtab 2 versty v anglijskom dyujme 1:84000. Moskva. [Topographic map of Moscow province, engraved in 1860 on 40 sheets of military Topographic depot. A scale of 2 miles in English inches, 1:84000. Moscow] (in Russian).

Chapter 25
Application of Fractal Theory Methods for Seismogram Analysis

Nikolay A. Makhutov, Rasim S. Akhmetkhanov, and Dmitry O. Reznikov

Abstract A variety of quantitative and qualitative methods and approaches are being used in risk assessment. One of these approaches is based on the application of the toolkit of nonlinear dynamics, theory of fractals, and multifractals.

The theory of fractals and multifractals is now widely used to describe properties of self-similarity and complex scaling properties of various processes. These studies showed that not all these processes can be characterized by a single value of the fractal dimension. Only one value of the Hausdorff dimension or scaling index is needed for quantitative description of fractals. This value describes the persistence of the geometry or statistical characteristics when the scale is being changed.

The studies on application of fractal analysis for assessing data sets of different nature have shown good results. The possibility to receive data that characterize the dynamic processes in terms of their scale properties gives us a hope that these methods combined with the other ones will allow us to explore the seismic processes more comprehensively and to estimate quantitatively the probability of seismic events within a short-term forecast.

About a million earthquakes occur on the Earth every year; most of them are insignificant, but some of the powerful ones are known for their devastating consequences. Since the beginning of the twenty-first century, there have been several catastrophic earthquakes, including the 2004 Sumatra earthquake accompanied by a tsunami with 227,898 victims and the 2010 Haiti earthquake with a magnitude of 7.0 that claimed the lives of more than 316,000 people.

In the year of 2011 alone, 180 earthquakes with a magnitude of 6.0 Richter scale or more were registered. Among them were several devastating earthquakes in Japan on March 11 with a magnitude from 7.6 to 9.0, the most powerful of which was accompanied by a tsunami. The total economic losses are apparently impossible to estimate. For example the March 2011 earthquakes caused damage at the Fukushima

---

N. A. Makhutov · R. S. Akhmetkhanov · D. O. Reznikov (✉)
Mechanical Engineering Research Institute of the RAS, Moscow, Russia

V. Svalova (ed.), *Natural Hazards and Risk Research in Russia*,
Innovation and Discovery in Russian Science and Engineering,
https://doi.org/10.1007/978-3-319-91833-4_25

nuclear power plant. Most of casualties and the maximum damage were recorded in the provinces of Iwate, Miyagi, and Fukushima due to a tsunami with the maximum run-up heights. The total economic loss in Japan is estimated at over 300 billion US dollars.

On November 13, 2017, an earthquake with the magnitude of 7.6 occurred at 204 km northeast of Baghdad and 104 km west of the Iranian city of Kermanshah where the number of victims was up to 530. In addition, more than 7.8 thousand people were injured.

Russia is not the most earthquake-prone place on Earth, but earthquakes here can bring a lot of troubles: over the past 25 years, 27 major earthquakes with a magnitude exceeding 7.0 on the Richter scale occurred in the country. The fact that population in many earthquake-prone areas such as Sakhalin, the Kuril Islands, Kamchatka, Altai, Yakutia, and the Baikal region is sparse – which is not true of the Caucasia – is in a sense helpful. The fact is, however, that more than 20 million people in total live in seismically active regions of Russia.

Prospects for the coming decades do not allow one to hope that the world seismic situation improves. To date this problem is considered unsolvable as the causes of seismic activity growth are not fully understood. Seismology mostly explains seismic activity growth through endogenous factors in the Earth's interior. The theory of plate tectonics is however generally accepted. It is the starting point for understanding the processes inside the Earth that lead to seismic and volcanic activity.

Most seismologists do not deny that the process of seismic activity is characterized by the existence of a particular clustering: usually earthquakes are localized at the boundaries of tectonic plates and the earthquake clustering is also observed in time domain. Seismologists however have substantial disagreement regarding the possibility for earthquake prediction.

To solve the problem of earthquake prediction through the assessment of seismic monitoring data, it is necessary that the study of quantitative and qualitative characteristics of the dynamics of seismic activity be carried out.

In recent years, alongside some traditional and new methods for the analysis of the distribution of risks of critical events [1, 2], approaches based on application of the patterns of self-similarity or fractals theories are becoming more and more popular [3, 4]. These approaches allow us to reveal the internal structure of data (time series) on the sequences of critical events and on the distribution of relevant risks along with time intervals. That is why the aim of the paper was to investigate the earthquake seismograms using fractal theory methods and to obtain qualitative and quantitative characteristics of dynamical processes described by the seismograms.

This fractal-based approach allows us to better understand the nature and properties of a sequence or set of critical events of different scale using universal self-similarity patterns inherent to different systems of various origin.

Self-similarity implies that copying and scaling of some reference sample allows nature to easily create a complex and multiscale structure.

B. Mandelbrot who was the founder of the fractal theory defined fractal as a set for which the Hausdorff–Besicovitch (or fractal) dimension D strictly exceeds the topological dimension D_T. In the Euclidean space R^E the dimension values D_T and D lie in the interval from 0 to E. Dimension D_T is always an integer number which is not necessarily the case with D.

Many observations of natural processes lead to temporal dependencies or series of measurements. The temporal sequence of measurements can be investigated using the method of normalized range, or Hurst method. Such sequences of measurements are characterized by the index H (Hurst exponent). Hurst investigated many natural processes such as river runoff, silt deposit, and the growth of tree rings. For this purpose he introduced the dimensionless ratio R/S, where R is a range and S is a standard deviation. Using this dimensionless ratio, it is possible to compare the ranges of different phenomena. According to Hurst, for many time series the observed normalized range R/S is well described by the empirical power law relation:

$$R/S = (\tau/2)^H.$$

Mandelbrot showed that the Hurst exponent and the fractal dimension are interrelated. For self-affine curves there is a relationship between the Hurst exponent H and the local fractal dimension D of the time series:

$$D = 2 - H.$$

This relationship is valid when the structure of the curve that describes the fractal function is investigated with high resolution, i.e., in the local limit.

The correlation function $C(t)$ between the future increments of the process and the past ones depends on the exponent of self-similarity H:

$$C(t) = 2^{2H-1} - 1 \quad (25.1)$$

It follows from formula (25.1) that when $H = 0.5$, there is no correlation between past and future increments for all t. The fractal dimension D of a random process with independent increments equals to 1.5.

Many of the signals that are to be analyzed in practice can be considered as representatives of a special class of multifractal processes. If simple (or monofractal) signals are homogeneous, i.e., their large-scale properties remain the same in all frequency ranges, then multifractal processes allow decomposition into subsets (segments) with different local-scale properties.

Monofractal signals are homogeneous, that is, their scaling characteristics remain the same in all scale ranges. The spectrum of power density for such signals can be expressed as $S(f) \sim f^{\beta}$ and does not change in a wide frequency range. Here the exponent β is constant, and such dependencies are referred to as power laws. The exponent β = const characterizes the homogeneity of the function; it is scale invariant and is defined in the range of $1 < \beta < 3$. Monofractal processes are characterized by a normal form of a probability density function.

As shown above, the fractal dimension of the time series can be evaluated by the Hurst exponent H which is determined in the range of $0 < H < 1$. If $0 < H < 0.5$, then the range corresponds to anti-persistent series. The process of this type has a short-range memory and correlation. When $H = 0.5$, no specific tendency of the process is detected, nor is there any reason to believe that it will appear in the future. In this case, the Hurst exponent indicates a random series. If $0.5 < H < 1.0$, then the time series studied is persistent or trend-resilient(there is a long-term memory). If in the previous period the series increases (or decreases), it is likely to keep the trend for some time in the future. Thus, the Hurst exponent can be considered as a predictive criterion to determine the likelihood of hazardous events.

The fractal seismogram analysis was conducted. It was based on data on four earthquakes:

- Kronotskoe earthquake (December 4, 1997), magnitude 8.4, recorded by Petropavlovsk-Kamchatsky station (PKS)
- Olyutorsky earthquake (April 20, 2006), magnitude 7.5, recorded by PKS
- Simushirskoe earthquake (November 15, 2006), magnitude 8.3, recorded by PKS
- Kuril earthquake (January 15, 2009), magnitude 7.4, recorded by Yuzhno-Sakhalinsk station (YSS)

Simushirskoe earthquake occurred on November 15, 2006, and became one of the most powerful in the history of the modern Kurils. It was caused by the earth's crust compression. The island of Matua proved the closest land to the epicenter. Fifteen minutes after the Matua earthquake, a tsunami came with the run-up height of 20 m in some places of the coast. Another earthquake occurred to the East of Simushir Isle. The magnitude of the earthquake exceeded 8.3 Richter scale.

Its seismogram $X(t_i)$ ($t_i = t_{i-1} + \Delta t$, $i = 1,..,l$) is shown in Fig. 25.1. And the results of the calculation of the Hurst exponent H and the histogram of its distribution are shown on Fig. 25.2. The length of the window considered during the Hurst exponent calculation was 10,000 values of the seismogram. The distribution of the Hurst exponent is characterized by a maximum density near $H = 0.9$.

Given the fact that the Hurst exponent estimates the distribution of the frequency spectrum (the coefficient β), the graph shows that the frequency spectrum of the seismogram is undergoing significant time changes. The changes of the Hurst exponent are of quasiregular character. Prior to the earthquake one can observe spikes of the Hurst exponent that is decreased from 0.95 to 0.7.

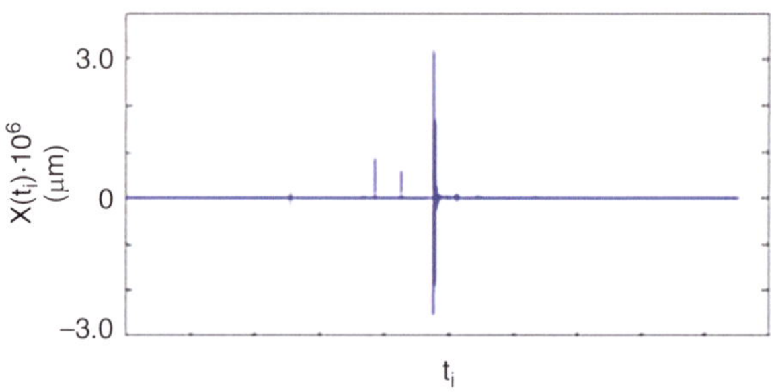

Fig. 25.1 Simushirskoe earthquake's seismogram: $l = 19085602$, duration 11 days

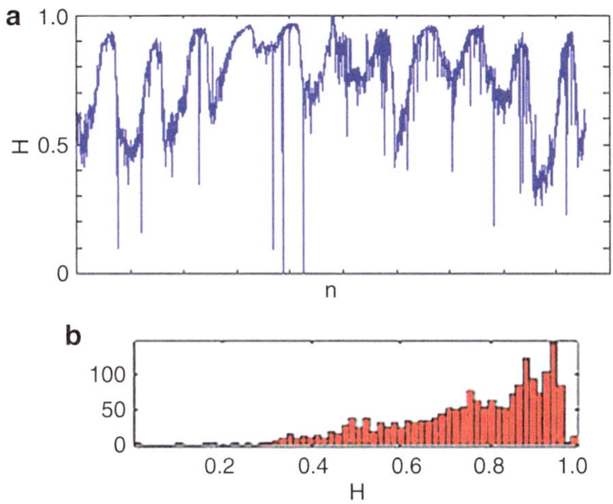

Fig. 25.2 (**a**) The pattern of change of the Hurst exponent: $H_{avr} = 0.7591$, std(H)=0.1689. (**b**) Hurst exponent histogram

Figure 25.3 shows the wavelet decomposition of the Hurst exponent dynamics into frequency components through wavelet transformation (MatLab software package was used) [3]. The original signal H is approximated by the coefficient a_5 and specified coefficients d_1, d_2, d_3, d_4, and d_5 [5]. The signal is decomposed into five levels.

Such decomposition allows us to estimate the frequency features of the Hurst exponent temporal dynamics. The nature of these frequency components is nonstationary. A quasiregular character of the Hurst exponent change is determined by the low-frequency components (coefficients) a_5, d_4, and d_5, whereas high-frequency components are described by the coefficients d_1, d_2, and d_3.

The data presented (Figs. 25.1, 25.2, and 25.3) give a detailed description of the specific features of the seismogram considered.

1997 Kronotskoe earthquake was the strongest over the past 40 years having the largest magnitude (7.9 on Richter scale) and size of the aftershock area. It occurred far away from the settlements on the Eastern coast of Kamchatka. The macroseismic effect of the earthquake accompanied by multiple ripping along the focal zone stretching over 200 km was lower than that for other Kamchatka earthquakes of similar magnitude and did not result in any destruction of the settlements in the region. The main impact was at the distance of 360 km from the city of Petropavlovsk-Kamchatsky, where the intensity of the shocks were in the range from 5 to 6 points. The seismogram of the earthquake is shown in Fig. 25.4, and the dynamics of the Hurst exponent is presented in Fig. 25.5.

The value of the Hurst exponent was about of 0.5 before the earthquake (Fig. 25.6a). The histogram of the exponent H distribution shows that the largest values of the density lie in the region from 0.8 to 0.95. The average value of the Hurst exponent H_{avr} and its standard deviation std(H) are close to the corresponding values of the previous earthquake.

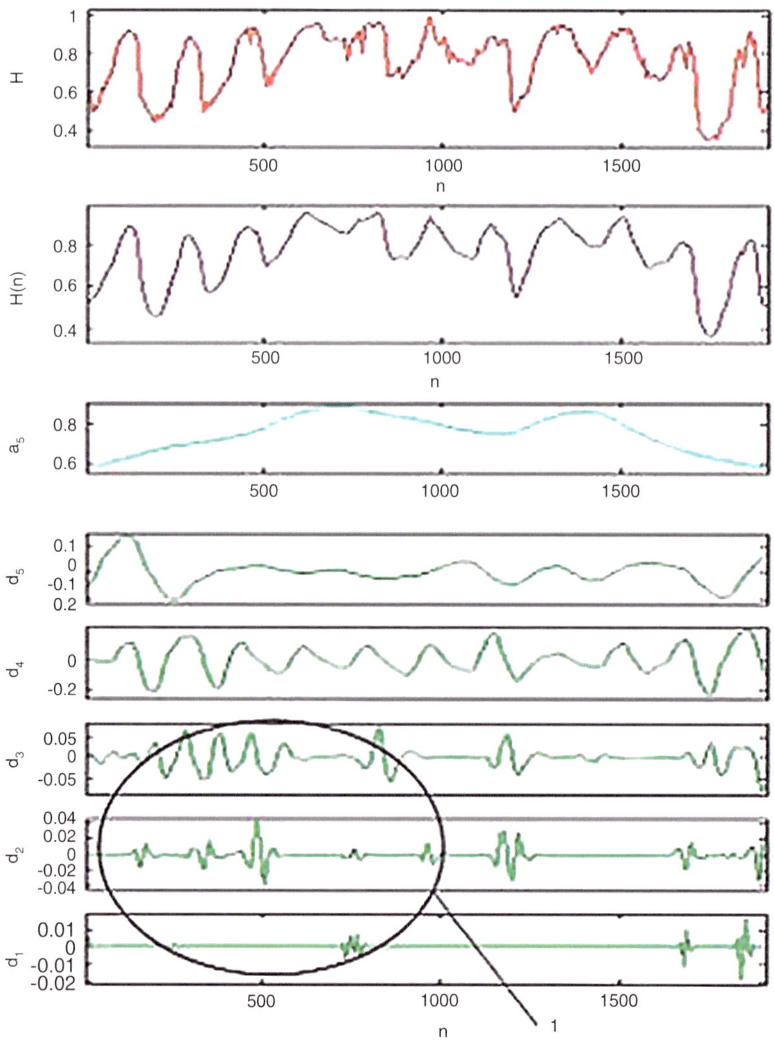

Fig. 25.3 Decomposition of the Hurst exponent into frequency components 1, the range of variation of frequency components d_1, d_2, and d_3

Changes in the dynamics of the Hurst exponent before the earthquake came to the appearance of high-frequency components of the wavelet decomposition (coefficients d_1, d_2, and d_3 in Fig. 25.6).

Olyutorsky earthquake with the magnitude of 7.8 on Richter scale occurred in 2006 in the North of the Kamchatka region. The earthquake left about 500 people homeless. The village of Tilichiki was completely destroyed. Korf airport runway was damaged and did not function. Liquefaction of the soil, formation of breaches, and cracks on the surface were the main damaging factors. The earthquake

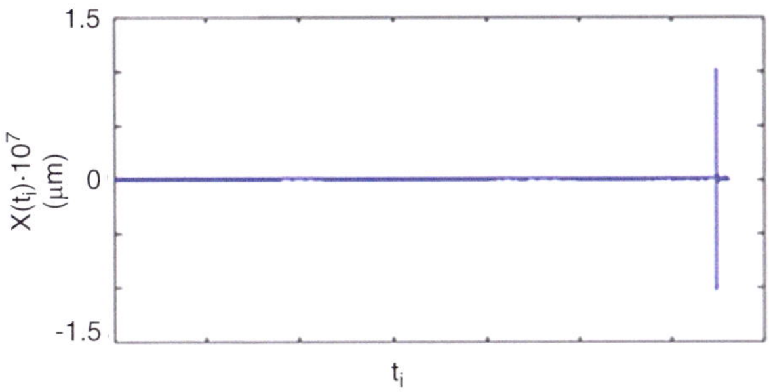

Fig. 25.4 Kronotsky earthquake seismogram: $l = 13234771$, duration 7 days 15 h 40 min

Fig. 25.5 (**a**) The pattern of change of the Hurst exponent: $H_{avr} = 0.8244$, std(H) =0.1391. (**b**) Hurst exponent histogram

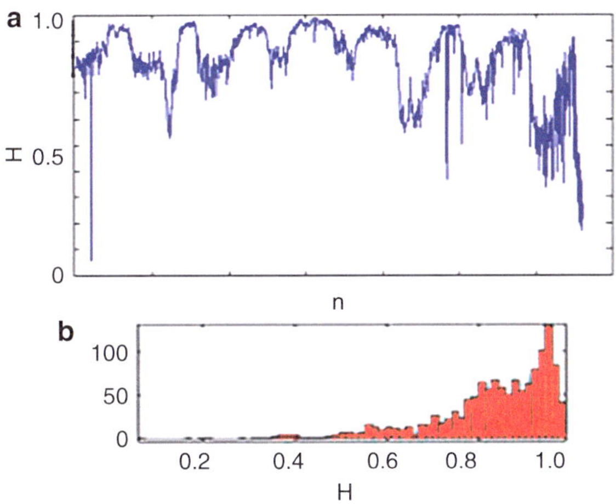

seismogram is represented in Fig. 25.7, and histogram of the Hurst exponent dynamics can be seen in Fig. 25.8.

In this case, the Hurst exponent H varies within broad limits from 0.1 to 1.0 (Fig. 25.8a). The histogram shape has four characteristic modes (Fig. 25.8b) and is determined by frequency wavelet decompositions: approximating coefficient a_5; specifying coefficients d_2, d_3, d_4, and d_5; and an additional coefficient d_1 in earthquake zone (Fig. 25.9).

On January 15, 2009, at 17 h, 49 min Greenwich time (20 h 49 min Moscow time) in the Central part of Kuril Ridge in the southern part of the North Kuril area, a strong earthquake with a magnitude of 7.6 points at a depth of 35 km occurred. The epicenter was located 428 km from the North-Kurilsk and 725 km from Petropavlovsk-Kamchatsky. The earthquake was felt with a magnitude of 3 points in Petropavlovsk-Kamchatsky, Severo-Kurilsk, Yuzhno-Kurilsk, and Malokurilskoe.

Let us consider on the seismogram the total length of which is about 23 days. Because of its large duration, the seismogram was divided into three stages.

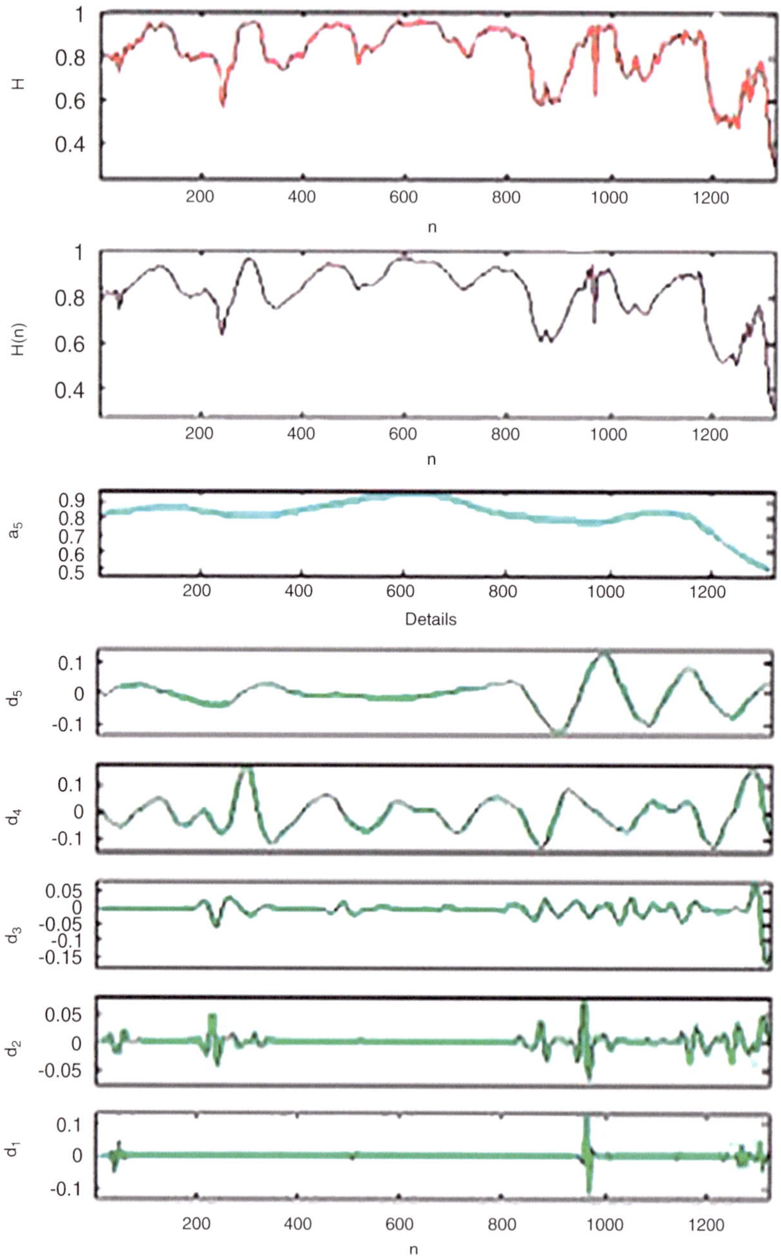

Fig. 25.6 Decomposition of the Hurst exponent into frequency components

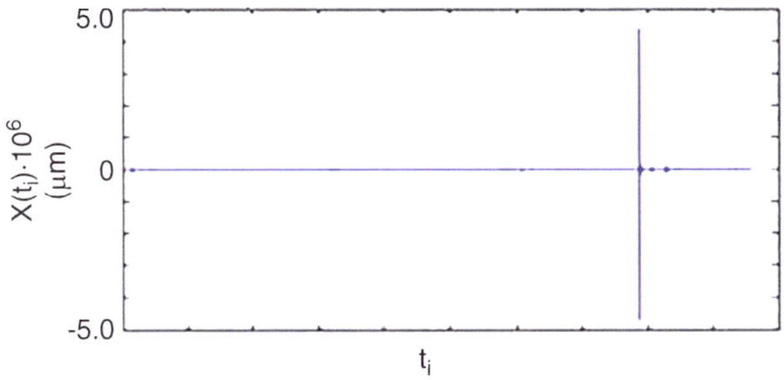

Fig. 25.7 Olyutorsk earthquake seismogram: $l = 19225000$; duration, 11 days 2 h 40 min

Fig. 25.8 (**a**) The pattern of change of the Hurst exponent: $H_{avr} = 0.6204$, std $(H) = 0.2524$). (**b**) Hurst exponent histogram

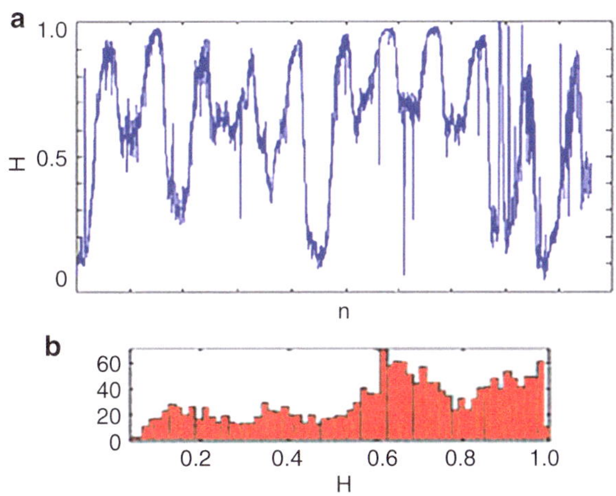

Figure 25.10 shows the seismogram of the first stage (8-day and 21 h time interval), and Fig. 25.11 presents the change of the Hurst exponent in this time range. The average value of the Hurst exponent is $H_{avr} = 0.6708$, and its standard deviation equals to std(H)=0.193. In the interval b (Fig. 25.11) a significant change in the Hurst exponent from 0.85 to 0.1 can be seen, and the characteristic region is observed on the seismogram (region a) with a significant decrease of the oscillation amplitude $X(t_i)$.

Figure 25.12 shows the wavelet decomposition of the Hurst exponent dynamics into its components; the dynamics of changes of the Hurst exponent is determined by the coefficients of a_5, d_5, d_4, and d_3. For the initial part of the seismogram, the additional coefficients d_1 and d_2 should also be taken into account.

The second stage lasting 8 days, 9 h, and 20 min is shown in Fig. 25.13. The seismogram contains two foreshocks. The Hurst exponent is characterized by changes in the range from 0 to 0.8 is shown in Fig. 25.14. And its average value is equal to $H_{avr} = 0.3486$, which is well below the previous values for the first stage.

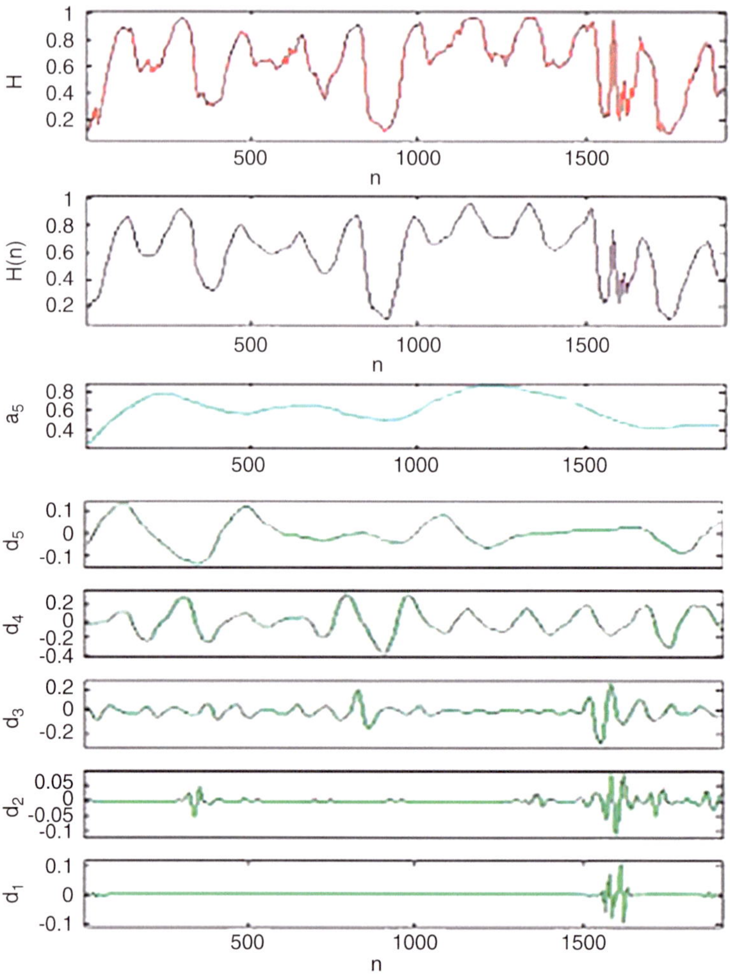

Fig. 25.9 Decomposition of the Hurst exponent dynamics into frequency components

Fig. 25.10 Seismogram of Kuril earthquake: stage 1, $l = 15358609$, the duration is 8 days and 21 h

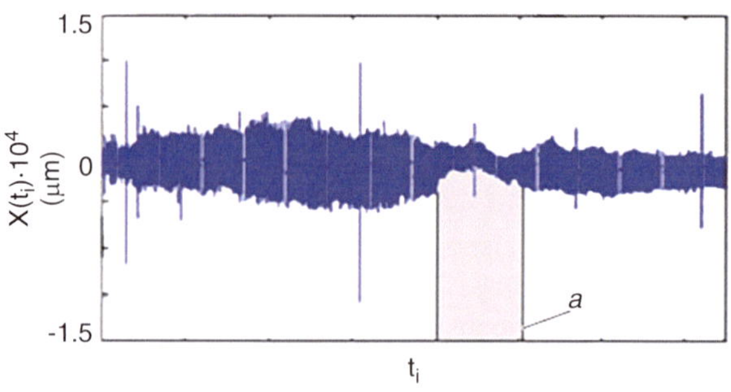

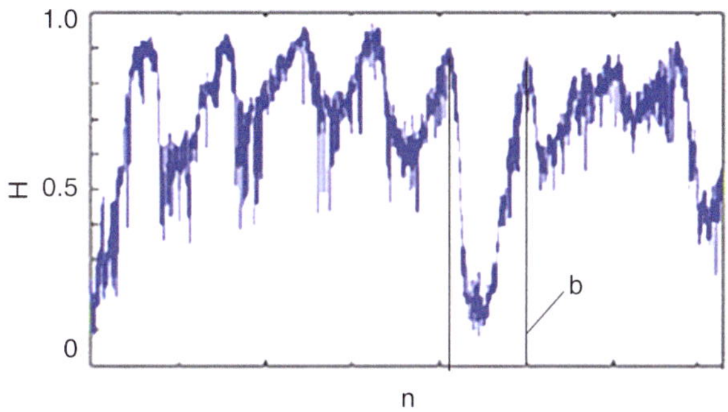

Fig. 25.11 The pattern of change of the Hurst exponent: $H_{\text{avr}} = 0.6708$, $\text{std}(H) = 0.1930$

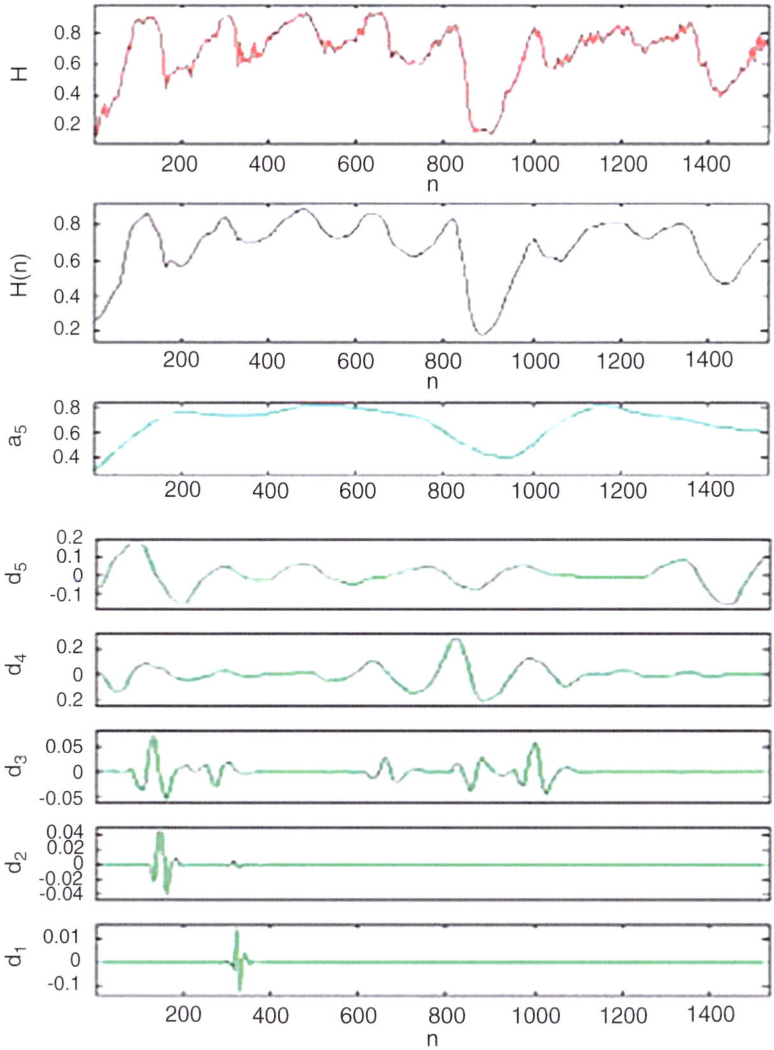

Fig. 25.12 Decomposition of the dynamics of the Hurst exponent into frequency components

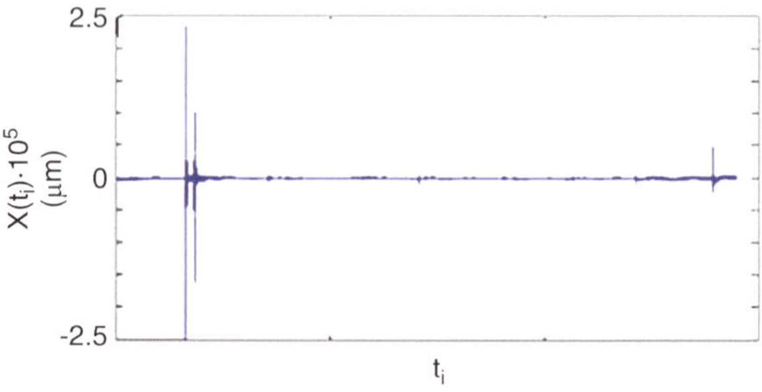

Fig. 25.13 Kuril earthquake's seismogram: stage 2, $l = 14511866$, duration – 8 days 9 h 20 min

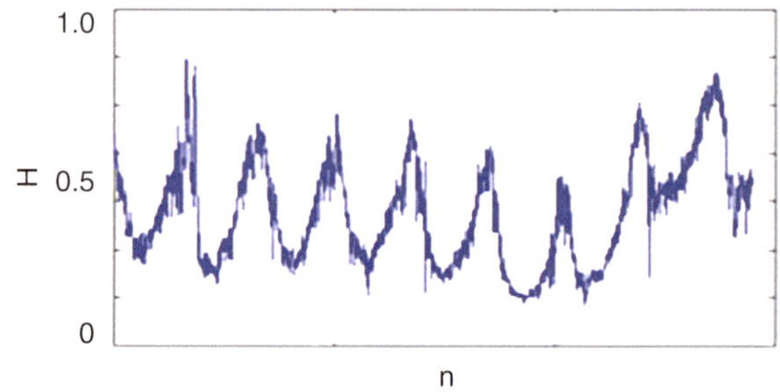

Fig. 25.14 The pattern of change of the Hurst exponent: $H_{avr} = 0.3486$, std(H) = 0.2200

Figure 25.15 shows the decomposition of the Hurst exponent dynamics into its components. The dynamics of changes of the Hurst exponent is determined by the decompositions a_5, d_3, d_4, and d_5, at the initial part of the seismogram and additionally by components d_1 and d_2 in the time domain associated with the foreshocks.

The second stage with duration of 5 days 17 h is shown in Fig. 25.16.

The dynamics of the Hurst exponent before the earthquake did not change substantially; it was varied within the range from 0 to 0.9 (Fig. 25.17). There is a characteristic increase in the values of the Hurst exponent before the earthquake, where it was varied in the range of 0.4–0.9.

Figure 25.18 shows the wavelet decomposition of the dynamics of the Hurst exponent into its components. The dynamics of changes of the Hurst exponent was determined by the coefficients a_5, d_5, and d_4 and additionally by the coefficient d_5 prior the earthquake and by the coefficients d_2 and d_1 after the earthquake.

We obtained the following average values of the Hurst exponent on the considered stages (1–3), H_{avr}, 0.6708, 0.3486, and 0.4477, and standard deviations, std*(H)*, 0.1930, 0.2200, and 0.2787. These average values and histograms of the distributions of the values of the Hurst exponent (Fig. 25.19) show that the system in its first stage has mainly persistent properties, while at stage 2 it has anti-persistent properties. The third stage is characterized by both processes. And the average values of the Hurst exponents are less than those previously seen in other seismograms.

All data shown above confirm that the dynamic processes described by seismograms are multifractal.

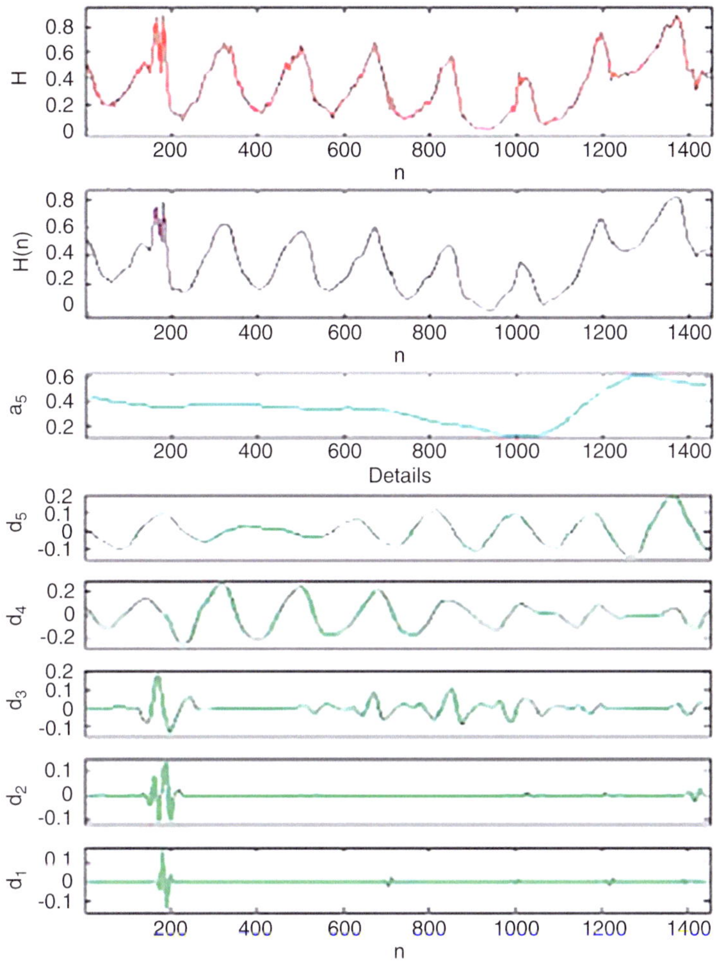

Fig. 25.15 Decomposition of the Hurst exponent into frequency components

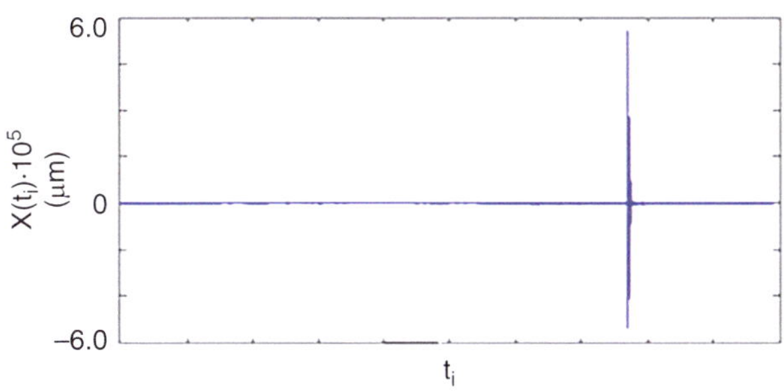

Fig. 25.16 Kuril earthquake's seismogram: stage 3, $l = 9874526$, duration is 5 days 17 h

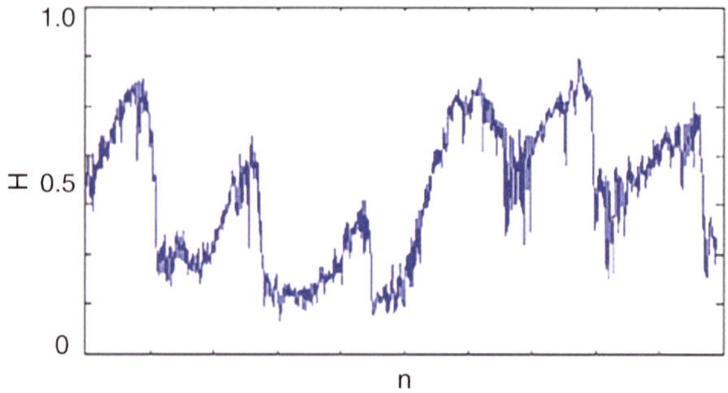

Fig. 25.17 The pattern of change of the Hurst exponent: $H_{avr} = 0.4477$, std(H)=0.2787

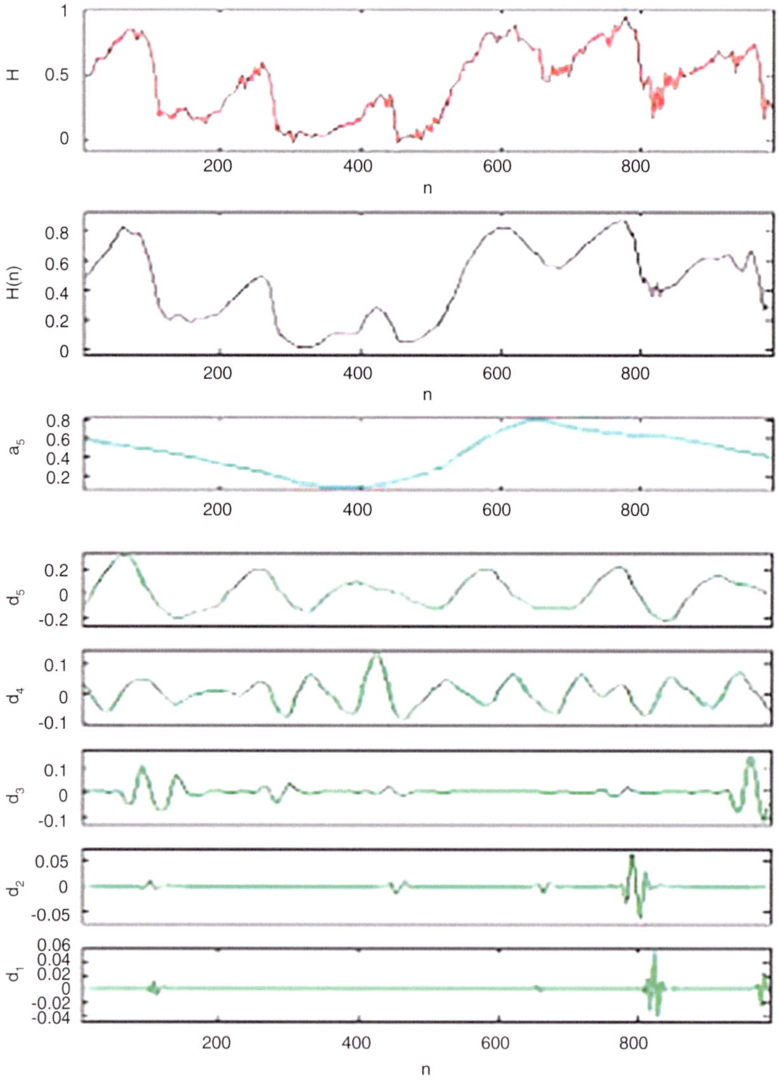

Fig. 25.18 Decomposition of the Hurst exponent into frequency components

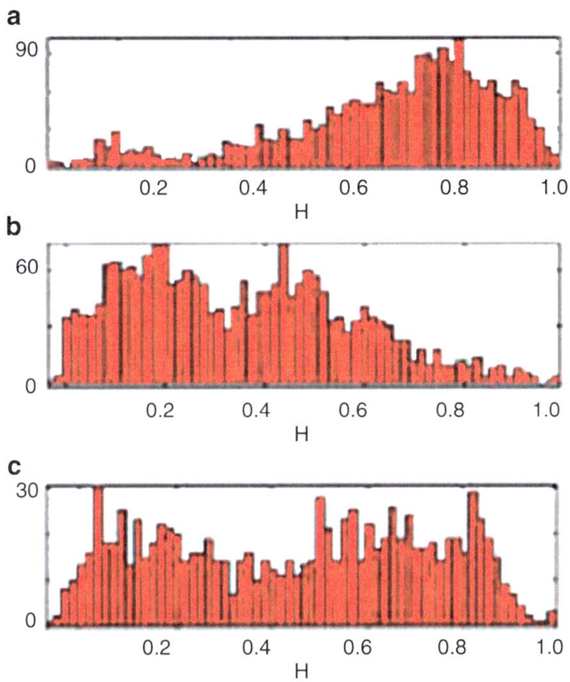

Fig. 25.19 Histograms of distribution of the Hurst exponent for the stages: (**a**) stage 1, (**b**) stage 2, and (**c**) stage 3

Multifractals are heterogeneous fractal objects, a full description of which, unlike that of the usual fractals, requires not only the introduction of one value (its fractal dimension *D*) but a whole range of dimensions whose number is, in general, infinite [4]. The reason for this is that, along with purely geometric characteristics defined by dimension *D*, they possess another important property, i.e., the distribution of some measure p_i along its geometric medium. Almost anything can be such a measure, i.e., population density, substance concentration, magnetization, energy, or pixel luminance. It is important that the measure is unevenly, but self-similarly, distributed along the multifractal. And the process spectrum differs significantly from that of the monofractal.

Multifractality of the process is usually presented as a multifractal spectrum (or singularity spectrum) $f(\alpha)$ [3, 4]. Multifractal spectra are characterized by a spectral width, asymmetry, and curvature. An increase of the width of the spectrum corresponds to the nonuniformity measures.

If dimension D_0 is used as a quantitative characteristics of complex fractal geometry, then the mathematical description of fractal measures is carried out in terms of spectrum of singularities $f(\alpha)$.The spectrum of singularities $f(\alpha)$ characterizes the dependence $N_\alpha(\varepsilon)$ of the number of elements of coverage corresponding to the points with the exponent singularity which is equal to a certain value of α [4]

$$N_\alpha(\varepsilon) \sim \varepsilon^{-f(\alpha)},$$

where ε – scaling factor.

If α = const the value of $f(\alpha)$ corresponds to the Hausdorff dimension. If the distribution of measure on a set is even, the spectrum of singularities is a unique point on the (α, f) plane. With an uneven distribution of measure, function $f(\alpha)$ has a more complex bell-shaped form. In the absence of multifractality, we have $D_q = D_1 = D_o = \alpha_{\max} = \alpha_{\min} = f(\alpha)$.

Application of multifractal formalism allows us to obtain the greatest deal of information on the following indicators:

1. Hausdorff–Besicovitch dimension D_0 that characterizes homogeneous fractals. The value of D_0 is determined by the maximum value of $f(\alpha)$, which corresponds to D_q at $q = 0$.
2. Information dimension D_1 that characterizes the speed of growth of the information quantity and corresponds to D_q at $q = 1$.
3. Correlation dimension D_2 that characterizes the probability of finding two points of the set in one and the same cell of coverage. It is determined by the value of D_q at $q = 2$.
4. Extreme values of the fractal dimension D_{+q} and D_{-q} that correspond to the degree of sparseness of a multifractal set. The parameter q is a value that makes for the increase in contribution of cells having relatively large values of measure μ with positive values of q and relatively low values of measure μ with negative values of q.

Let's consider the multifractal spectra of the seismogram's diagnostic data obtained for different time periods that are shown in Fig. 25.20. Each seismogram was divided into three stages, the third stage being right before the earthquake.

The maximum value of the fractal dimension of the subsets $f(\alpha)$ is equal to the fractal dimension of the measure. In the examples presented, the measure is equal to 1. As for measures defined on fractals with fractal dimension D, we obtain:

$$f_{\max}(\alpha) = D.$$

A multifractal analysis of seismograms showed that only one out of the four seismograms can be characterized by a substantial change of the width of the multifractal spectrum and its shift toward the increase of parameter α before the earthquake.

The comparison of histograms of the Hurst exponent distribution during Kuril earthquake (Fig. 25.19) reviles that on the third stage, the exponent has the most even distribution over the entire possible range (from 0 to 1.0), and consequently, the multifractal spectrum is wider.

The analysis of the Hurst exponent changes for three earthquakes showed that H_{avr} varies in the range from 0.6204 to 0.8244, while the value of the std(H) varies in the range from 0,1391 to 0,2524. And the same characteristics of Kuril earthquake vary within the following limits: H_{avr} varies from 0.3486 to 0.6708 and std(H) in the range of 0,1930–0,2787. These differences correspond well to the data of the analysis (see Fig. 25.20d), where multifractal spectra of three stages differ one from another by the width of the multifractal spectrum and its asymmetry.

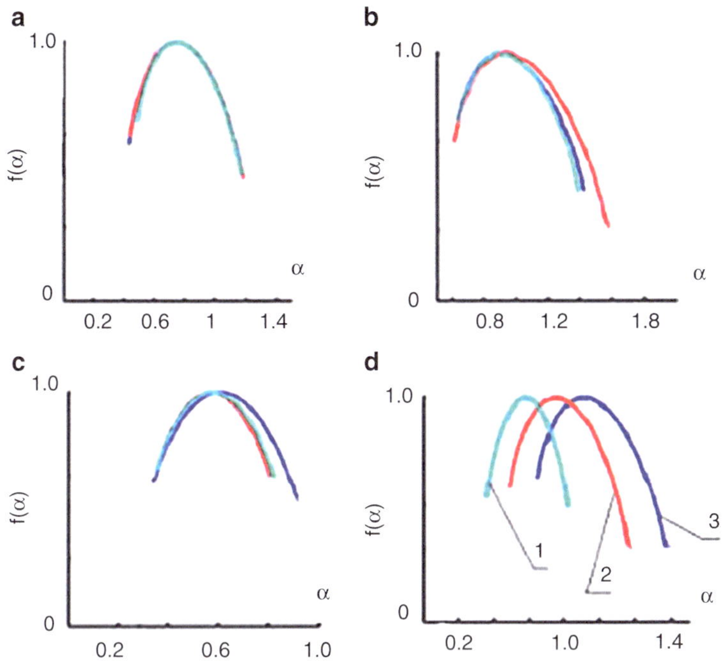

Fig. 25.20 Multifractal seismogram spectra at time $t_1(1)$ $t_2(2)$, $t_3(3)$: (**a**) the Samusenko, (**b**) Kronotsky earthquake,(**c**) Olyutorsky earthquake, and (**d**) Kuril earthquake (1, 2, 3 – the order of the stages)

Thus, regardless of multifractal dynamics of seismograms, it is obvious that monofractal analysis of seismograms provides a substantial amount of additional information about the current seismic situation.

The assessment of potential of the fractal analysis of time series shows its high sensitivity to the changes in the nature of the signal, which allows one to apply this method for analyzing the seismic activity. Different dimensions and quantitative and qualitative characteristics obtained through the method can be used as additional criteria for the assessment of seismic activity of the region.

References

1. Makhutov, N. A., Petrov, V. P., & Akhmetkhanov, R. S. (2004). *Natural-technogenic-social systems and risks. Problems of safety and emergency situations* (pp. 3–28). No 3. (in Russian).
2. Akhmetkhanov, R. S. (2006). *Fractal theory and risks. Problems of safety and emergency situations* (pp. 22–36). No 4. (in Russian).
3. Akhmetkhanov R. S. (2009). *Application of fractal theory and wavelet analysis for identification of singularities of time series in system diagnostics. Herald of scientific and technical development* (pp. 26–31). No 1. (in Russian).

4. Pavlov, A. N., & Anishchenko, V. S. (2007). Multifractal. Analysis of complex signals. *Advances in Physical Sciences, 177*(8), 859–876 (in Russian).
5. Dikonov, V. P. (2002). *Wavelets. From theory to practice*. Moscow: SALON-R Publ 448 p. (in Russian).

Chapter 26
The New Method of the Potentially Hazardous Geodynamic Active Zone Mapping

Ekaterina Karfidova and Vladimir Makeev

Abstract Authors developed GIS methods to detect the azimuthal disagreements of zones at surfaces of different ages of key horizons for the mapping of the subhorizontal geodynamic active zones at the platforms. The method is based on the analysis of digital elevation models to detect areas of an azimuthal and angle falling disagreement zones. Graphics of these zones indicate a heterogeneous structure of the key horizons, which is the reason for the development of subhorizontal geodynamic active zones. It is assumed that the greater the disagreement different age deformation, the higher the probability of the development of these zones which are hazardous to intensify geological processes.

26.1 Introduction

The concept and principles of the allocation of geodynamic active zones were first developed by V.I. Makarov in the mid-1990s. The 21 century and developed in the works of his followers [1–4, 7, 8]. The research is based on the concept of development of neotectonic deformations in geodynamic active zones of the subvertical type. They studied, as an alternative to active faults allocated to low activity platform conditions. In our opinion, conditions for the formation of active faults on ancient platforms are absent [2].

In our case, the geodynamic active zones are two-dimensional linear or three-dimensional volumes of the geological environment in which conditions arise for the concentration of additional tectonic stresses and their discharge. In applied research, the study of geodynamic active zones has acquired particular relevance in connection with the provision of geodynamic safety of the projected nuclear power plants, radioactive waste disposal sites, the buffer strip of main pipelines, and other urban territories [4].

E. Karfidova (✉) · V. Makeev
IEG RAS, Moscow, Russia

V. Svalova (ed.), *Natural Hazards and Risk Research in Russia*,
Innovation and Discovery in Russian Science and Engineering,
https://doi.org/10.1007/978-3-319-91833-4_26

The engineering-geological zoning of the territory of the Moscow megapolis showed that the rocks of the sedimentary cover in some areas have an increased layer-by-layer material-structural misalignment relative to each other, expressed by different angles of incidence of rocks and a sharp change in the material composition [3–5, 7, 8].

The study of this discrepancy has shown that it can be one of the reasons for the formation of stratified zones of increased stresses and deformations – subhorizontal geodynamic active zones.

It is assumed that the degree of mismatch of uneven-age deformations determines their differentiation with respect to the strained and deformed state: the greater the difference between the azimuths of deformations, including the angles of falling, the higher the tectonic tension in the sedimentary cover.

The East European Platform, central part (Moscow region), is investigated in order to detect the tectonic plicate and disjunctive deformations, including the geodynamic active zones (GdAZ). GdAZ associate with an abnormally concentration of high stresses and deformations. GdAZ are characterized by high fragmentation and fracturing of rocks, the local manifestations of karst, hydrogeological "window," and some other exogenous processes.

These researches are part of the project, which is realized at the Sergeev Institute of Environmental Geoscience RAS (IEG RAS) based on GIS technologies in order to create the scientific informational geo-environmental fund as a large geodatabase for Moscow megapolis. The surfaces of the key horizons are created with GIS technology of the processing borehole database of IEG RAS [6]. It should be noted that the calculation of the horizons as digital elevation model (DEM) with regular GRID has a cell size of 100 m. The results of the geostatistical analysis of the researching key horizons are presented in Table 26.1.

Advanced analysis of digital elevation models key horizons is based on the software ESRI – ArcView-ArcGIS – plus different additional modules of ESRI users (directional slope, curvature, the hydrological modeling, the evaluation elevation variety, vertical and horizontal partitioning/ruggedness of relief). The histogram of range Z, calculated by the method Neighborhood Statistics of Surface for GRID Z in circle $R = 1000$ m for Vr (Verey) clays and C-O (Callovian-Oxfordian) clays, is presented in Fig. 26.1.

Then the aspect and slope GRID are calculated in analysis of digital elevation model (DEM) of horizon surfaces. Geodynamic active zones and their dislocations are extending in the subvertical and subhorizontal directions. If subvertical (classic)

Table 26.1 The results of the geostatistical analysis of the researching key horizons

Statistics	Grd_VR	Grd_C-O
Minimum value	−79.00	95.00
Maximum value	22.58	142.00
Mean value	−29.46	119.94
Standard deviation	24.16	10.49
Variance	584.03	110.10
Range	101.58	47.00

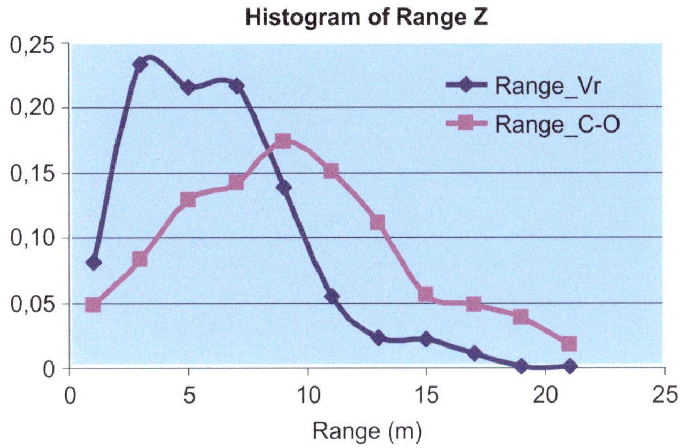

Fig. 26.1 The histogram of range *Z*, calculated by the method Neighborhood Statistics of Surface for GRID *Z* in circle $R = 1000$ m for Vr Verey clays, C-O Callovian-Oxfordian clays

GdAZ more or less successfully are identified using the structural geomorphological methods, then the subhorizontal zones with layered dislocations cannot be detected in this way. Their cartography is based on the paleotectonic analysis of ancient (Precambrian, Paleozoic, Mesozoic) and newest (Neogene-Quaternary) key horizons with the detection of an azimuthal disagreement zones.

In this work we studied two key horizons:

1. The roof of Verey clays of the Moscovian stage of Pennsylvanian.
2. The roof of Callovian-Oxfordian clays of Middle-Upper Jurassic . The character of their coherence with each other is defined using borehole data.

Then horizon surfaces are defined by the character of their coherence with each other through the detection of an azimuthal disagreement zones.

26.2 Methods

Advanced analysis of digital elevation models key horizons is based on the software ESRI – ArcView-ArcGIS – plus different additional modules of ESRI users (directional slope, curvature, the hydrological modeling, the evaluation elevation variety, vertical and horizontal partitioning/ruggedness of relief). The histogram of range *Z*, calculated by the method Neighborhood Statistics of Surface for GRID *Z* in circle $R = 1000$ m for Vr (Verey) clays and C-O (Callovian-Oxfordian) clays, is presented in Fig. 26.1.

Then the aspect and slope GRID are calculated in analysis of digital elevation model (DEM) of horizon surfaces. Geodynamic active zones and their dislocations are extending in the subvertical and subhorizontal directions. If subvertical (classic) GdAZ more or less successfully are identified using the structural geomorphological

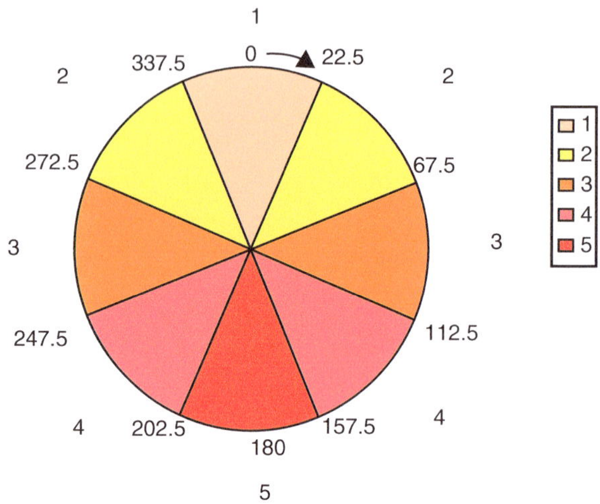

Fig. 26.2 Classification of dip angle (angle falling) by five classes

methods, then the subhorizontal zones with layered dislocations cannot be detected in this way. Their cartography is based on the paleotectonic analysis of ancient (Precambrian, Paleozoic, Mesozoic) and newest (Neogene-Quaternary) key horizons with the detection of an azimuthal disagreement zones. In this work we studied two key horizons: (1) the roof of Verey (Vr) clays of the Moscovian stage of Pennsylvanian and (2) the roof of Callovian-Oxfordian (C-O) clays of Middle-Upper Jurassic. The character of their coherence with each other is defined using borehole data.

Then horizon surfaces are defined by the character of their coherence with each other through the detection of an azimuthal disagreement zones.

The azimuthal disagreement is calculated based on the difference between azimuthal directions of key horizon surfaces. GdAZ are classified by five classes (Fig. 26.2).

The first and second classes have weak stage of an azimuthal disagreement.

The third class represents the type of an azimuthal disagreement – normal crossing – when angle 90° is between the considered horizons.

The fourth and fifth classes are strong degree of an azimuth disagreement, when the gradients of the angle falling have a different signs (Fig. 26.2). Model for detection of azimuthal disagreement zones is created using Model Builder (ESRI) (Fig. 26.3).

When the researcher has DEM surface, he can use famous cartographic methods and try to make a better cartographic presentation: gradients, profiles, histograms, and others to understand the problem essence.

The proposed method is based on the use of new cartographic layers, simple calculations, and the drawing of a special azimuthal disagreement graph, which is a principal innovation.

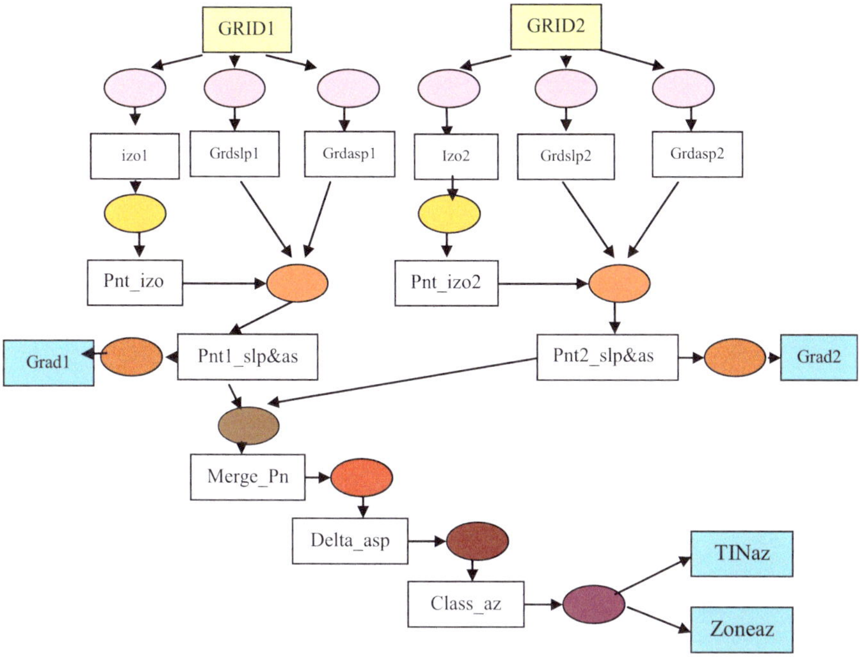

Fig. 26.3 Model for detection of the azimuthal disagreement zones of two surfaces of different ages. GRID1 и GRID2 digital elevation model of key horizons (DEM1 and DEM2), izo1 и izo2 calculated isolines of DEM, Grdslp1 и Grdslp2 calculated grids of slope, Grdasp1 и Grdasp2 calculated grids of aspect, Pnt_izo1 и Pnt_izo2 point layer for each surface, Grad1 и Grad2 gradients for each surface, Merge_pnt merge of point layers of two surfaces, Delta_asp difference between aspects of surfaces, Class_az classification by Delta_asp in five classes, TIN_az triangulated irregular net (TIN) by Class_az, Zone_az zones bounded by isolines of the azimuthal disagreement of class 3 and more

Each surface of the key horizons carries out the following steps (Fig. 26.4):

1. A circle with a radius of 1 km is built in the center of the disagreement zone.
2. On the circle, starting from the direction to the north, points are drawn through 45° in the clockwise direction (total of eight points).
3. The absolute elevation is determined in the center of circle.
4. The absolute elevation of horizon is referenced for circle points.
5. For each point of the circle, the differences in elevation between the center of the circle and points on the circle are calculated.
6. The graph is constructed. The x-axis is the azimuth direction (0–360°); the y-axis is the difference in elevation marks.

   The first and the last tenth point is assigned a zero value (for a symmetric view). A smoothing curve is plotted along the points.
7. The azimuthal disagreement zones are allocated on the plotted graph.

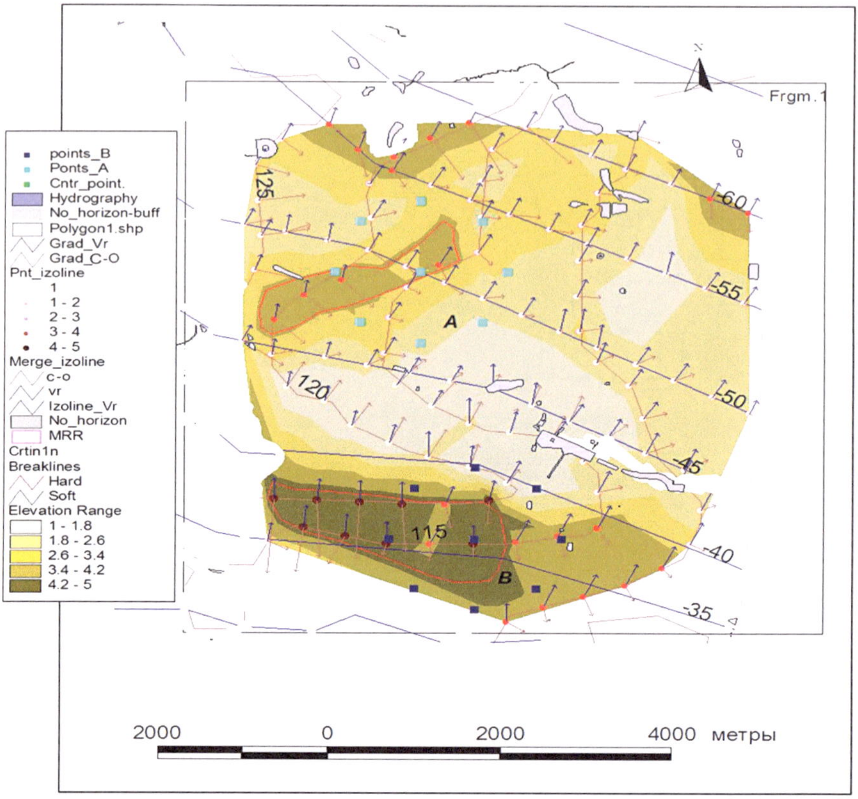

Fig. 26.4 The cartographic presentation of new method. Graph Delta Z vs. φ is calculated for circle A (top) and circle B (bottom). Cntr_point central point of the circle with radius 1000 m. Points A and points B points located on the circle, divided in eight segments (45°). AD azimuthal disagreement; Delta $Z = Z_i - Z_c$; Z_i value Z of DEM; i, 0, 1, 2. . .10; φ, 0, 45, 90 . . . 360; z_c, value Z at central point of circle. Circle A has partial azimuthal disagreement, class 3, circle B, entirely Azimuthal Disagreement, class 5

Therefore, the new digital cartographic method makes it possible to detect GdAZ and to determine the degree of hazard for the territories of especially important object design.

The cartographic results of new method are presented in Fig. 26.3. The Azimuth Disagreement TIN with gradients dip angle (falling).

A and B – center of circle with radius 1000 m at GdAZ

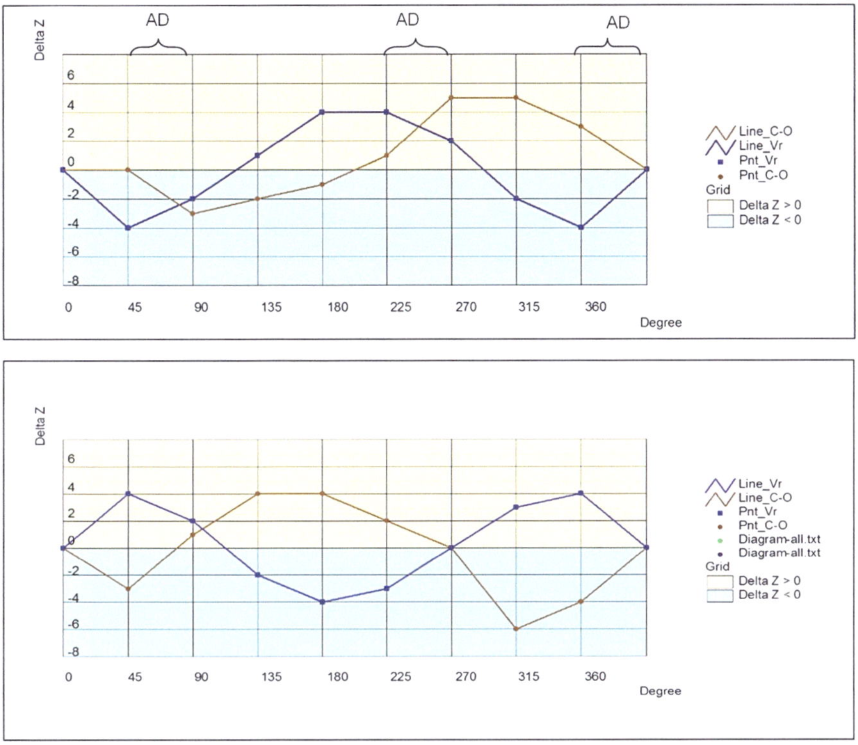

Fig. 26.5 Graph delta *Z* vs. φ is calculated for circle A (top) and circle B (bottom). According to the structural-geodynamic map of Moscow [3, 8], the geodynamic active zones – GdAZ – A is located within the boundary separating the lifting and the lowering. The GdAZ B is within the lowland with a stable and long-lasting tendency of relative subsidence. Fragment of the structural geomorphologic map of Moscow [3] (Fig. 26.5). The color shows various structural forms: yellow, elevations with a tendency to rise, green and blue, lowlands with a steady tendency to descend. The black lines of different thickness, the boundaries between blocks of different orders, lines of different colors, lineaments. A rectangle shows the parts of detailed studies corresponding to GdAZ A and B, respectively (Fig. 26.6)

26.3 Conclusion

The proposed geoinformation methods allow to identify, evaluate, and analyze subhorizontal disagreement zones of different age deformations. These zones are considered by us as a new category – subhorizontal geodynamic active zones. The detection of subhorizontal geodynamic active zones is relevant from the point of view of solving applied problems aimed at assessing geoecological safety in spatial planning schemes for the placement of complex engineering structures: nuclear power plants, hydroelectric power plants, radioactive waste disposal, etc.

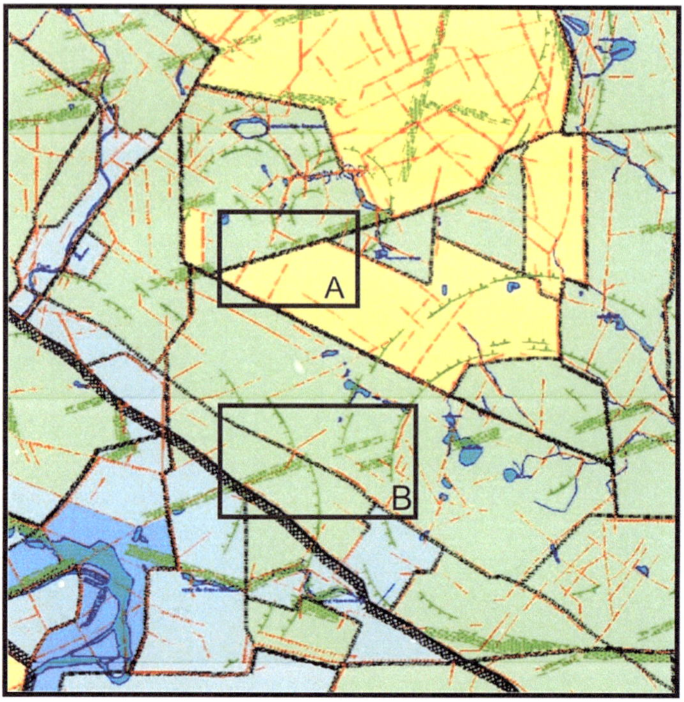

Fig. 26.6 Fragment of the structural-geomorphologic map of Moscow [3]

References

1. Makarov, V. I., Dorozhko, A. L., Makarova, N. V., & Makeev, V. M. (2007). Geodynamic active zones of platforms. *Geoecology. Engineering geology. Hydrogeology. Geocryology*, (2), 99–110 (on the Russian).
2. Makarov, V. I., Grigorieva, S. V., Makeev, V. M., Korobova, I. V., & Krylova, V. S. (2013). Structural-geodynamic map of Moscow, scale 1:25 000. Small encyclopedic atlas of Moscow. M. (on the Russian).
3. Makeev, V. M., Makarova, N. V., Ledenev, V. N., Dorozhko, A. L., Sukhanova, N. V., Karfidova, E. A., & Korobova, I. V. (2014). Fundamentals of geodynamic safety concept areas of environmentally hazardous facilities. *Geoecology*, (2), 99–110 (on the Russian).
4. Makeev, V. M., Karfidova, E. A., Miklyaev, P. S., Avkhimovich, A. N., Dorozhko, A. L., & Korobova, A., IV. (2013). *Method for studying layered structural disagreements in geodynamic studies. Proceedings of the VII University Geological Readings: Problems of regional geology and mineral exploration* (pp. 20–22). Minsk: BSU Publishing Center (on the Russian).
5. Makarov, V. I., Grigorieva, S. V., & Makeev, V. M. (2011). Structural and geodynamic researches in engineering geology surveys and environment study. Environmental Geosciences and Engineering Survey for Territory Protection and Population Safety. International Conference under the aegis of IAEG. EngeoPro. Moscow. Abstracts to Proceeding/Moscow/ 06-08/09/2011 (on the Russian).

6. Mironov, O. K. (2011). GIS technology to produce the large-scale geological mapping the territory of Moscow. *Environmental Geoscience. Engineering Geology. Hydrogeology. Geocryology,* (3), Moscow, 200–216 (on the Russian).
7. Karfidova, E. A., & Sizov, A. P. (2008). Evaluation study of the conditions of the urban area for engineering and geological mapping. Use and protection of natural resources NIA-NATURE, (1). Moscow, pp. 65–72 (on the Russian).
8. Makeev, V. M., Karfidova, E. A., & Dorozhko, A. L. (2014). *Structural-geodynamic mapping including the geostatistical analysis methods of the platform areas on the example of Moscow Megalopolis*. Torino: Springer, IAEG XII Congress.

Index

V. Svalova (ed.), *Natural Hazards and Risk Research in Russia*,
Innovation and Discovery in Russian Science and Engineering,
https://doi.org/10.1007/978-3-319-91833-4

H

MIX
Papier aus verantwortungsvollen Quellen
Paper from responsible sources
FSC® C105338

FSC
www.fsc.org

If you have any concerns about our products,
you can contact us on
ProductSafety@springernature.com

In case Publisher is established outside the EU,
the EU authorized representative is:
Springer Nature Customer Service Center GmbH
Europaplatz 3, 69115 Heidelberg, Germany

Printed by Libri Plureos GmbH
in Hamburg, Germany